Sixth Edition

Student Solutions Guide

Chemistry

Steven S. Zumdahl
Susan Arena Zumdahl

Thomas J. Hummel
Steven S. Zumdahl
Susan Arena Zumdahl

University of Illinois at Urbana-Champaign

HOUGHTON MIFFLIN COMPANY   Boston   New York

Editor-in-Chief:	Charles Hartford
Executive Editor:	Richard Stratton
Development Editor:	Sara Wise
Senior Project Editor:	Cathy Brooks
Editorial Assistant:	Rosemary Mack
Senior Marketing Manager:	Katherine Greig

Printed in the U.S.A.

ISBN: 0-618-221638
        6789—CRS—06 05 04

TABLE OF CONTENTS

TO THE STUDENT: HOW TO USE THIS GUIDE

Solutions to odd-numbered end of chapter questions and exercises are in this manual. This "Solutions Guide" can be very valuable if you use it properly. The way NOT to use it is to look at an exercise in the book and then immediately check the solution, often saying to yourself, "That's easy, I can do it." Developing problem solving skills takes practice. Don't look up a solution to a problem until you have tried to work it on your own. If you are completely stuck, see if you can find a similar problem in the Sample Exercises in the chapter. Only look up the solution as a last resort. If you do this for a problem, look for a similar problem in the end of chapter exercises and try working it. The more problems you do, the easier chemistry becomes. It is also in your self interest to try to work as many problems as possible. Most exams that you will take in chemistry will involve a lot of problem solving. If you have worked several problems similar to the ones on an exam, you will do much better than if the exam is the first time you try to solve a particular type of problem. No matter how much you read and study the text, or how well you think you understand the material, you don't really understand it until you have taken the information in the text and applied the principles to problem solving. You will make mistakes, but the good students learn from their mistakes.

In this manual we have worked problems as in the textbook. We have shown intermediate answers to the correct number of significant figures and used the rounded answer in later calculations. Thus, some of your answers may differ slightly from ours. When we have not followed this convention, we have usually noted this in the solution. The most common exception is when working with the natural logarithm (ln) function, where we usually carried extra significant figures in order to reduce round-off error. In addition, we tried to use constants and conversion factors reported to at least one more significant figure as compared to numbers given in the problem. The practice of carrying one extra significant figure in constants helps minimize round-off error.

We are grateful to Claire O. Szoke for her outstanding effort in preparing the manuscript for this manual. We also thank Marjorie M. Needham and Linda C. Bush for their careful and thorough accuracy review of the Solutions Guide.

TJH
SSZ
SAZ

CHAPTER ONE

CHEMICAL FOUNDATIONS

Questions

17. No, it is useful whenever a systematic approach of observation and hypothesis testing can be used.

19. Volume readings are estimated to one decimal place past the markings on the glassware. The assumed uncertainty is ±1 in the estimated digit. For glassware a, the volume would be estimated to the tenths place since the markings are to the ones place. A sample reading would be 4.2 with an uncertainty of ±0.1. This reading has two significant figures. For glassware b, 10.52 ±0.01 would be a sample reading and the uncertainty; this reading has four significant figures. For glassware c, 18 ±1 would be a sample reading and uncertainty, with the reading having two significant figures.

21. Chemical changes involve the making and breaking of chemical forces (bonds). Physical changes do not. The identity of a substance changes after a chemical change, but not after a physical change.

Exercises

Significant Figures and Unit Conversions

23. a. inexact   b. exact   c. exact

   For c, $\dfrac{36 \text{ in}}{\text{yd}} \times \dfrac{2.54 \text{ cm}}{\text{in}} \times \dfrac{1 \text{ m}}{100 \text{ cm}} = \dfrac{0.9144 \text{ m}}{\text{yd}}$ (All conversion factors used are exact.)

   d. inexact; Although this number appears to be exact, it probably isn't. The announced attendance may be tickets sold but not the number who were actually in the stadium. Some people who paid may not have gone, some may leave early or arrive late, some may sneak in without paying, etc.

   e. exact   f. inexact

25. a. 12; 2 significant figures (S.F.); Nonzero integers always count as significant figures.

   b. 1098; 4 S.F.; Captive zeros always count as significant figures.

   c. 2001; 4 S.F.   d. 2.001×10^3; 4 S.F.

   e. 0.0000101; 3 S.F.; Leading zeros never count as significant figures.

f. $\underline{1.01} \times 10^{-5}$; 3 S.F.

g. $\underline{1000.}$; 4 S.F.; Trailing zeros are only counted as significant figures if the number contains a decimal point.

h. $\underline{22.04030}$; 7 S.F.; The trailing zero is a significant figure since the number contains a decimal point.

27.   a. 3.13×10^2 b. 3.13×10^{-4} c. 3.13×10^7

d. 3.13×10^{-1} e. 3.13×10^{-2}

29.   For addition and/or subtraction, the result has the same number of decimal places as the number in the calculation with the fewest decimal places. When the result is rounded to the correct number of significant figures, the last significant figure stays the same if the number after this significant figure is less than 5 and increases by one if the number is greater than or equal to 5.

a. $97.381 + 4.2502 + 0.99195 = \underline{102.623}15 = 102.623$; Since 97.381 has only three decimal places, the result should only have three decimal places.

b. $171.5 + 72.915 - 8.23 = \underline{236.1}85 = 236.2$

c. $1.00914 + 0.87104 + 1.2012 = \underline{3.0813}8 = 3.0814$

d. $21.901 - 13.21 - 4.0215 = \underline{4.669}5 = 4.67$

31.   a. 467; The difference of $25.27 - 24.16 = 1.11$ has only three significant figures. The answer will only have three significant figures since we have a four significant figure number multiplied by a five significant figure number multiplied by a three significant figure number. For this problem and for subsequent problems, the addition/subtraction rule must be applied separately from the multiplication/division rule.

b. 0.24; The difference of $8.925 - 8.904 = 0.021$ has only 2 significant figures. When a two significant figure number is divided by a four significant figure number, the result is reported to two significant figures (division rule).

c. $(9.04 - 8.23 + 21.954 + 81.0) \div 3.1416 = 103.8 \div 3.1416 = 33.04$

Here, apply the addition/subtraction rule first; then apply the multiplication/division rule to arrive at the four significant figure answer. We will generally round off at intermediate steps in order to show the correct number of significant figures. However, you should round off at the end of all the mathematical operations in order to avoid round-off error. Make sure you keep track of the correct number of significant figures during intermediate steps, but round off at the end.

d. $\dfrac{9.2 \times 100.65}{8.321 + 4.026} = \dfrac{9.2 \times 100.65}{12.347} = 75$

e. $0.1654 + 2.07 - 2.114 = 0.12$

Uncertainty begins to appear in the second decimal place. Numbers were added as written and the answer was rounded off to 2 decimal places at the end. If you round to 2 decimal places and add you get 0.13. Always round off at the end of the operation to avoid round-off error.

f. $8.27(4.987 - 4.962) = 8.27(0.025) = 0.21$

g. $\dfrac{9.5 + 4.1 + 2.8 + 3.175}{4} = \dfrac{19.6}{4} = 4.90 = 4.9$

Uncertainty appears in the first decimal place. The average of several numbers can only be as precise as the least precise number. Averages can be exceptions to the significant figure rules.

h. $\dfrac{9.025 - 9.024}{9.025} \times 100 = \dfrac{0.001}{9.025} \times 100 = 0.01$

33. a. $8.43 \text{ cm} \times \dfrac{1 \text{ m}}{100 \text{ cm}} \times \dfrac{1000 \text{ mm}}{\text{m}} = 84.3 \text{ mm}$ b. $2.41 \times 10^2 \text{ cm} \times \dfrac{1 \text{ m}}{100 \text{ cm}} = 2.41 \text{ m}$

c. $294.5 \text{ nm} \times \dfrac{1 \text{ m}}{1 \times 10^9 \text{ nm}} \times \dfrac{100 \text{ cm}}{\text{m}} = 2.945 \times 10^{-5} \text{ cm}$ d. $1.445 \times 10^4 \text{ m} \times \dfrac{1 \text{ km}}{1000 \text{ m}} = 14.45 \text{ km}$

e. $235.3 \text{ m} \times \dfrac{1000 \text{ mm}}{\text{m}} = 2.353 \times 10^5 \text{ mm}$

f. $903.3 \text{ nm} \times \dfrac{1 \text{ m}}{1 \times 10^9 \text{ nm}} \times \dfrac{1 \times 10^6 \, \mu\text{m}}{\text{m}} = 0.9033 \, \mu\text{m}$

35. a. Appropriate conversion factors are found in Appendix 6. In general, the number of significant figures we use in the conversion factors will be one more than the number of significant figures from the numbers given in the problem. This is usually sufficient to avoid round-off error.

$3.91 \text{ kg} \times \dfrac{1 \text{ lb}}{0.4536 \text{ kg}} = 8.62 \text{ lb};\quad 0.62 \text{ lb} \times \dfrac{16 \text{ oz}}{\text{lb}} = 9.9 \text{ oz}$

Baby's weight = 8 lb and 9.9 oz or to the nearest ounce, 8 lb and 10 oz.

$51.4 \text{ cm} \times \dfrac{1 \text{ in}}{2.54 \text{ cm}} = 20.2 \text{ in} \approx 20 \ 1/4 \text{ in} = \text{baby's height}$

b. $25{,}000 \text{ mi} \times \dfrac{1.61 \text{ km}}{\text{mi}} = 4.0 \times 10^4 \text{ km};\quad 4.0 \times 10^4 \text{ km} \times \dfrac{1000 \text{ m}}{\text{km}} = 4.0 \times 10^7 \text{ m}$

c. $V = l \times w \times h = 1.0 \text{ m} \times \left(5.6 \text{ cm} \times \dfrac{1 \text{ m}}{100 \text{ cm}} \right) \times \left(2.1 \text{ dm} \times \dfrac{1 \text{ m}}{10 \text{ dm}} \right) = 1.2 \times 10^{-2} \text{ m}^3$

$$1.2 \times 10^{-2} \text{ m}^3 \times \left(\dfrac{10 \text{ dm}}{\text{m}} \right)^3 \times \dfrac{1 \text{ L}}{\text{dm}^3} = 12 \text{ L}$$

$$12 \text{ L} \times \dfrac{1000 \text{ cm}^3}{\text{L}} \times \left(\dfrac{1 \text{ in}}{2.54 \text{ cm}} \right)^3 = 730 \text{ in}^3; \quad 730 \text{ in}^3 \times \left(\dfrac{1 \text{ ft}}{12 \text{ in}} \right)^3 = 0.42 \text{ ft}^3$$

37. a. $1.25 \text{ mi} \times \dfrac{8 \text{ furlongs}}{\text{mi}} = 10.0 \text{ furlongs}; \quad 10.0 \text{ furlongs} \times \dfrac{40 \text{ rods}}{\text{furlong}} = 4.00 \times 10^2 \text{ rods}$

$$4.00 \times 10^2 \text{ rods} \times \dfrac{5.5 \text{ yd}}{\text{rod}} \times \dfrac{36 \text{ in}}{\text{yd}} \times \dfrac{2.54 \text{ cm}}{\text{in}} \times \dfrac{1 \text{ m}}{100 \text{ cm}} = 2.01 \times 10^3 \text{ m}$$

$$2.01 \times 10^3 \text{ m} \times \dfrac{1 \text{ km}}{1000 \text{ m}} = 2.01 \text{ km}$$

b. Let's assume we know this distance to ± 1 yard. First convert 26 miles to yards.

$$26 \text{ mi} \times \dfrac{5280 \text{ ft}}{\text{mi}} \times \dfrac{1 \text{ yd}}{3 \text{ ft}} = 45,760. \text{ yd}$$

$$26 \text{ mi} + 385 \text{ yd} = 45,760. \text{ yd} + 385 \text{ yd} = 46,145 \text{ yards}$$

$$46,145 \text{ yard} \times \dfrac{1 \text{ rod}}{5.5 \text{ yd}} = 8390.0 \text{ rods}; \quad 8390.0 \text{ rods} \times \dfrac{1 \text{ furlong}}{40 \text{ rods}} = 209.75 \text{ furlongs}$$

$$46,145 \text{ yard} \times \dfrac{36 \text{ in}}{\text{yd}} \times \dfrac{2.54 \text{ cm}}{\text{in}} \times \dfrac{1 \text{ m}}{100 \text{ cm}} = 42,195 \text{ m}; \quad 42,195 \text{ m} \times \dfrac{1 \text{ km}}{1000 \text{ m}} = 42.195 \text{ km}$$

39. a. $1 \text{ troy lb} \times \dfrac{12 \text{ troy oz}}{\text{troy lb}} \times \dfrac{20 \text{ pw}}{\text{troy oz}} \times \dfrac{24 \text{ grains}}{\text{pw}} \times \dfrac{0.0648 \text{ g}}{\text{grain}} \times \dfrac{1 \text{ kg}}{1000 \text{ g}} = 0.373 \text{ kg}$

$$1 \text{ troy lb} = 0.373 \text{ kg} \times \dfrac{2.205 \text{ lb}}{\text{kg}} = 0.822 \text{ lb}$$

b. $1 \text{ troy oz} \times \dfrac{20 \text{ pw}}{\text{troy oz}} \times \dfrac{24 \text{ grains}}{\text{pw}} \times \dfrac{0.0648 \text{ g}}{\text{grain}} = 31.1 \text{ g}$

$$1 \text{ troy oz} = 31.1 \text{ g} \times \dfrac{1 \text{ carat}}{0.200 \text{ g}} = 156 \text{ carats}$$

c. $1 \text{ troy lb} = 0.373 \text{ kg}; \quad 0.373 \text{ kg} \times \dfrac{1000 \text{ g}}{\text{kg}} \times \dfrac{1 \text{ cm}^3}{19.3 \text{ g}} = 19.3 \text{ cm}^3$

41. $1.71 \text{ warp factor} = \left(5.00 \times \dfrac{3.00 \times 10^8 \text{ m}}{\text{s}} \right) \times \dfrac{1.094 \text{ yd}}{\text{m}} \times \dfrac{60 \text{ s}}{\text{min}} \times \dfrac{60 \text{ min}}{\text{hr}}$

$$\times \dfrac{1 \text{ knot}}{2000 \text{ yd/hr}} = 2.95 \times 10^9 \text{ knots}$$

43. $\dfrac{14 \text{ km}}{\text{L}} \times \dfrac{1 \text{ mi}}{1.61 \text{ km}} \times \dfrac{3.79 \text{ L}}{\text{gal}} = 33 \text{ mi/gal};$ The spouse's car has the better gas mileage.

45. $1.00 \text{ lb} \times \dfrac{0.4536 \text{ kg}}{\text{lb}} \times \dfrac{4.00 \text{ euros}}{\text{kg}} \times \dfrac{\$1.00}{1.14 \text{ euros}} = \$1.59$

Temperature

47. $T_C = \dfrac{5}{9}(T_F - 32) = \dfrac{5}{9}(102.5 - 32) = 39.2°C; \; T_K = T_C + 273.2 = 312.4 \text{ K}$ (Note: 32 is exact)

49.   a. $T_F = \dfrac{9}{5} \times T_C + 32 = \dfrac{9}{5} \times 78.1°C + 32 = 173°F$

$T_K = T_C + 273.2 = 78.1 + 273.2 = 351.3 \text{ K}$

b. $T_F = \dfrac{9}{5} \times (-25) + 32 = -13°F; \; T_K = -25 + 273 = 248 \text{ K}$

c. $T_F = \dfrac{9}{5} \times (-273) + 32 = -459°F; \; T_K = -273 + 273 = 0 \text{ K}$

d. $T_F = \dfrac{9}{5} \times 801 + 32 = 1470°F; \; T_K = 801 + 273 = 1074 \text{ K}$

51.   We can do this two ways. One way is to calculate the high and low temperature and get the uncertainty from the range. $20.6°C \pm 0.1°C$ means the temperature can range from $20.5°C$ to $20.7°C$.

$T_F = \dfrac{9}{5} \times T_c + 32 \leftarrow$ (exact); $\; T_F = \dfrac{9}{5} \times 20.6 + 32 = 69.1°F$

$T_F (\text{min}) = \dfrac{9}{5} \times 20.5 + 32 = 68.9°F; \; T_F (\text{max}) = \dfrac{9}{5} \times 20.7 + 32 = 69.3°F$

So the temperature ranges from $68.9°F$ to $69.3°F$ which we can express as $69.1 \pm 0.2°F$.

An alternative way is to treat the uncertainty and the temperature in °C separately.

$$T_F = \frac{9}{5} \times T_C + 32 = \frac{9}{5} \times 20.6 + 32 = 69.1°F; \quad \pm 0.1°C \times \frac{9°F}{5°C} = \pm 0.18°F \approx \pm 0.2°F$$

Combining the two calculations: $T_F = 69.1 \pm 0.2°F$

Density

53.
$$\frac{2.70 \text{ g}}{\text{cm}^3} \times \frac{1 \text{ kg}}{1000 \text{ g}} \times \left(\frac{100 \text{ cm}}{\text{m}}\right)^3 = \frac{2.70 \times 10^3 \text{ kg}}{\text{m}^3}$$

$$\frac{2.70 \text{ g}}{\text{cm}^3} \times \frac{1 \text{ lb}}{453.6 \text{ g}} \times \left(\frac{2.54 \text{ cm}}{\text{in}}\right)^3 \times \left(\frac{12 \text{ in}}{\text{ft}}\right)^3 = \frac{169 \text{ lb}}{\text{ft}^3}$$

55.
$$V = \frac{4}{3}\pi r^3 = \frac{4}{3} \times 3.14 \times \left(7.0 \times 10^5 \text{ km} \times \frac{1000 \text{ m}}{\text{km}} \times \frac{100 \text{ cm}}{\text{m}}\right)^3 = 1.4 \times 10^{33} \text{ cm}^3$$

$$\text{density} = \frac{\text{mass}}{\text{volume}} = \frac{2 \times 10^{36} \text{ kg} \times \dfrac{1000 \text{ g}}{\text{kg}}}{1.4 \times 10^{33} \text{ cm}^3} = 1.4 \times 10^6 \text{ g/cm}^3 = 1 \times 10^6 \text{ g/cm}^3$$

57.
$$5.0 \text{ carat} \times \frac{0.200 \text{ g}}{\text{carat}} \times \frac{1 \text{ cm}^3}{3.51 \text{ g}} = 0.28 \text{ cm}^3$$

59.
$$V = 21.6 \text{ mL} - 12.7 \text{ mL} = 8.9 \text{ mL}; \quad \text{density} = \frac{33.42 \text{ g}}{8.9 \text{ mL}} = 3.8 \text{ g/mL} = 3.8 \text{ g/cm}^3$$

61.   a. Both have the same mass of 1.0 kg.

    b. 1.0 mL of mercury; Mercury has a greater density than water. Note: 1 mL = 1 cm³

$$1.0 \text{ mL} \times \frac{13.6 \text{ g}}{\text{mL}} = 14 \text{ g of mercury}; \quad 1.0 \text{ mL} \times \frac{0.998 \text{ g}}{\text{mL}} = 1.0 \text{ g of water}$$

    c. Same; Both represent 19.3 g of substance.

$$19.3 \text{ mL} \times \frac{0.9982 \text{ g}}{\text{mL}} = 19.3 \text{ g of water}; \quad 1.00 \text{ mL} \times \frac{19.32 \text{ g}}{\text{mL}} = 19.3 \text{ g of gold}$$

d.   1.0 L of benzene (880 g vs 670 g)

$$75 \text{ mL} \times \frac{8.96 \text{ g}}{\text{mL}} = 670 \text{ g of copper; } 1.0 \text{ L} \times \frac{1000 \text{ mL}}{\text{L}} \times \frac{0.880 \text{ g}}{\text{mL}} = 880 \text{ g of benzene}$$

63. $V = 1.00 \times 10^3 \text{ g} \times \frac{1 \text{ cm}^3}{22.57 \text{ g}} = 44.3 \text{ cm}^3$

44.3 cm^3 = 1 × w × h = 4.00 cm × 4.00 cm × h, h = 2.77 cm

Classification and Separation of Matter

65.   Solid: own volume, own shape, does not flow; Liquid: own volume, takes shape of container, flows; Gas: takes volume and shape of container, flows

67.   A gas has molecules that are very far apart from each other while a solid or liquid has molecules that are very close together. An element has the same type of atom, whereas a compound contains two or more different elements. Picture i represents an element that exists as two atoms bonded together (like H$_2$ or O$_2$ or N$_2$). Picture iv represents a compound (like CO, NO, or HF). Pictures iii and iv contain representations of elements that exist as individual atoms (like Ar, Ne, or He).

a.   Picture iv represents a gaseous compound. Note that pictures ii and iii also contain a gaseous compound, but they also both have a gaseous element present.

b.   Picture vi represents a mixture of two gaseous elements.

c.   Picture v represents a solid element.

d.   Pictures ii and iii both represent a mixture of a gaseous element and a gaseous compound.

69.   A physical change is a change in the state of a substance (solid, liquid and gas are the three states of matter); a physical change does not change the chemical composition of the substance. A chemical change is a change in which a given substance is converted into another substance having a different formula (composition).

a.   Vaporization refers to a liquid converting to a gas, so this is a physical change. The formula (composition) of the moth ball does not change.

b.   This is a chemical change since hydrofluoric acid (HF) is reacting with glass (SiO$_2$) to form new compounds which wash away.

c.   This is a physical change since all that is happening is the conversion of liquid alcohol to gaseous alcohol. The alcohol formula (C$_2$H$_5$OH) does not change.

d.   This is a chemical change since the acid is reacting with cotton to form new compounds.

Additional Exercises

71. $1 \; \mu mol \times \dfrac{1 \; mol}{1 \times 10^6 \; \mu mol} \times \dfrac{6.02 \times 10^{23} \; atoms}{mol} = 6.02 \times 10^{17} \; atoms \; of \; helium$

$1.25 \times 10^{20} \; atoms \times \dfrac{1 \; mol}{6.02 \times 10^{23} \; atoms} = 2.08 \times 10^{-4} \; mol \; helium$

73. $Total \; volume = \left(200. \; m \times \dfrac{100 \; cm}{m} \right) \times \left(300. \; m \times \dfrac{100 \; cm}{m} \right) \times 4.0 \; cm = 2.4 \times 10^9 \; cm^3$

$Vol. \; of \; topsoil \; covered \; by \; 1 \; bag = \left[10. \; ft^2 \times \left(\dfrac{12 \; in}{ft} \right)^2 \times \left(\dfrac{2.54 \; cm}{in} \right)^2 \right] \times \left(1.0 \; in \times \dfrac{2.54 \; cm}{in} \right)$

$$= 2.4 \times 10^4 \; cm^3$$

$2.4 \times 10^9 \; cm^3 \times \dfrac{1 \; bag}{2.4 \times 10^4 \; cm^3} = 1.0 \times 10^5 \; bags \; topsoil$

75. $Volume \; of \; lake = 100 \; mi^2 \times \left(\dfrac{5280 \; ft}{mi} \right)^2 \times 20 \; ft = 6 \times 10^{10} \; ft^3$

$6 \times 10^{10} \; ft^3 \times \left(\dfrac{12 \; in}{ft} \times \dfrac{2.54 \; cm}{in} \right)^3 \times \dfrac{1 \; mL}{cm^3} \times \dfrac{0.4 \; \mu g}{mL} = 7 \times 10^{14} \; \mu g \; mercury$

$7 \times 10^{14} \; \mu g \times \dfrac{1 \; g}{10^6 \; \mu g} \times \dfrac{1 \; kg}{10^3 \; g} = 7 \times 10^5 \; kg \; of \; mercury$

77.   a.   Volume × density = mass; the orange block is more dense. Since mass (orange) > mass (blue) and since volume (orange) < volume (blue), the density of the orange block must be greater to account for the larger mass of the orange block.

b.   Which block is more dense cannot be determined. Since mass (orange) > mass (blue) and since volume (orange) > volume (blue), the density of the orange block may or may not be larger than the blue block. If the blue block is more dense, its density cannot be so large that its mass is larger than the orange block's mass.

c.   The blue block is more dense. Since mass (blue) = mass (orange) and since volume (blue) < volume (orange), the density of the blue block must be larger in order to equate the masses.

d.   The blue block is more dense. Since mass (blue) > mass (orange) and since the volumes are equal, the density of the blue block must be larger in order to give the blue block the larger mass.

79. $V = V_{final} - V_{initial}; \; d = \dfrac{28.90 \; g}{9.8 \; cm^3 - 6.4 \; cm^3} = \dfrac{28.90 \; g}{3.4 \; cm^3} = 8.5 \; g/cm^3$

$$d_{max} = \frac{mass_{max}}{V_{min}}; \quad \text{We get } V_{min} \text{ from } 9.7 \text{ cm}^3 - 6.5 \text{ cm}^3 = 3.2 \text{ cm}^3.$$

$$d_{max} = \frac{28.93 \text{ g}}{3.2 \text{ cm}^3} = \frac{9.0 \text{ g}}{\text{cm}^3}; \quad d_{min} = \frac{mass_{min}}{V_{max}} = \frac{28.87 \text{ g}}{9.9 \text{ cm}^3 - 6.3 \text{ cm}^3} = \frac{8.0 \text{ g}}{\text{cm}^3}$$

The density is: 8.5 ± 0.5 g/cm^3.

Challenge Problems

81.   In subtraction, the result gets smaller but the uncertainties add. If the two numbers are very close together, the uncertainty may be larger than the result. For example, let us assume we want to take the difference of the following two measured quantities: $999,999 \pm 2$ and $999,996 \pm 2$. The difference is 3 ± 4. Because of the uncertainty, subtracting two similar numbers is bad practice.

83.   Heavy pennies (old): mean mass = 3.08 ± 0.05 g

Light pennies (new): mean mass $= \dfrac{(2.467 + 2.545 + 2.518)}{3} = 2.51 \pm 0.04$ g

Since we are assuming that the volume is additive, let's calculate the volume of 100. g of each type of penny then calculate the density of the alloy. For 100. g of the old pennies, 95 g will be Cu and 5 g will be Zn.

$$V = 95 \text{ g Cu} \times \frac{1 \text{ cm}^3}{8.96 \text{ g}} + 5 \text{ g Zn} \times \frac{1 \text{ cm}^3}{7.14 \text{ g}} = 11.3 \text{ cm}^3 \text{ (carrying one extra sig. fig.)}$$

$$\text{Density of old pennies} = \frac{100. \text{ g}}{11.3 \text{ cm}^3} = 8.8 \text{ g/cm}^3$$

For 100. g of new pennies, 97.6 g will be Zn and 2.4 g will be copper.

$$V = 2.4 \text{ g Cu} \times \frac{1 \text{ cm}^3}{8.96 \text{ g}} + 97.6 \text{ g Zn} \times \frac{1 \text{ cm}^3}{7.14 \text{ g}} = 13.94 \text{ cm}^3 \text{ (carrying one extra sig. fig.)}$$

$$\text{Density of new pennies} = \frac{100. \text{ g}}{13.94 \text{ cm}^3} = 7.17 \text{ g/cm}^3$$

Since $d = \dfrac{mass}{volume}$ and since the volume of both types of pennies are assumed equal, then:

$$\frac{d_{new}}{d_{old}} = \frac{mass_{new}}{mass_{old}} = \frac{7.17 \text{ g/cm}^3}{8.8 \text{ g/cm}^3} = 0.81$$

The calculated average mass ratio is: $\dfrac{\text{mass}_{new}}{\text{mass}_{old}} = \dfrac{2.51\ g}{3.08\ g} = 0.815$

To the first two decimal places, the ratios are the same. If the assumptions are correct, then we can reasonably conclude that the difference in mass is accounted for by the difference in alloy used.

85.   a.   One possibility is that rope B is not attached to anything and rope A and rope C are connected via a pair of pulleys and/or gears.

       b.   Try to pull rope B out of the box. Measure the distance moved by C for a given movement of A. Hold either A or C firmly while pulling on the other.

CHAPTER TWO

ATOMS, MOLECULES, AND IONS

Questions

15. a. Atoms have mass and are neither destroyed nor created by chemical reactions. Therefore, mass is neither created nor destroyed by chemical reactions. Mass is conserved.

   b. The composition of a substance depends on the number and kinds of atoms that form it.

   c. Compounds of the same elements differ only in the numbers of atoms of the elements forming them, i.e., NO, N_2O, NO_2.

17. Deflection of cathode rays by magnetic and electric fields led to the conclusion that they were negatively charged. The cathode ray was produced at the negative electrode and repelled by the negative pole of the applied electric field.

19. The proton and neutron have similar mass with the mass of the neutron slightly larger than that of the proton. Each of these particles has a mass approximately 1800 times greater than that of an electron. The combination of the protons and the neutrons in the nucleus makes up the bulk of the mass of an atom, but the electrons make the greatest contribution to the chemical properties of the atom.

21. The atomic number of an element is equal to the number of protons in the nucleus of an atom of that element. The mass number is the sum of the number of protons plus neutrons in the nucleus. The atomic mass is the actual mass of a particular isotope (including electrons). As we will see in Chapter Three, the average mass of an atom is taken from a measurement made on a large number of atoms. The average atomic mass value is listed in the periodic table.

23. A compound will always contain the same numbers (and types) of atoms. A given amount of hydrogen will react only with a specific amount of oxygen. Any excess oxygen will remain unreacted.

Exercises

Development of the Atomic Theory

25. a. The composition of a substance depends on the numbers of atoms of each element making up the compound (i.e., on the formula of the compound) and not on the composition of the mixture from which it was formed.

b. Avogadro's hypothesis implies that volume ratios are equal to molecule ratios at constant temperature and pressure. $H_2(g) + Cl_2(g) \rightarrow 2\ HCl(g)$. From the balanced equation (2 molecules of HCl are produced per molecule of H_2 or Cl_2 reacted), the volume of HCl produced will be twice the volume of H_2 (or Cl_2) reacted.

27. $\dfrac{1.188}{1.188} = 1.000;\quad \dfrac{2.375}{1.188} = 1.999;\quad \dfrac{3.563}{1.188} = 2.999$

The masses of fluorine are simple ratios of whole numbers to each other, 1:2:3.

29.   To get the atomic mass of H to be 1.00, we divide the mass of hydrogen that reacts with 1.00 g of oxygen by 0.126, i.e., $\dfrac{0.126}{0.126} = 1.00$. To get Na, Mg and O on the same scale, we do the same division.

Na: $\dfrac{2.875}{0.126} = 22.8;\quad$ Mg: $\dfrac{1.500}{0.126} = 11.9;\quad$ O: $\dfrac{1.00}{0.126} = 7.94$

	H	O	Na	Mg
Relative Value	1.00	7.94	22.8	11.9
Accepted Value	1.008	16.00	22.99	24.31

The atomic masses of O and Mg are incorrect; the atomic masses of H and Na are close to the values in the periodic table. Something must be wrong about the assumed formulas of the compounds. It turns out the correct formulas are H_2O, Na_2O, and MgO. The smaller discrepancies result from the error in the atomic mass of H.

The Nature of the Atom

31.   Density of hydrogen nucleus (contains one proton only):

$$V_{nucleus} = \frac{4}{3}\pi r^3 = \frac{4}{3}(3.14)(5 \times 10^{-14}\ cm)^3 = 5 \times 10^{-40}\ cm^3$$

$$d = \frac{1.67 \times 10^{-24}\ g}{5 \times 10^{-40}\ cm^3} = 3 \times 10^{15}\ g/cm^3$$

Density of H-atom (contains one proton and one electron):

$$V_{atom} = \frac{4}{3}(3.14)(1 \times 10^{-8}\ cm)^3 = 4 \times 10^{-24}\ cm^3$$

$$d = \frac{1.67 \times 10^{-24} + 9 \times 10^{-28}\ g}{4 \times 10^{-24}\ cm^3} = 0.4\ g/cm^3$$

33. $5.93 \times 10^{-18}\ C \times \dfrac{1\ electron\ charge}{1.602 \times 10^{-19}\ C} = 37$ negative (electron) charges on the oil drop

35.   sodium - Na; beryllium - Be; manganese - Mn; chromium - Cr; uranium - U

37.   Sn - tin; Pt - platinum; Co - cobalt; Ni - nickel; Mg - magnesium; Ba - barium; K - potassium

39.   The noble gases are He, Ne, Ar, Kr, Xe, and Rn (helium, neon, argon, krypton, xenon, and radon).
       Radon has only radioactive isotopes. In the periodic table, the whole number enclosed in parentheses
       is the mass number of the longest-lived isotope of the element.

41.   a.   Eight; Li to Ne   b.   Eight; Na to Ar

       c.   Eighteen; K to Kr   d.   Five; N, P, As, Sb, Bi

43.   a. $^{238}_{94}$Pu: 94 protons, 238 - 94 = 144 neutrons   b. $^{65}_{29}$Cu: 29 protons, 65 - 29 = 36 neutrons

       c. $^{52}_{24}$Cr: 24 protons, 28 neutrons   d. $^{4}_{2}$He: 2 protons, 2 neutrons

       e. $^{60}_{27}$Co: 27 protons, 33 neutrons   f. $^{54}_{24}$Cr: 24 protons, 30 neutrons

45.   a.   Element #5 is boron. $^{12}_{5}$B   b.   Z = 7; A = 7 + 8 = 15; $^{15}_{7}$N

       c.   Z = 17; A = 17 + 18 = 35; $^{35}_{17}$Cl   d.   A = 92 + 143 = 235; $^{235}_{92}$U

       e.   Z = 6; A = 14; $^{14}_{6}$C   f.   Z = 15; A = 31; $^{31}_{15}$P

47.   Atomic number = 63 (Eu); Charge = +63 - 60 = +3; Mass number = 63 + 88 = 151;
       Symbol: $^{151}_{63}$Eu^{3+}

       Atomic number = 50 (Sn); Mass number = 50 + 68 = 118; Net charge = +50 - 48 = +2; The symbol
       is $^{118}_{50}$Sn^{2+}.

49.

Symbol	Number of protons in nucleus	Number of neutrons in nucleus	Number of electrons	Net charge
$^{75}_{33}$As^{3+}	33	42	30	3+
$^{128}_{52}$Te^{2-}	52	76	54	2-
$^{32}_{16}$S	16	16	16	0

Symbol	Number of protons in nucleus	Number of neutrons in nucleus	Number of electrons	Net charge
$^{204}_{81}Tl^+$	81	123	80	1+
$^{195}_{78}Pt$	78	117	78	0

51.   Metals: Mg, Ti, Au, Bi, Ge, Eu, Am.   Nonmetals: Si, B, At, Rn, Br.

53.   a and d. A group is a vertical column of elements in the periodic table. Elements in the same family (group) have similar chemical properties.

55.   Carbon is a nonmetal. Silicon and germanium are metalloids. Tin and lead are metals. Thus, metallic character increases as one goes down a family in the periodic table.

57.   Metals lose electrons to form cations, and nonmetals gain electrons to form anions. Group 1A, 2A and 3A metals form stable +1,+2 and +3 charged cations, respectively. Group 5A, 6A and 7A nonmetals form -3, -2 and -1 charged anions, respectively.

a. Lose 1 e⁻ to form Na^+.   b. Lose 2 e⁻ to form Sr^{2+}.   c. Lose 2 e⁻ to form Ba^{2+}.

d. Gain 1 e⁻ to form I^-.   e. Lose 3 e⁻ to form Al^{3+}.   f. Gain 2 e⁻ to form S^{2-}.

Nomenclature

59.   a. sodium chloride   b. rubidium oxide

c. calcium sulfide   d. aluminum iodide

61.   a. chromium(VI) oxide   b. chromium(III) oxide   c. aluminum oxide

d. sodium hydride   e. calcium bromide

f. zinc chloride (Zinc only forms +2 ions so no Roman numerals are needed for zinc compounds.)

63.   a. potassium perchlorate   b. calcium phosphate

c. aluminum sulfate   d. lead(II) nitrate

65.   a. nitrogen triiodide   b. sulfur difluoride

c. phosphorus trichloride   d. dinitrogen tetrafluoride

67. a. copper(I) iodide   b. copper(II) iodide   c. cobalt(II) iodide

     d. sodium carbonate   e. sodium hydrogen carbonate or sodium bicarbonate

     f. tetrasulfur tetranitride g. sulfur hexafluoride   h. sodium hypochlorite

     i. barium chromate   j. ammonium nitrate

69. a. CsBr   b. $BaSO_4$ c. NH_4Cl d. ClO

     e. $SiCl_4$ f. ClF_3 g. BeO   h. MgF_2

71. a. Na_2O b. Na_2O_2 c. KCN

     d. $Cu(NO_3)_2$ e. $SeBr_4$ f. PbS

     g. PbS_2 h. CuCl   i. GaAs (Predict Ga^{3+} and As^{3-} ions.)

     j. CdSe (Cadmium only forms +2 charged ions in compounds.)

     k. ZnS (Zinc only forms +2 charged ions in compounds.)

     l. HNO_2 m. P_2O_5

Additional Exercises

73. Yes, 1.0 g H would react with 37.0 g ^{37}Cl and 1.0 g H would react with 35.0 g ^{35}Cl.

     No, the mass ratio of H/Cl would always be 1 g H/37 g Cl for ^{37}Cl and 1 g H/35 g Cl for ^{35}Cl.
     As long as we had pure ^{37}Cl or pure ^{35}Cl, the above ratios will always hold. If we have a mixture
     (such as the natural abundance of chlorine), the ratio will also be constant as long as the composition
     of the mixture of the two isotopes does not change.

75. a. nitric acid, HNO_3 b. perchloric acid, $HClO_4$ c. acetic acid, $HC_2H_3O_2$

     d. sulfuric acid, H_2SO_4 e. phosphoric acid, H_3PO_4

77. a. $Pb(C_2H_3O_2)_2$: lead(II) acetate   b. $CuSO_4$: copper(II) sulfate

     c. CaO: calcium oxide   d. $MgSO_4$: magnesium sulfate

     e. $Mg(OH)_2$: magnesium hydroxide   f. $CaSO_4$: calcium sulfate

     g. N_2O: dinitrogen monoxide or nitrous oxide

79. A chemical formula gives the actual number and kind of atoms in a compound. In all cases, 12
     hydrogen atoms are present. For example:

$$4 \text{ molecules } H_3PO_4 \times \frac{3 \text{ atoms } H}{\text{molecule } H_3PO_4} = 12 \text{ atoms } H$$

81.   a. Ca^{2+} and N^{3-}: Ca_3N_2, calcium nitride   b. K^+ and O^{2-}: K_2O, potassium oxide

      a. Rb^+ and F^-: RbF, rubidium fluoride   d. Mg^{2+} and S^{2-}: MgS, magnesium sulfide

      e. Ba^{2+} and I^-: BaI_2, barium iodide   f. Al^{3+} and Se^{2-}: Al_2Se_3, aluminum selenide

      g. Cs^+ and P^{3-}: Cs_3P, cesium phosphide

      h. In^{3+} and Br^-: $InBr_3$, indium(III) bromide. In also forms In^+ ions, but one would predict In^{3+} ions from its position in the periodic table.

Challenge Problems

83.   Copper(Cu), silver (Ag) and gold(Au) make up the coinage metals.

85.   Compound I: $\dfrac{14.0 \text{ g R}}{3.00 \text{ g Q}} = \dfrac{4.67 \text{ g R}}{1.00 \text{ g Q}}$;   Compound II: $\dfrac{7.00 \text{ g R}}{4.50 \text{ g Q}} = \dfrac{1.56 \text{ g R}}{1.00 \text{ g Q}}$

      The ratio of the masses of R that combine with 1.00 g Q is: $\dfrac{4.67}{1.56} = 2.99 \approx 3$

      As expected from the law of multiple proportions, this ratio is a small whole number.

      Since Compound I contains three times the mass of R per gram of Q as compared to Compound II (RQ), then the formula of Compound I should be R_3Q.

87.   a.   Both compounds have C_2H_6O as the formula. Because they have the same formula, their mass percent composition will be identical. However, these are different compounds with different properties since the atoms are bonded together differently. These compounds are called isomers of each other.

      b.   When wood burns, most of the solid material in wood is converted to gases, which escape. The gases produced are most likely CO_2 and H_2O.

      c.   The atom is not an indivisible particle, but is instead composed of other smaller particles, e.g., electrons, neutrons, protons.

      d.   The two hydride samples contain different isotopes of either hydrogen and/or lithium. Although the compounds are composed of different isotopes, their properties are similar because different isotopes of the same element have similar properties (except, of course, their mass).

CHAPTER THREE

STOICHIOMETRY

Questions

19. The molecular formula tells us the actual number of atoms of each element in a molecule (or formula unit) of a compound. The empirical formula tells only the simplest whole number ratio of atoms of each element in a molecule. The molecular formula is a whole number multiple of the empirical formula. If that multiplier is one, the molecular and empirical formulas are the same. For example, both the molecular and empirical formulas of water are H_2O. They are the same. For hydrogen peroxide, the empirical formula is OH; the molecular formula is H_2O_2.

Exercises

Atomic Masses and the Mass Spectrometer

21. A = atomic mass = 0.7899(23.9850 amu) + 0.1000(24.9858 amu) + 0.1101(25.9826 amu)

    A = 18.95 amu + 2.499 amu + 2.861 amu = 24.31 amu

23. Let x = % of ^{151}Eu and y = % of ^{153}Eu, then $x + y = 100$ and $y = 100 - x$.

    $$151.96 = \frac{x(150.9196) + (100 - x)(152.9209)}{100}$$

    $15196 = 150.9196\,x + 15292.09 - 152.9209\,x, \quad -96 = -2.0013\,x$

    $x = 48\%$; 48% ^{151}Eu and 100 - 48 = 52% ^{153}Eu

25. There are three peaks in the mass spectrum, each 2 mass units apart. This is consistent with two isotopes, differing in mass by two mass units. The peak at 157.84 corresponds to a Br_2 molecule composed of two atoms of the lighter isotope. This isotope has mass equal to 157.84/2 or 78.92. This corresponds to ^{79}Br. The second isotope is ^{81}Br with mass equal to 161.84/2 = 80.92. The peaks in the mass spectrum correspond to $^{79}Br_2$, $^{79}Br^{81}Br$, and $^{81}Br_2$ in order of increasing mass. The intensities of the highest and lowest mass tell us the two isotopes are present in about equal abundance. The actual abundance is 50.69% ^{79}Br and 49.31% ^{81}Br. The calculation of the abundance from the mass spectrum is beyond the scope of this text.

Moles and Molar Masses

27.   When more than one conversion factor is necessary to determine the answer, we will usually put all the conversion factors into one calculation instead of determining intermediate answers. This method reduces round-off error and is a time saver.

$$500. \text{ atoms Fe} \times \frac{1 \text{ mol Fe}}{6.022 \times 10^{23} \text{ atoms Fe}} \times \frac{55.85 \text{ g Fe}}{\text{mol Fe}} = 4.64 \times 10^{-20} \text{ g Fe}$$

29. $$1.00 \text{ carat} \times \frac{0.200 \text{ g C}}{\text{carat}} \times \frac{1 \text{ mol C}}{12.01 \text{ g C}} \times \frac{6.022 \times 10^{23} \text{ atoms C}}{\text{mol C}} = 1.00 \times 10^{22} \text{ atoms C}$$

31. Al_2O_3: $2(26.98) + 3(16.00) = 101.96$ g/mol

        Na_3AlF_6: $3(22.99) + 1(26.98) + 6(19.00) = 209.95$ g/mol

33.   a.   The formula is NH_3. 14.01 g/mol $+ 3(1.008$ g/mol$) = 17.03$ g/mol

        b.   The formula is N_2H_4. $2(14.01) + 4(1.008) = 32.05$ g/mol

        c. $(NH_4)_2Cr_2O_7$: $2(14.01) + 8(1.008) + 2(52.00) + 7(16.00) = 252.08$ g/mol

35.   a. $$1.00 \text{ g NH}_3 \times \frac{1 \text{ mol NH}_3}{17.03 \text{ g NH}_3} = 0.0587 \text{ mol NH}_3$$

        b. $$1.00 \text{ g N}_2\text{H}_4 \times \frac{1 \text{ mol N}_2\text{H}_4}{32.05 \text{ g N}_2\text{H}_4} = 0.0312 \text{ mol N}_2\text{H}_4$$

        c. $$1.00 \text{ g (NH}_4)_2\text{Cr}_2\text{O}_7 \times \frac{1 \text{ mol (NH}_4)_2\text{Cr}_2\text{O}_7}{252.08 \text{ g (NH}_4)_2\text{Cr}_2\text{O}_7} = 3.97 \times 10^{-3} \text{ mol (NH}_4)_2\text{Cr}_2\text{O}_7$$

37.   a. $$5.00 \text{ mol NH}_3 \times \frac{17.03 \text{ g NH}_4}{\text{mol NH}_3} = 85.2 \text{ g NH}_3$$

        b. $$5.00 \text{ mol N}_2\text{H}_4 \times \frac{32.05 \text{ g N}_2\text{H}_4}{\text{mol N}_2\text{H}_4} = 160. \text{ g N}_2\text{H}_4$$

        c. $$5.00 \text{ mol (NH}_4)_2\text{Cr}_2\text{O}_7 \times \frac{252.08 \text{ g (NH}_4)_2\text{Cr}_2\text{O}_7}{\text{mol (NH}_4)_2\text{Cr}_2\text{O}_7} = 1260 \text{ g (NH}_4)_2\text{Cr}_2\text{O}_7$$

39.   Chemical formulas give atom ratios as well as mol ratios.

        a. $$5.00 \text{ mol NH}_3 \times \frac{1 \text{ mol N}}{\text{mol NH}_3} \times \frac{14.01 \text{ g N}}{\text{mol N}} = 70.1 \text{ g N}$$

b. $5.00 \text{ mol N}_2\text{H}_4 \times \dfrac{2 \text{ mol N}}{\text{mol N}_2\text{H}_4} \times \dfrac{14.01 \text{ g N}}{\text{mol N}} = 140. \text{ g N}$

c. $5.00 \text{ mol (NH}_4)_2\text{Cr}_2\text{O}_7 \times \dfrac{2 \text{ mol N}}{\text{mol (NH}_4)_2\text{Cr}_2\text{O}_7} \times \dfrac{14.01 \text{ g N}}{\text{mol N}} = 140. \text{ g N}$

41. a. $1.00 \text{ g NH}_3 \times \dfrac{1 \text{ mol NH}_3}{17.03 \text{ g NH}_3} \times \dfrac{6.022 \times 10^{23} \text{ molecules NH}_3}{\text{mol NH}_3} = 3.54 \times 10^{22} \text{ molecules NH}_3$

  b. $1.00 \text{ g N}_2\text{H}_4 \times \dfrac{1 \text{ mol N}_2\text{H}_4}{32.05 \text{ g N}_2\text{H}_4} \times \dfrac{6.022 \times 10^{23} \text{ molecules N}_2\text{H}_4}{\text{mol N}_2\text{H}_4} = 1.88 \times 10^{22} \text{ molecules N}_2\text{H}_4$

  c. $1.00 \text{ g (NH}_4)_2\text{Cr}_2\text{O}_7 \times \dfrac{1 \text{ mol (NH}_4)_2\text{Cr}_2\text{O}_7}{252.08 \text{ g (NH}_4)_2\text{Cr}_2\text{O}_7} \times \dfrac{6.022 \times 10^{23} \text{ formula units (NH}_4)_2\text{Cr}_2\text{O}_7}{\text{mol (NH}_4)_2\text{Cr}_2\text{O}_7}$

  $$= 2.39 \times 10^{21} \text{ formula units (NH}_4)_2\text{Cr}_2\text{O}_7$$

43. Using answers from Exercise 41:

  a. $3.54 \times 10^{22} \text{ molecules NH}_3 \times \dfrac{1 \text{ atom N}}{\text{molecule NH}_3} = 3.54 \times 10^{22} \text{ atoms N}$

  b. $1.88 \times 10^{22} \text{ molecules N}_2\text{H}_4 \times \dfrac{2 \text{ atoms N}}{\text{molecule N}_2\text{H}_4} = 3.76 \times 10^{22} \text{ atoms N}$

  c. $2.39 \times 10^{21} \text{ formula units (NH}_4)_2\text{Cr}_2\text{O}_7 \times \dfrac{2 \text{ atoms N}}{\text{formula unit (NH}_4)_2\text{Cr}_2\text{O}_7} = 4.78 \times 10^{21} \text{ atoms N}$

45. Molar mass of $\text{C}_6\text{H}_8\text{O}_6 = 6(12.01) + 8(1.008) + 6(16.00) = 176.12 \text{ g/mol}$

  $500.0 \text{ mg} \times \dfrac{1 \text{ g}}{1000 \text{ mg}} \times \dfrac{1 \text{ mol}}{176.12 \text{ g}} = 2.839 \times 10^{-3} \text{ mol}$

  $2.839 \times 10^{-3} \text{ mol} \times \dfrac{6.022 \times 10^{23} \text{ molecules}}{\text{mol}} = 1.710 \times 10^{21} \text{ molecules}$

47. a. $2.49 \times 10^{20} \text{ molecules CO} \times \dfrac{1 \text{ mol CO}}{6.022 \times 10^{23} \text{ molecules CO}} = 4.13 \times 10^{-4} \text{ mol CO}$

  b. $15.0 \text{ g CuSO}_4 \times \dfrac{1 \text{ mol CuSO}_4}{159.62 \text{ g CuSO}_4} = 9.40 \times 10^{-2} \text{ mol CuSO}_4$

  c. $100 \text{ molecules H}_2\text{SO}_4 \times \dfrac{1 \text{ mol H}_2\text{SO}_4}{6.022 \times 10^{23} \text{ molecules H}_2\text{SO}_4} = 1.661 \times 10^{-22} \text{ mol H}_2\text{SO}_4$

  d. $6.210 \text{ mg K}_2\text{O} \times \dfrac{1 \text{ g}}{1000 \text{ mg}} \times \dfrac{1 \text{ mol K}_2\text{O}}{94.20 \text{ g K}_2\text{O}} = 6.592 \times 10^{-5} \text{ mol K}_2\text{O}$

49.   a. $1.27 \text{ mmol } CO_2 \times \dfrac{1 \text{ mol}}{1000 \text{ mmol}} \times \dfrac{44.01 \text{ g } CO_2}{\text{mol } CO_2} = 5.59 \times 10^{-2} \text{ g } CO_2$

      b. $2.00 \times 10^{22} \text{ molecules } NCl_3 \times \dfrac{1 \text{ mol } NCl_3}{6.022 \times 10^{23} \text{ molecules}} \times \dfrac{120.36 \text{ g}}{\text{mol } NCl_3} = 4.00 \text{ g } NCl_3$

      c. $0.00451 \text{ mol } (NH_4)_2CO_3 \times \dfrac{96.09 \text{ g } (NH_4)_2CO_3}{\text{mol } (NH_4)_2CO_3} = 0.433 \text{ g } (NH_4)_2CO_3$

      d. $1 \text{ molecule } N_2 \times \dfrac{1 \text{ mol } N_2}{6.022 \times 10^{23} \text{ molecules } N_2} \times \dfrac{28.02 \text{ g } N_2}{\text{mol } N_2} = 4.653 \times 10^{-23} \text{ g } N_2$

      e. $62.7 \text{ mol } CuSO_4 \times \dfrac{159.62 \text{ g } CuSO_4}{\text{mol } CuSO_4} = 1.00 \times 10^{4} \text{ g } CuSO_4$

51.   a. $14 \text{ mol } C \left(\dfrac{12.01 \text{ g}}{\text{mol } C} \right) + 18 \text{ mol } H \left(\dfrac{1.008 \text{ g}}{\text{mol } H} \right) + 2 \text{ mol } N \left(\dfrac{14.01 \text{ g}}{\text{mol } N} \right) + 5 \text{ mol } O \left(\dfrac{16.00 \text{ g}}{\text{mol } O} \right)$

$$= 294.30 \text{ g/mol}$$

      b. $10.0 \text{ g aspartame} \times \dfrac{1 \text{ mol}}{294.30 \text{ g}} = 3.40 \times 10^{-2} \text{ mol}$

      c. $1.56 \text{ mol} \times \dfrac{294.30 \text{ g}}{\text{mol}} = 459 \text{ g}$

      d. $5.0 \text{ mg} \times \dfrac{1 \text{ g}}{1000 \text{ mg}} \times \dfrac{1 \text{ mol}}{294.30 \text{ g}} \times \dfrac{6.02 \times 10^{23} \text{ molecules}}{\text{mol}} = 1.0 \times 10^{19} \text{ molecules}$

      e.   The chemical formula tells us that 1 molecule of aspartame contains two atoms of N. The chemical formula also says that 1 mol of aspartame contains two mol of N.

         $1.2 \text{ g aspartame} \times \dfrac{1 \text{ mol aspartame}}{294.30 \text{ g aspartame}} \times \dfrac{2 \text{ mol } N}{\text{mol aspartame}} \times \dfrac{6.02 \times 10^{23} \text{ atoms } N}{\text{mol } N}$

$$= 4.9 \times 10^{21} \text{ atoms of nitrogen}$$

      f. $1.0 \times 10^{9} \text{ molecules} \times \dfrac{1 \text{ mol}}{6.02 \times 10^{23} \text{ molecules}} \times \dfrac{294.30 \text{ g}}{\text{mol}} = 4.9 \times 10^{-13} \text{ g or } 490 \text{ fg}$

      g. $1 \text{ molecule aspartame} \times \dfrac{1 \text{ mol}}{6.022 \times 10^{23} \text{ molecules}} \times \dfrac{294.30 \text{ g}}{\text{mol}} = 4.887 \times 10^{-22} \text{ g}$

Percent Composition

53.   In 1 mole of $YBa_2Cu_3O_7$, there are 1 mole of Y, 2 moles of Ba, 3 moles of Cu and 7 moles of O.

$$\text{Molar mass} = 1 \text{ mol Y} \left(\frac{88.91 \text{ g Y}}{\text{mol Y}} \right) + 2 \text{ mol Ba} \left(\frac{137.3 \text{ g Ba}}{\text{mol Ba}} \right)$$

$$+ 3 \text{ mol Cu} \left(\frac{63.55 \text{ g Cu}}{\text{mol Cu}} \right) + 7 \text{ mol O} \left(\frac{16.00 \text{ g O}}{\text{mol O}} \right)$$

Molar mass = 88.91 + 274.6 + 190.65 + 112.00 = 666.2 g/mol

$$\%Y = \frac{88.91 \text{ g}}{666.2 \text{ g}} \times 100 = 13.35\% \text{ Y}; \quad \%Ba = \frac{274.6 \text{ g}}{666.2 \text{ g}} \times 100 = 41.22\% \text{ Ba}$$

$$\%Cu = \frac{190.65 \text{ g}}{666.2 \text{ g}} \times 100 = 28.62\% \text{ Cu}; \quad \%O = \frac{112.00 \text{ g}}{666.2 \text{ g}} \times 100 = 16.81\% \text{ O}$$

55.   a. NO: $\%N = \dfrac{14.01 \text{ g N}}{30.01 \text{ g NO}} \times 100 = 46.68\% \text{ N}$

b. NO_2: $\%N = \dfrac{14.01 \text{ g N}}{46.01 \text{ g NO}_2} \times 100 = 30.45\% \text{ N}$

c. N_2O_4: $\%N = \dfrac{28.02 \text{ g N}}{92.02 \text{ g N}_2O_4} \times 100 = 30.45\% \text{ N}$

d. N_2O: $\%N = \dfrac{28.02 \text{ g N}}{44.02 \text{ g N}_2O} \times 100 = 63.65\% \text{ N}$

The order from lowest to highest mass percentage of nitrogen is: $NO_2 = N_2O_4 < NO < N_2O$.

57.   There are many valid methods to solve this problem. We will assume 100.00 g of compound; then determine from the information in the problem how many mol of compound equals 100.00 g of compound. From this information, we can determine the mass of one mol of compound (the molar mass) by setting up a ratio. Assuming 100.00 g cyanocobalamin:

$$\text{mol cyanocobalamin} = 4.34 \text{ g Co} \times \frac{1 \text{ mol Co}}{58.93 \text{ g Co}} \times \frac{1 \text{ mol cyanocobalamin}}{\text{mol Co}}$$

$$= 7.36 \times 10^{-2} \text{ mol cyanocobalamin}$$

$$\frac{x \text{ g cyanocobalamin}}{1 \text{ mol cyanocobalamin}} = \frac{100.00 \text{ g}}{7.36 \times 10^{-2} \text{ mol}}, \quad x = \text{molar mass} = 1360 \text{ g/mol}$$

Empirical and Molecular Formulas

59.   a.   Molar mass of $CH_2O = 1 \text{ mol C} \left(\dfrac{12.01 \text{ g}}{\text{mol C}} \right) + 2 \text{ mol H} \left(\dfrac{1.008 \text{ g H}}{\text{mol H}} \right)$

$$+ 1 \text{ mol O} \left(\dfrac{16.00 \text{ g}}{\text{mol O}} \right) = 30.03 \text{ g/mol}$$

$\%C = \dfrac{12.01 \text{ g C}}{30.03 \text{ g } CH_2O} \times 100 = 39.99\% \text{ C}; \quad \%H = \dfrac{2.016 \text{ g H}}{30.03 \text{ g } CH_2O} \times 100 = 6.713\% \text{ H}$

$\%O = \dfrac{16.00 \text{ g O}}{30.03 \text{ g } CH_2O} \times 100 = 53.28\% \text{ O} \quad$ or $\%O = 100.00 - (39.99 + 6.713) = 53.30\%$

b.   Molar Mass of $C_6H_{12}O_6 = 6(12.01) + 12(1.008) + 6(16.00) = 180.16 \text{ g/mol}$

$\%C = \dfrac{72.06 \text{ g C}}{180.16 \text{ g } C_6H_{12}O_6} \times 100 = 40.00\%; \quad \%H = \dfrac{12(1.008) \text{ g}}{180.16 \text{ g}} \times 100 = 6.714\%$

$\%O = 100.00 - (40.00 + 6.714) = 53.29\%$

c.   Molar mass of $HC_2H_3O_2 = 2(12.01) + 4(1.008) + 2(16.00) = 60.05 \text{ g/mol}$

$\%C = \dfrac{24.02 \text{ g}}{60.05 \text{ g}} \times 100 = 40.00\%; \quad \%H = \dfrac{4.032 \text{ g}}{60.05 \text{ g}} \times 100 = 6.714\%$

$\%O = 100.00 - (40.00 + 6.714) = 53.29\%$

61.   a.   The molecular formula is N_2O_4. The smallest whole number ratio of the atoms (the empirical formula) is NO_2.

b.   Molecular formula: C_3H_6; empirical formula $= CH_2$

c.   Molecular formula: P_4O_{10}; empirical formula $= P_2O_5$

d.   Molecular formula: $C_6H_{12}O_6$; empirical formula $= CH_2O$

63.   Out of 100.0 g of the pigment, there are:

$59.9 \text{ g Ti} \times \dfrac{1 \text{ mol Ti}}{47.88 \text{ g Ti}} = 1.25 \text{ mol Ti}; \quad 40.1 \text{ g O} \times \dfrac{1 \text{ mol O}}{16.00 \text{ g O}} = 2.51 \text{ mol O}$

Empirical formula $= TiO_2$ since mol O to mol Ti are in a 2:1 mol ratio $(2.51/1.25 = 2.01)$.

65.   Compound I: mass O $= 0.6498 \text{ g } Hg_xO_y - 0.6018 \text{ g Hg} = 0.0480 \text{ g O}$

$0.6018 \text{ g Hg} \times \dfrac{1 \text{ mol Hg}}{200.6 \text{ g Hg}} = 3.000 \times 10^{-3} \text{ mol Hg}$

$$0.0480 \text{ g O} \times \frac{1 \text{ mol O}}{16.00 \text{ g O}} = 3.00 \times 10^{-3} \text{ mol O}$$

The mol ratio between Hg and O is 1:1, so the empirical formula of compound I is HgO.

Compound II: mass Hg = 0.4172 g Hg_xO_y - 0.016 g O = 0.401 g Hg

$$0.401 \text{ g Hg} \times \frac{1 \text{ mol Hg}}{200.6 \text{ g Hg}} = 2.00 \times 10^{-3} \text{ mol Hg}; \quad 0.016 \text{ g O} \times \frac{1 \text{ mol O}}{16.00 \text{ g O}} = 1.0 \times 10^{-3} \text{ mol O}$$

The mol ratio between Hg and O is 2:1, so the empirical formula is Hg_2O.

67.   Out of 100.0 g compound: $30.4 \text{ g N} \times \dfrac{1 \text{ mol N}}{14.01 \text{ g N}} = 2.17 \text{ mol N}$

%O = 100.0 - 30.4 = 69.6% O; $69.6 \text{ g O} \times \dfrac{1 \text{ mol O}}{16.00 \text{ g O}} = 4.35 \text{ mol O}$

$\dfrac{2.17}{2.17} = 1.00; \quad \dfrac{4.35}{2.17} = 2.00;$ Empirical formula is NO_2.

The empirical formula mass of $NO_2 \approx 14 + 2(16) = 46$ g/mol.

$\dfrac{92 \text{ g}}{46 \text{ g}} = 2.0;$ Therefore, the molecular formula is N_2O_4.

69.   Assuming 100.00 g of compound (mass hydrogen = 100.00 g - 49.31 g C - 43.79 g O = 6.90 g H):

$$49.31 \text{ g C} \times \frac{1 \text{ mol C}}{12.01 \text{ g C}} = 4.106 \text{ mol C}; \quad 6.90 \text{ g H} \times \frac{1 \text{ mol H}}{1.008 \text{ g H}} = 6.85 \text{ mol H}$$

$$43.79 \text{ g O} \times \frac{1 \text{ mol O}}{16.00 \text{ g O}} = 2.737 \text{ mol O}$$

Dividing all mole values by 2.737 gives:

$$\frac{4.106}{2.737} = 1.500; \quad \frac{6.85}{2.737} = 2.50; \quad \frac{2.737}{2.737} = 1.000$$

Since a whole number ratio is required, the empirical formula is $C_3H_5O_2$.

The empirical formula mass is: 3(12.01) + 5(1.008) + 2(16.00) = 73.07 g/mol

$$\frac{\text{molar mass}}{\text{empirical formula mass}} = \frac{146.1}{73.07} = 1.999; \quad \text{molecular formula} = (C_3H_5O_2)_2 = C_6H_{10}O_4$$

71.   When combustion data are given, it is assumed that all the carbon in the compound ends up as carbon in CO_2 and all the hydrogen in the compound ends up as hydrogen in H_2O. In the sample of propane combusted, the moles of C and H are:

$$\text{mol C} = 2.641 \text{ g CO}_2 \times \frac{1 \text{ mol CO}_2}{44.01 \text{ g CO}_2} \times \frac{1 \text{ mol C}}{\text{mol CO}_2} = 0.06001 \text{ mol C}$$

$$\text{mol H} = 1.442 \text{ g H}_2\text{O} \times \frac{1 \text{ mol H}_2\text{O}}{18.02 \text{ g H}_2\text{O}} \times \frac{2 \text{ mol H}}{\text{mol H}_2\text{O}} = 0.1600 \text{ mol H}$$

$$\frac{\text{mol H}}{\text{mol C}} = \frac{0.1600}{0.06001} = 2.666$$

Multiplying this ratio by three gives the empirical formula of C_3H_8.

73.   The combustion data allow determination of the amount of hydrogen in cumene. One way to determine the amount of carbon in cumene is to determine the mass percent of hydrogen in the compound from the data in the problem; then determine the mass percent of carbon by difference (100.0 - mass %H = mass %C).

$$42.8 \text{ mg H}_2\text{O} \times \frac{1 \text{ g}}{1000 \text{ mg}} \times \frac{2.016 \text{ g H}}{18.02 \text{ g H}_2\text{O}} \times \frac{1000 \text{ mg}}{\text{g}} = 4.79 \text{ mg H}$$

$$\%\text{H} = \frac{4.79 \text{ mg H}}{47.6 \text{ mg cumene}} \times 100 = 10.1\% \text{ H}; \ \%\text{C} = 100.0 - 10.1 = 89.9\% \text{ C}$$

Now solve this empirical formula problem. Out of 100.0 g cumene, we have:

$$89.9 \text{ g C} \times \frac{1 \text{ mol C}}{12.01 \text{ g C}} = 7.49 \text{ mol C}; \ 10.1 \text{ g H} \times \frac{1 \text{ mol H}}{1.008 \text{ g H}} = 10.0 \text{ mol H}$$

$\frac{10.0}{7.49} = 1.34 \approx \frac{4}{3}$, i.e., mol H to mol C are in a 4:3 ratio. Empirical formula = C_3H_4

Empirical formula mass $\approx 3(12) + 4(1) = 40$ g/mol

The molecular formula is $(C_3H_4)_3$ or C_9H_{12} since the molar mass will be between 115 and 125 g/mol (molar mass $\approx 3 \times 40$ g/mol = 120 g/mol).

Balancing Chemical Equations

75.   When balancing reactions, start with elements that appear in only one of the reactants and one of the products, then go on to balance the remaining elements.

a. $Fe + O_2 \rightarrow Fe_2O_3$. Balancing Fe first, then O, gives: $2 Fe + 3/2 O_2 \rightarrow Fe_2O_3$. The best balanced equation contains the smallest whole numbers. To convert to whole numbers, multiply each coefficient by two, which gives: $4 Fe(s) + 3 O_2(g) \rightarrow 2 Fe_2O_3(s)$

b. $Ca + H_2O \rightarrow Ca(OH)_2 + H_2$; Calcium is already balanced, so concentrate on oxygen next. Balancing O gives: $Ca(s) + 2 H_2O(l) \rightarrow Ca(OH)_2(aq) + H_2(g)$. The equation is balanced.

Note: Hydrogen is the most difficult element to balance since it appears in both products. It is generally easiest to save these atoms for last when balancing an equation.

c. $Ba(OH)_2 + H_2SO_4 \rightarrow BaSO_4 + H_2O$; Ba and S are already balanced. There are 6 O atoms on the reactant side and, in order to get 6 O atoms on the product side, we will need 2 H_2O molecules. The balanced equation is: $Ba(OH)_2(aq) + H_2SO_4(aq) \rightarrow BaSO_4(s) + 2\ H_2O(l)$.

77.   a. $Cu(s) + 2\ AgNO_3(aq) \rightarrow 2\ Ag(s) + Cu(NO_3)_2(aq)$

b. $Zn(s) + 2\ HCl(aq) \rightarrow ZnCl_2(aq) + H_2(g)$

c. $Au_2S_3(s) + 3\ H_2(g) \rightarrow 2\ Au(s) + 3\ H_2S(g)$

79.   a.   The formulas of the reactants and products are $C_6H_6(l) + O_2(g) \rightarrow CO_2(g) + H_2O(g)$. To balance this combustion reaction, notice that all of the carbon in C_6H_6 has to end up as carbon in CO_2 and all of the hydrogen in C_6H_6 has to end up as hydrogen in H_2O. To balance C and H, we need 6 CO_2 molecules and 3 H_2O molecules for every 1 molecule of C_6H_6. We do oxygen last. Since we have 15 oxygen atoms in 6 CO_2 molecules and 3 H_2O molecules, we need 15/2 O_2 molecules in order to have 15 oxygen atoms on the reactant side.

$C_6H_6(l) + \dfrac{15}{2}\ O_2(g) \rightarrow 6\ CO_2(g) + 3\ H_2O(g)$; Multiply by two to give whole numbers.

$2\ C_6H_6(l) + 15\ O_2(g) \rightarrow 12\ CO_2(g) + 6\ H_2O(g)$

b.   The formulas of the reactants and products are $C_4H_{10}(g) + O_2(g) \rightarrow CO_2(g) + H_2O(g)$.

$C_4H_{10}(g) + \dfrac{13}{2}\ O_2(g) \rightarrow 4\ CO_2(g) + 5\ H_2O(g)$; Multiply by two to give whole numbers.

$2\ C_4H_{10}(g) + 13\ O_2(g) \rightarrow 8\ CO_2(g) + 10\ H_2O(g)$

c. $C_{12}H_{22}O_{11}(s) + 12\ O_2(g) \rightarrow 12\ CO_2(g) + 11\ H_2O(g)$

d. $2\ Fe(s) + \dfrac{3}{2}\ O_2(g) \rightarrow Fe_2O_3(s)$;   For whole numbers: $4\ Fe(s) + 3\ O_2(g) \rightarrow 2\ Fe_2O_3(s)$

e. $2\ FeO(s) + \dfrac{1}{2}\ O_2(g) \rightarrow Fe_2O_3(s)$;   For whole numbers, multiply by two.

$4\ FeO(s) + O_2(g) \rightarrow 2\ Fe_2O_3(s)$

81.   a. $SiO_2(s) + C(s) \rightarrow Si(s) + CO(g)$

Balance oxygen atoms: $SiO_2 + C \rightarrow Si + 2\ CO$

Balance carbon atoms: $SiO_2(s) + 2\ C(s) \rightarrow Si(s) + 2\ CO(g)$

b. $SiCl_4(l) + Mg(s) \rightarrow Si(s) + MgCl_2(s)$

Balance Cl atoms: $SiCl_4 + Mg \rightarrow Si + 2\ MgCl_2$

Balance Mg atoms: $SiCl_4(l) + 2\ Mg(s) \rightarrow Si(s) + 2\ MgCl_2(s)$

c. $Na_2SiF_6(s) + Na(s) \rightarrow Si(s) + NaF(s)$

Balance F atoms: $Na_2SiF_6 + Na \rightarrow Si + 6\ NaF$

Balance Na atoms: $Na_2SiF_6(s) + 4\ Na(s) \rightarrow Si(s) + 6\ NaF(s)$

83. $C_{12}H_{22}O_{11}(aq) + H_2O(l) \rightarrow 4\ C_2H_5OH(aq) + 4\ CO_2(g)$

Reaction Stoichiometry

85.   The stepwise method to solve stoichiometry problems is outlined in the text. Instead of calculating intermediate answers for each step, we will combine conversion factors into one calculation. This practice reduces round-off error and saves time.

The balanced reaction is: $(NH_4)_2Cr_2O_7(s) \rightarrow Cr_2O_3(s) + N_2(g) + 4\ H_2O(g)$

$$10.8\text{ g }(NH_4)_2Cr_2O_7 \times \frac{1\text{ mol }(NH_4)_2Cr_2O_7}{252.08\text{ g}} = 4.28 \times 10^{-2}\text{ mol }(NH_4)_2Cr_2O_7$$

$$4.28 \times 10^{-2}\text{ mol }(NH_4)_2Cr_2O_7 \times \frac{1\text{ mol }Cr_2O_3}{\text{mol }(NH_4)_2Cr_2O_7} \times \frac{152.00\text{ g }Cr_2O_3}{\text{mol }Cr_2O_3} = 6.51\text{ g }Cr_2O_3$$

$$4.28 \times 10^{-2}\text{ mol }(NH_4)_2Cr_2O_7 \times \frac{1\text{ mol }N_2}{\text{mol }(NH_4)_2Cr_2O_7} \times \frac{28.02\text{ g }N_2}{\text{mol }N_2} = 1.20\text{ g }N_2$$

$$4.28 \times 10^{-2}\text{ mol }(NH_4)_2Cr_2O_7 \times \frac{4\text{ mol }H_2O}{\text{mol }(NH_4)_2Cr_2O_7} \times \frac{18.02\text{ g }H_2O}{\text{mol }H_2O} = 3.09\text{ g }H_2O$$

87. $$1.000\text{ kg Al} \times \frac{1000\text{ g Al}}{\text{kg Al}} \times \frac{1\text{ mol Al}}{26.98\text{ g Al}} \times \frac{3\text{ mol }NH_4ClO_4}{3\text{ mol Al}} \times \frac{117.49\text{ g }NH_4ClO_4}{\text{mol }NH_4ClO_4} = 4355\text{ g}$$

89. $$1.0\text{ ton CuO} \times \frac{907\text{ kg}}{\text{ton}} \times \frac{1000\text{ g}}{\text{kg}} \times \frac{1\text{ mol CuO}}{79.55\text{ g CuO}} \times \frac{1\text{ mol C}}{2\text{ mol CuO}} \times \frac{12.01\text{ g C}}{\text{mol C}} \times \frac{100.\text{ g coke}}{95\text{ g C}}$$

$$= 7.2 \times 10^4\text{ g coke}$$

91.   a. Molar mass = $195.1 + 2(14.01) + 6(1.008) + 2(35.45) = 300.1$ g/mol

% Pt = $\frac{195.1\text{ g}}{300.1\text{ g}} \times 100 = 65.01\%$ Pt; % N = $\frac{28.02\text{ g}}{300.1\text{ g}} \times 100 = 9.337\%$ N

$$\% \text{ H} = \frac{6.048 \text{ g}}{300.1 \text{ g}} \times 100 = 2.015\% \text{ H}; \quad \% \text{ Cl} = \frac{70.90 \text{ g}}{300.1 \text{ g}} \times 100 = 23.63\% \text{ Cl}$$

65.01% Pt; 9.337% N; 2.015% H; 23.63% Cl

What mass can be made of 100 g

b. $100. \text{ g K}_2\text{PtCl}_4 \times \dfrac{1 \text{ mol K}_2\text{PtCl}_4}{415.1 \text{ g K}_2\text{PtCl}_4} \times \dfrac{1 \text{ mol Pt(NH}_3)_2\text{Cl}_2}{\text{mol K}_2\text{PtCl}_4} \times \dfrac{300.1 \text{ g Pt(NH}_3)_2\text{Cl}_2}{\text{mol Pt(NH}_3)_2\text{Cl}_2}$

$$= 72.3 \text{ g Pt(NH}_3)_2\text{Cl}_2$$

$100. \text{ g K}_2\text{PtCl}_4 \times \dfrac{1 \text{ mol K}_2\text{PtCl}_4}{415.1 \text{ g K}_2\text{PtCl}_4} \times \dfrac{2 \text{ mol KCl}}{\text{mol K}_2\text{PtCl}_4} \times \dfrac{74.55 \text{ g KCl}}{\text{mol KCl}} = 35.9 \text{ g KCl}$

Limiting Reactants and Percent Yield

93. a. $\text{Mg(s)} + \text{I}_2\text{(s)} \rightarrow \text{MgI}_2\text{(s)}$

From the balanced equation, 100 molecules of I_2 reacts completely with 100 atoms of Mg. We have a stoichiometric mixture. Neither is limiting.

b. $150 \text{ atoms Mg} \times \dfrac{1 \text{ molecule I}_2}{1 \text{ atom Mg}} = 150 \text{ molecules I}_2 \text{ needed}$

We need 150 molecules I_2 to react completely with 150 atoms Mg; we only have 100 molecules I_2. Therefore, I_2 is limiting.

c. $200 \text{ atoms Mg} \times \dfrac{1 \text{ molecule I}_2}{1 \text{ atom Mg}} = 200 \text{ molecules I}_2\text{;}$ Mg is limiting since 300 molecules I_2 are present.

d. $0.16 \text{ mol Mg} \times \dfrac{1 \text{ mol I}_2}{1 \text{ mol Mg}} = 0.16 \text{ mol I}_2\text{;}$ Mg is limiting since 0.25 mol I_2 are present.

e. $0.14 \text{ mol Mg} \times \dfrac{1 \text{ mol I}_2}{1 \text{ mol Mg}} = 0.14 \text{ mol I}_2 \text{ needed;}$ Stoichiometric mixture. Neither is limiting.

f. $0.12 \text{ mol Mg} \times \dfrac{1 \text{ mol I}_2}{1 \text{ mol Mg}} = 0.12 \text{ mol I}_2 \text{ needed;}$ I_2 is limiting since only 0.08 mol I_2 are present.

g. $6.078 \text{ g Mg} \times \dfrac{1 \text{ mol Mg}}{24.31 \text{ g Mg}} \times \dfrac{1 \text{ mol I}_2}{1 \text{ mol Mg}} \times \dfrac{253.8 \text{ g I}_2}{\text{mol I}_2} = 63.46 \text{ g I}_2$

Stoichiometric mixture. Neither is limiting.

h. $1.00 \text{ g Mg} \times \dfrac{1 \text{ mol Mg}}{24.31 \text{ g Mg}} \times \dfrac{1 \text{ mol I}_2}{1 \text{ mol Mg}} \times \dfrac{253.8 \text{ g I}_2}{\text{mol I}_2} = 10.4 \text{ g I}_2$

10.4 g I_2 needed, but we only have 2.00 g. I_2 is limiting.

i.   From h above, we calculated that 10.4 g I_2 will react completely with 1.00 g Mg. We have 20.00 g I_2. I_2 is in excess. Mg is limiting.

95. a. $10.0 \text{ g Hg} \times \dfrac{1 \text{ mol Hg}}{200.6 \text{ g Hg}} = 4.99 \times 10^{-2} \text{ mol Hg}$

$9.00 \text{ g Br}_2 \times \dfrac{1 \text{ mol Br}_2}{159.80 \text{ g Br}_2} = 5.63 \times 10^{-2} \text{ mol Br}_2$

The required mol ratio from the balanced equation is 1 mol Br_2 to 1 mol Hg. The actual mol ratio is:

$\dfrac{5.63 \times 10^{-2} \text{ mol Br}}{4.99 \times 10^{-2} \text{ mol Hg}} = 1.13$

This is higher than the required ratio, so Hg is the limiting reagent.

$4.99 \times 10^{-2} \text{ mol Hg} \times \dfrac{1 \text{ mol HgBr}_2}{\text{mol Hg}} \times \dfrac{360.4 \text{ g HgBr}_2}{\text{mol HgBr}_2} = 18.0 \text{ g HgBr}_2 \text{ produced}$

$4.99 \times 10^{-2} \text{ mol Hg} \times \dfrac{1 \text{ mol Br}_2}{1 \text{ mol Hg}} \times \dfrac{159.80 \text{ g Br}_2}{\text{mol Br}_2} = 7.97 \text{ g Br}_2 \text{ reacted}$

excess Br_2 = 9.00 g Br_2 - 7.97 g Br_2 = 1.03 g Br_2

b. $5.00 \text{ mL Hg} \times \dfrac{13.6 \text{ g Hg}}{\text{mL Hg}} \times \dfrac{1 \text{ mol Hg}}{200.6 \text{ g Hg}} = 0.339 \text{ mol Hg}$

$5.00 \text{ mL Br}_2 \times \dfrac{3.10 \text{ g Br}_2}{\text{mL Br}_2} \times \dfrac{1 \text{ mol Br}_2}{159.80 \text{ g Br}_2} = 0.0970 \text{ mol Br}_2$

Br_2 is limiting since the actual moles of Br_2 present is well below the required 1:1 mol ratio.

$0.0970 \text{ mol Br}_2 \times \dfrac{1 \text{ mol HgBr}_2}{\text{mol Br}_2} \times \dfrac{360.4 \text{ g HgBr}_2}{\text{mol HgBr}_2} = 35.0 \text{ g HgBr}_2 \text{ produced}$

97. $Ca_3(PO_4)_2 + 3 \text{ H}_2SO_4 \rightarrow 3 \text{ CaSO}_4 + 2 \text{ H}_3PO_4$

$1.0 \times 10^3 \text{ g Ca}_3(PO_4)_2 \times \dfrac{1 \text{ mol Ca}_3(PO_4)_2}{310.18 \text{ g Ca}_3(PO_4)_2} = 3.2 \text{ mol Ca}_3(PO_4)_2$

$1.0 \times 10^3 \text{ g conc. H}_2SO_4 \times \dfrac{98 \text{ g H}_2SO_4}{100 \text{ g conc. H}_2SO_4} \times \dfrac{1 \text{ mol H}_2SO_4}{98.09 \text{ g H}_2SO_4} = 10. \text{ mol H}_2SO_4$

The required mol ratio from the balanced equation is 3 mol H_2SO_4 to 1 mol $Ca_3(PO_4)_2$. The actual ratio is: $\dfrac{10.\ mol\ H_2SO_4}{3.2\ mol\ Ca_3(PO_4)_2} = 3.1$

This is higher than the required mol ratio, so $Ca_3(PO_4)_2$ is the limiting reagent.

$$3.2\ mol\ Ca_3(PO_4)_2 \times \frac{3\ mol\ CaSO_4}{mol\ Ca_3(PO_4)_2} \times \frac{136.15\ g\ CaSO_4}{mol\ CaSO_4} = 1300\ g\ CaSO_4\ produced$$

$$3.2\ mol\ Ca_3(PO_4)_2 \times \frac{2\ mol\ H_3PO_4}{mol\ Ca_3(PO_4)_2} \times \frac{97.99\ g\ H_3PO_4}{mol\ H_3PO_4} = 630\ g\ H_3PO_4\ produced$$

99. $C_2H_6(g) + Cl_2(g) \rightarrow C_2H_5Cl(g) + HCl(g)$

$$300.\ g\ C_2H_6 \times \frac{1\ mol\ C_2H_6}{30.07\ g\ C_2H_6} = 9.98\ mol\ C_2H_6;\quad 650.\ g\ Cl_2 \times \frac{1\ mol\ Cl_2}{70.90\ g\ Cl_2} = 9.17\ mol\ Cl_2$$

The balanced equation requires a 1:1 mol ratio between reactants. 9.17 mol of C_2H_6 will react with all of the Cl_2 present (9.17 mol). Since 9.98 mol C_2H_6 is present, Cl_2 is the limiting reagent.

The theoretical yield of C_2H_5Cl is:

$$9.17\ mol\ Cl_2 \times \frac{1\ mol\ C_2H_5Cl}{mol\ Cl_2} \times \frac{64.51\ g\ C_2H_5Cl}{mol\ C_2H_5Cl} - 592\ g\ C_2H_5Cl$$

$$Percent\ yield = \frac{actual}{theoretical} \times 100 = \frac{490.\ g}{592\ g} \times 100 = 82.8\%$$

101. $2.50\ metric\ tons\ Cu_3FeS_3 \times \dfrac{1000\ kg}{metric\ ton} \times \dfrac{1000\ g}{kg} \times \dfrac{1\ mol\ Cu_3FeS_3}{342.71\ g} \times \dfrac{3\ mol\ Cu}{1\ mol\ Cu_3FeS_3} \times \dfrac{63.55\ g}{mol\ Cu}$

$$= 1.39 \times 10^6\ g\ Cu\ (theoretical)$$

$$1.39 \times 10^6\ g\ Cu\ (theoretical) \times \frac{86.3\ g\ Cu\ (actual)}{100.\ g\ Cu\ (theoretical)} = 1.20 \times 10^6\ g\ Cu = 1.20 \times 10^3\ kg\ Cu$$

$$= 1.20\ metric\ tons\ Cu\ (actual)$$

Additional Exercises

103. $\dfrac{9.123 \times 10^{-23}\ g}{atom} \times \dfrac{6.022 \times 10^{23}\ atom}{mol} = \dfrac{54.94\ g}{mol}$

The atomic mass is 54.94 amu. From the periodic table, the element is manganese (Mn).

105.   Empirical formula mass = 12.01 + 1.008 = 13.02 g/mol; Since 104.14/13.02 = 7.998 ≈ 8, the molecular formula for styrene is $(CH)_8 = C_8H_8$.

$$2.00 \text{ g C}_8\text{H}_8 \times \frac{1 \text{ mol C}_8\text{H}_8}{104.14 \text{ g C}_8\text{H}_8} \times \frac{8 \text{ mol H}}{\text{mol C}_8\text{H}_8} \times \frac{6.022 \times 10^{23} \text{ atoms H}}{\text{mol H}} = 9.25 \times 10^{22} \text{ atoms H}$$

107. $17.3 \text{ g H} \times \dfrac{1 \text{ mol H}}{1.008 \text{ g H}} = 17.2 \text{ mol H};$ $82.7 \text{ g C} \times \dfrac{1 \text{ mol C}}{12.01 \text{ g C}} = 6.89 \text{ mol C}$

$\dfrac{17.2}{6.89} = 2.50;$ The empirical formula is C_2H_5.

The empirical formula mass is ~29 g, so two times the empirical formula would put the compound in the correct range of the molar mass. Molecular formula = $(C_2H_5)_2 = C_4H_{10}$

$$2.59 \times 10^{23} \text{ atoms H} \times \frac{1 \text{ molecule C}_4\text{H}_{10}}{10 \text{ atoms H}} \times \frac{1 \text{ mol C}_4\text{H}_{10}}{6.022 \times 10^{23} \text{ molecules}} = 4.30 \times 10^{-2} \text{ mol C}_4\text{H}_{10}$$

$$4.30 \times 10^{-2} \text{ mol C}_4\text{H}_{10} \times \frac{58.12 \text{ g}}{\text{mol C}_4\text{H}_{10}} = 2.50 \text{ g C}_4\text{H}_{10}$$

109.   Mass of H_2O = 0.755 g $CuSO_4 \cdot xH_2O$ - 0.483 g $CuSO_4$ = 0.272 g H_2O

$$0.483 \text{ g CuSO}_4 \times \frac{1 \text{ mol CuSO}_4}{159.62 \text{ g CuSO}_4} = 0.00303 \text{ mol CuSO}_4$$

$$0.272 \text{ g H}_2\text{O} \times \frac{1 \text{ mol H}_2\text{O}}{18.02 \text{ g H}_2\text{O}} = 0.0151 \text{ mol H}_2\text{O}$$

$$\frac{0.0151 \text{ mol H}_2\text{O}}{0.00303 \text{ mol CuSO}_4} = \frac{4.98 \text{ mol H}_2\text{O}}{1 \text{ mol CuSO}_4};$$ Compound formula = $CuSO_4 \cdot 5H_2O$, $x = 5$

111. $$1.20 \text{ g CO}_2 \times \frac{1 \text{ mol CO}_2}{44.01 \text{ g}} \times \frac{1 \text{ mol C}}{\text{mol CO}_2} \times \frac{1 \text{ mol C}_{24}\text{H}_{30}\text{N}_3\text{O}}{24 \text{ mol C}} \times \frac{376.51 \text{ g}}{\text{mol C}_{24}\text{H}_{30}\text{N}_3\text{O}} = 0.428 \text{ g C}_{24}\text{H}_{30}\text{N}_3\text{O}$$

$$\frac{0.428 \text{ g C}_{24}\text{H}_{30}\text{N}_3\text{O}}{1.00 \text{ g sample}} \times 100 = 42.8\% \text{ C}_{24}\text{H}_{30}\text{N}_3\text{O}$$

113. $$453 \text{ g Fe} \times \frac{1 \text{ mol Fe}}{55.85 \text{ g Fe}} \times \frac{1 \text{ mol Fe}_2\text{O}_3}{2 \text{ mol Fe}} \times \frac{159.70 \text{ g Fe}_2\text{O}_3}{\text{mol Fe}_2\text{O}_3} = 648 \text{ g Fe}_2\text{O}_3$$

mass % Fe_2O_3 = $\dfrac{648 \text{ g Fe}_2\text{O}_3}{752 \text{ g ore}} \times 100 = 86.2\%$

115.   Assuming one mol of vitamin A (286.4 g Vitamin A):

$$\text{mol C} = 286.4 \text{ g Vitamin A} \times \frac{0.8386 \text{ g C}}{\text{g Vitamin A}} \times \frac{1 \text{ mol C}}{12.01 \text{ g C}} = 20.00 \text{ mol C}$$

$$\text{mol H} = 286.4 \text{ g Vitamin A} \times \frac{0.1056 \text{ g H}}{\text{g Vitamin A}} \times \frac{1 \text{ mol H}}{1.008 \text{ g H}} = 30.00 \text{ mol H}$$

Since one mol of Vitamin A contains 20 mol C and 30 mol H, the molecular formula of Vitamin A is $C_{20}H_{30}E$. To determine E, let's calculate the molar mass of E.

$$286.4 \text{ g} = 20(12.01) + 30(1.008) + \text{molar mass E}, \text{molar mass E} = 16.0 \text{ g/mol}$$

From the periodic table, E = oxygen and the molecular formula of Vitamin A is $C_{20}H_{30}O$.

Challenge Problems

117.   First, we will determine composition in mass percent. We assume all the carbon in the 0.213 g CO_2 came from 0.157 g of the compound and that all the hydrogen in the 0.0310 g H_2O came from the 0.157 g of the compound.

$$0.213 \text{ g } CO_2 \times \frac{12.01 \text{ g C}}{44.01 \text{ g } CO_2} = 0.0581 \text{ g C}; \%C = \frac{0.0581 \text{ g C}}{0.157 \text{ g compound}} \times 100 = 37.0\% \text{ C}$$

$$0.0310 \text{ g } H_2O \times \frac{2.016 \text{ g H}}{18.02 \text{ g } H_2O} = 3.47 \times 10^{-3} \text{ g H}; \%H = \frac{3.47 \times 10^{-3} \text{ g}}{0.157 \text{ g}} = 2.21\% \text{ H}$$

We get %N from the second experiment:

$$0.0230 \text{ g } NH_3 \times \frac{14.01 \text{ g N}}{17.03 \text{ g } NH_3} = 1.89 \times 10^{-2} \text{ g N}$$

$$\%N = \frac{1.89 \times 10^{-2} \text{ g}}{0.103 \text{ g}} \times 100 = 18.3\% \text{ N}$$

The mass percent of oxygen is obtained by difference:

$$\%O = 100.00 - (37.0 + 2.21 + 18.3) = 42.5\%$$

So out of 100.00 g of compound, there are:

$$37.0 \text{ g C} \times \frac{1 \text{ mol C}}{12.01 \text{ g C}} = 3.08 \text{ mol C}; 2.21 \text{ g H} \times \frac{1 \text{ mol H}}{1.008 \text{ g H}} = 2.19 \text{ mol H}$$

$$18.3 \text{ g N} \times \frac{1 \text{ mol N}}{14.01 \text{ g N}} = 1.31 \text{ mol N}; 42.5 \text{ g O} \times \frac{1 \text{ mol O}}{16.00 \text{ g O}} = 2.66 \text{ mol O}$$

The last, and often the hardest part, is to find simple whole number ratios. Divide all mole values by the smallest number:

$$\frac{3.08}{1.31} = 2.35;\quad \frac{2.19}{1.31} = 1.67;\quad \frac{1.31}{1.31} = 1.00;\quad \frac{2.66}{1.31} = 2.03$$

Multiplying all these ratios by 3 gives an empirical formula of $C_7H_5N_3O_6$.

119.   Total mass of copper used:

$$10,000 \text{ boards} \times \frac{(8.0 \text{ cm} \times 16.0 \text{ cm} \times 0.060 \text{ cm})}{\text{board}} \times \frac{8.96 \text{ g}}{\text{cm}^3} = 6.9 \times 10^5 \text{ g Cu}$$

Amount of Cu removed = $0.80 \times 6.9 \times 10^5 \text{ g} = 5.5 \times 10^5 \text{ g Cu}$

$$5.5 \times 10^5 \text{ g Cu} \times \frac{1 \text{ mol Cu}}{63.55 \text{ g Cu}} \times \frac{1 \text{ mol Cu(NH}_3)_4\text{Cl}_2}{\text{mol Cu}} \times \frac{202.59 \text{ g Cu(NH}_3)_4\text{Cl}_2}{\text{mol Cu(NH}_3)_4\text{Cl}_2}$$

$$= 1.8 \times 10^6 \text{ g Cu(NH}_3)_4\text{Cl}_2$$

$$5.5 \times 10^5 \text{ g Cu} \times \frac{1 \text{ mol Cu}}{63.55 \text{ g Cu}} \times \frac{4 \text{ mol NH}_3}{\text{mol Cu}} \times \frac{17.03 \text{ g NH}_3}{\text{mol NH}_3} = 5.9 \times 10^5 \text{ g NH}_3$$

121. $10.00 \text{ g XCl}_2 + \text{excess Cl}_2 \rightarrow 12.55 \text{ g XCl}_4$; 2.55 g Cl reacted with XCl_2 to form XCl_4. XCl_4 contains 2.55 g Cl and 10.00 g XCl_2. From mol ratios, 10.00 g XCl_2 must also contain 2.55 g Cl; mass X in $XCl_2 = 10.00 - 2.55 = 7.45 \text{ g X}$.

$$2.55 \text{ g Cl} \times \frac{1 \text{ mol Cl}}{35.45 \text{ g Cl}} \times \frac{1 \text{ mol XCl}_2}{2 \text{ mol Cl}} \times \frac{1 \text{ mol X}}{\text{mol XCl}_2} = 3.60 \times 10^{-2} \text{ mol X}$$

So, $3.60 \times 10^{-2} \text{ mol X}$ must equal 7.45 g X. The molar mass of X is:

$$\frac{7.45 \text{ g X}}{3.60 \times 10^{-2} \text{ mol X}} = \frac{207 \text{ g}}{\text{mol X}}; \text{ Atomic mass} = 207 \text{ amu so X is Pb.}$$

123.   Consider the case of aluminum plus oxygen. Aluminum forms Al^{3+} ions; oxygen forms O^{2-} anions. The simplest compound of the two elements is Al_2O_3. Similarly, we would expect the formula of any group 6A element with Al to be Al_2X_3. Assuming this, out of 100.00 g of compound there are 18.56 g Al and 81.44 g of the unknown element, X. Let's use this information to determine the molar mass of X which will allow us to identify X from the periodic table.

$$18.56 \text{ g Al} \times \frac{1 \text{ mol Al}}{26.98 \text{ g Al}} \times \frac{3 \text{ mol X}}{2 \text{ mol Al}} = 1.032 \text{ mol X}$$

81.44 g of X must contain 1.032 mol of X.

The molar mass of X = $\dfrac{81.44 \text{ g X}}{1.032 \text{ mol X}}$ = 78.91 g/mol X.

From the periodic table, the unknown element is selenium and the formula is Al_2Se_3.

125.   The balanced equations are:

$$4 \text{ NH}_3(g) + 5 \text{ O}_2(g) \rightarrow 4 \text{ NO}(g) + 6 \text{ H}_2\text{O}(g) \text{ and } 4 \text{ NH}_3(g) + 7 \text{ O}_2(g) \rightarrow 4 \text{ NO}_2(g) + 6 \text{ H}_2\text{O}(g)$$

Let 4x = number of mol of NO formed, and let 4y = number of mol of NO_2 formed. Then:

$$4x \text{ NH}_3 + 5x \text{ O}_2 \rightarrow 4x \text{ NO} + 6x \text{ H}_2\text{O and } 4y \text{ NH}_3 + 7y \text{ O}_2 \rightarrow 4y \text{ NO}_2 + 6y \text{ H}_2\text{O}$$

All the NH_3 reacted, so 4x + 4y = 2.00. 10.00 - 6.75 = 3.25 mol O_2 reacted, so 5x + 7y = 3.25.

Solving by the method of simultaneous equations:

$$
\begin{aligned}
20x + 28y &= 13.0 \\
\underline{-20x - 20y} &= \underline{-10.0} \\
8y &= 3.0, \quad y = 0.38; \quad 4x + 4 \times 0.38 = 2.00, \quad x = 0.12
\end{aligned}
$$

mol NO = 4x = 4 × 0.12 = 0.48 mol NO formed

CHAPTER FOUR

TYPES OF CHEMICAL REACTIONS AND SOLUTION STOICHIOMETRY

Questions

9.   "Slightly soluble" refers to substances that dissolve only to a small extent. A slightly soluble salt may still dissociate completely to ions and, hence, be a strong electrolyte. An example of such a substance is $Mg(OH)_2$. It is a strong electrolyte, but not very soluble. A weak electrolyte is a substance that doesn't dissociate completely to produce ions. A weak electrolyte may be very soluble in water, or it may not be very soluble. Acetic acid is an example of a weak electrolyte that is very soluble in water.

Exercises

Aqueous Solutions: Strong and Weak Electrolytes

11.   a. $NaBr(s) \rightarrow Na^+(aq) + Br^-(aq)$ b. $MgCl_2(s) \rightarrow Mg^{2+}(aq) + 2\ Cl^-(aq)$

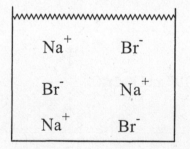   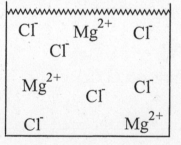

Your drawing should show equal   Your drawing should show twice the number
numbers of Na^+ and Br^- ions.   of Cl^- ions as Mg^{2+} ions.

c. $Al(NO_3)_3(s) \rightarrow Al^{3+}(aq) + 3\ NO_3^-(aq)$ d. $(NH_4)_2SO_4(s) \rightarrow 2\ NH_4^+(aq) + SO_4^{2-}(aq)$

Al^{3+}	NO_3^-	NO_3^-
NO_3^-	NO_3^-	Al^{3+}
NO_3^-	NO_3^-	NO_3^-
NO_3^-	Al^{3+}	NO_3^-

SO_4^{2-}		NH_4^+
NH_4^+	NH_4^+	SO_4^{2-}
SO_4^{2-}		NH_4^+
NH_4^+		NH_4^+

For e-i, your drawings should show equal numbers of the cations and anions present as each salt is a 1:1 salt. The ions present are listed in the following dissolution reactions.

e. $NaOH(s) \rightarrow Na^+(aq) + OH^-(aq)$ f. $FeSO_4(s) \rightarrow Fe^{2+}(aq) + SO_4^{2-}(aq)$

g. $KMnO_4(s) \rightarrow K^+(aq) + MnO_4^-(aq)$ h. $HClO_4(aq) \rightarrow H^+(aq) + ClO_4^-(aq)$

i. $NH_4C_2H_3O_2(s) \rightarrow NH_4^+(aq) + C_2H_3O_2^-(aq)$

13. $CaCl_2(s) \rightarrow Ca^{2+}(aq) + 2\ Cl^-(aq)$

Solution Concentration: Molarity

15. a. $5.623\ g\ NaHCO_3 \times \dfrac{1\ mol\ NaHCO_3}{84.01\ g\ NaHCO_3} = 6.693 \times 10^{-2}\ mol\ NaHCO_3$

$M = \dfrac{6.693 \times 10^{-2}\ mol}{250.0\ mL} \times \dfrac{1000\ mL}{L} = 0.2677\ M\ NaHCO_3$

b. $0.1846\ g\ K_2Cr_2O_7 \times \dfrac{1\ mol\ K_2Cr_2O_7}{294.20\ g\ K_2Cr_2O_7} = 6.275 \times 10^{-4}\ mol\ K_2Cr_2O_7$

$M = \dfrac{6.275 \times 10^{-4}\ mol}{500.0 \times 10^{-3}\ L} = 1.255 \times 10^{-3}\ M\ K_2Cr_2O_7$

c. $0.1025\ g\ Cu \times \dfrac{1\ mol\ Cu}{63.55\ g\ Cu} = 1.613 \times 10^{-3}\ mol\ Cu = 1.613 \times 10^{-3}\ mol\ Cu^{2+}$

$M = \dfrac{1.613 \times 10^{-3}\ mol\ Cu^{2+}}{200.0\ mL} \times \dfrac{1000\ mL}{L} = 8.065 \times 10^{-3}\ M\ Cu^{2+}$

17. a. $CaCl_2(s) \rightarrow Ca^{2+}(aq) + 2\ Cl^-(aq)$; $M_{Ca^{2+}} = 0.15\ M$; $M_{Cl^-} = 2(0.15) = 0.30\ M$

b. $Al(NO_3)_3(s) \rightarrow Al^{3+}(aq) + 3\ NO_3^-(aq)$; $M_{Al^{3+}} = 0.26\ M$; $M_{NO_3^-} = 3(0.26) = 0.78\ M$

c. $K_2Cr_2O_7(s) \rightarrow 2\ K^+(aq) + Cr_2O_7^{2-}(aq)$; $M_{K^+} = 2(0.25) = 0.50\ M$; $M_{Cr_2O_7^{2-}} = 0.25\ M$

d. $Al_2(SO_4)_3(s) \rightarrow 2\ Al^{3+}(aq) + 3\ SO_4^{2-}(aq)$

$$M_{Al^{3+}} = \frac{2.0 \times 10^{-3}\ \text{mol Al}_2(SO_4)_3}{L} \times \frac{2\ \text{mol Al}^{3+}}{\text{mol Al}_2(SO_4)_3} = 4.0 \times 10^{-3}\ M$$

$$M_{SO_4^{2-}} = \frac{2.0 \times 10^{-3}\ \text{mol Al}_2(SO_4)_3}{L} \times \frac{3\ \text{mol SO}_4^{2-}}{\text{mol Al}_2(SO_4)_3} = 6.0 \times 10^{-3}\ M$$

19. mol solute = volume (L) × molarity $\left(\dfrac{\text{mol}}{L}\right)$; $AlCl_3(s) \rightarrow Al^{3+}(aq) + 3\ Cl^-(aq)$

$$\text{mol Cl}^- = 0.1000\ L \times \frac{0.30\ \text{mol AlCl}_3}{L} \times \frac{3\ \text{mol Cl}^-}{\text{mol AlCl}_3} = 9.0 \times 10^{-2}\ \text{mol Cl}^-$$

$MgCl_2(s) \rightarrow Mg^{2+}(aq) + 2\ Cl^-(aq)$

$$\text{mol Cl}^- = 0.0500\ L \times \frac{0.60\ \text{mol MgCl}_2}{L} \times \frac{2\ \text{mol Cl}^-}{\text{mol MgCl}_2} = 6.0 \times 10^{-2}\ \text{mol Cl}^-$$

$NaCl(s) \rightarrow Na^+(aq) + Cl^-(aq)$

$$\text{mol Cl}^- = 0.2000\ L \times \frac{0.40\ \text{mol NaCl}}{L} \times \frac{1\ \text{mol Cl}^-}{\text{mol NaCl}} = 8.0 \times 10^{-2}\ \text{mol Cl}^-$$

100.0 mL of 0.30 M $AlCl_3$ contains the most moles of Cl^- ions.

21. Molar mass of $NaHCO_3$ = 22.99 + 1.008 + 12.01 + 3(16.00) = 84.01 g/mol

$$\text{Volume} = 0.350\ \text{g NaHCO}_3 \times \frac{1\ \text{mol NaHCO}_3}{84.01\ \text{g NaHCO}_3} \times \frac{1\ L}{0.100\ \text{mol NaHCO}_3} = 0.0417\ L = 41.7\ \text{mL}$$

41.7 mL of 0.100 M $NaHCO_3$ contains 0.350 g $NaHCO_3$.

23. a. $2.00\ L \times \dfrac{0.250\ \text{mol NaOH}}{L} \times \dfrac{40.00\ \text{g NaOH}}{\text{mol}} = 20.0\ \text{g NaOH}$

Place 20.0 g NaOH in a 2 L volumetric flask; add water to dissolve the NaOH, and fill to the mark with water, mixing several times along the way.

b. $2.00\ L \times \dfrac{0.250\ \text{mol NaOH}}{L} \times \dfrac{1\ L\ \text{stock}}{1.00\ \text{mol NaOH}} = 0.500\ L$

Add 500. mL of 1.00 M NaOH stock solution to a 2 L volumetric flask; fill to the mark with water, mixing several times along the way.

c. $2.00 \text{ L} \times \dfrac{0.100 \text{ mol K}_2\text{CrO}_4}{\text{L}} \times \dfrac{194.20 \text{ g K}_2\text{CrO}_4}{\text{mol K}_2\text{CrO}_4} = 38.8 \text{ g K}_2\text{CrO}_4$

Similar to the solution made in part a, instead using 38.8 g K_2CrO_4.

d. $2.00 \text{ L} \times \dfrac{0.100 \text{ mol K}_2\text{CrO}_4}{\text{L}} \times \dfrac{1 \text{ L stock}}{1.75 \text{ mol K}_2\text{CrO}_4} = 0.114 \text{ L}$

Similar to the solution made in part b, instead using 114 mL of the 1.75 M K_2CrO_4 stock solution.

25. $10.8 \text{ g (NH}_4)_2\text{SO}_4 \times \dfrac{1 \text{ mol}}{132.15 \text{ g}} = 8.17 \times 10^{-2} \text{ mol (NH}_4)_2\text{SO}_4$

Molarity $= \dfrac{8.17 \times 10^{-2} \text{ mol}}{100.0 \text{ mL}} \times \dfrac{1000 \text{ mL}}{\text{L}} = 0.817 \text{ M (NH}_4)_2\text{SO}_4$

Moles of $(NH_4)_2SO_4$ in final solution

$10.00 \times 10^{-3} \text{ L} \times \dfrac{0.817 \text{ mol}}{\text{L}} = 8.17 \times 10^{-3} \text{ mol}$

Molarity of final solution $= \dfrac{8.17 \times 10^{-3} \text{ mol}}{(10.00 + 50.00) \text{ mL}} \times \dfrac{1000 \text{ mL}}{\text{L}} = 0.136 \text{ M (NH}_4)_2\text{SO}_4$

$(NH_4)_2SO_4(s) \rightarrow 2 \text{ NH}_4^+(aq) + SO_4^{2-}(aq); \quad M_{NH_4^+} = 2(0.136) = 0.272 \text{ M}; \quad M_{SO_4^{2-}} = 0.136 \text{ M}$

27.   Stock solution $= \dfrac{10.0 \text{ mg}}{500.0 \text{ mL}} = \dfrac{10.0 \times 10^{-3} \text{ g}}{500.0 \text{ mL}} = \dfrac{2.00 \times 10^{-5} \text{ g steroid}}{\text{mL}}$

$100.0 \times 10^{-6} \text{ L stock} \times \dfrac{1000 \text{ mL}}{\text{L}} \times \dfrac{2.00 \times 10^{-5} \text{ g steroid}}{\text{mL}} = 2.00 \times 10^{-6} \text{ g steroid}$

This is diluted to a final volume of 100.0 mL.

$\dfrac{2.00 \times 10^{-6} \text{ g steroid}}{100.0 \text{ mL}} \times \dfrac{1000 \text{ mL}}{\text{L}} \times \dfrac{1 \text{ mol steroid}}{336.43 \text{ g steroid}} = 5.94 \times 10^{-8} \text{ M steroid}$

Precipitation Reactions

29.   In these reactions, soluble ionic compounds are mixed together. To predict the precipitate, switch the anions and cations in the two reactant compounds to predict possible products; then use the solubility rules in Table 4.1 to predict if any of these possible products are insoluble (are the precipitate).

a.   Possible products = $BaSO_4$ and NaCl; precipitate = $BaSO_4(s)$

b. Possible products = $PbCl_2$ and KNO_3; precipitate = $PbCl_2(s)$
c. Possible products = Ag_3PO_4 and $NaNO_3$; precipitate = $Ag_3PO_4(s)$
d. Possible products = $NaNO_3$ and $Fe(OH)_3$; precipitate = $Fe(OH)_3(s)$

31.   For the following answers, the balanced molecular equation is first, followed by the complete ionic equation, then the net ionic equation.

a. $BaCl_2(aq) + Na_2SO_4(aq) \rightarrow BaSO_4(s) + 2\ NaCl(aq)$

$Ba^{2+}(aq) + 2\ Cl^-(aq) + 2\ Na^+(aq) + SO_4^{2-}(aq) \rightarrow BaSO_4(s) + 2\ Na^+(aq) + 2\ Cl^-(aq)$

$Ba^{2+}(aq) + SO_4^{2-}(aq) \rightarrow BaSO_4(s)$

b. $Pb(NO_3)_2(aq) + 2\ KCl(aq) \rightarrow PbCl_2(s) + 2\ KNO_3(aq)$

$Pb^{2+}(aq) + 2\ NO_3^-(aq) + 2\ K^+(aq) + 2\ Cl^-(aq) \rightarrow PbCl_2(s) + 2\ K^+(aq) + 2\ NO_3^-(aq)$

$Pb^{2+}(aq) + 2\ Cl^-(aq) \rightarrow PbCl_2(s)$

c. $3\ AgNO_3(aq) + Na_3PO_4(aq) \rightarrow Ag_3PO_4(s) + 3\ NaNO_3(aq)$

$3\ Ag^+(aq) + 3\ NO_3^-(aq) + 3\ Na^+(aq) + PO_4^{3-}(aq) \rightarrow Ag_3PO_4(s) + 3\ Na^+(aq) + 3\ NO_3^-(aq)$

$3\ Ag^+(aq) + PO_4^{3-}(aq) \rightarrow Ag_3PO_4(s)$

d. $3\ NaOH(aq) + Fe(NO_3)_3(aq) \rightarrow Fe(OH)_3(s) + 3\ NaNO_3(aq)$

$3\ Na^+(aq) + 3\ OH^-(aq) + Fe^{3+}(aq) + 3\ NO_3^-(aq) \rightarrow Fe(OH)_3(s) + 3\ Na^+(aq) + 3\ NO_3^-(aq)$

$Fe^{3+}(aq) + 3\ OH^-(aq) \rightarrow Fe(OH)_3(s)$

33.   a. When $CuSO_4(aq)$ is added to $Na_2S(aq)$, the precipitate that forms is $CuS(s)$. Therefore, Na^+ (the grey spheres) and SO_4^{2-} (the blueish-green spheres) are the spectator ions.

$CuSO_4(aq) + Na_2S(aq) \rightarrow CuS(s) + Na_2SO_4(aq);\ Cu^{2+}(aq) + S^{2-}(aq) \rightarrow CuS(s)$

b. When $CoCl_2(aq)$ is added to $NaOH(aq)$, the precipitate that forms is $Co(OH)_2(s)$. Therefore, Na^+ (the grey spheres) and Cl^- (the green spheres) are the spectator ions.

$CoCl_2(aq) + 2\ NaOH(aq) \rightarrow Co(OH)_2(s) + 2\ NaCl(aq);\ Co^{2+}(aq) + 2\ OH^-(aq) \rightarrow Co(OH)_2(s)$

c. When $AgNO_3(aq)$ is added to $KI(aq)$, the precipitate that forms is $AgI(s)$. Therefore, K^+ (the red spheres) and NO_3^- (the blue spheres) are the spectator ions.

$AgNO_3(aq) + KI(aq) \rightarrow AgI(s) + KNO_3(aq);\ Ag^+(aq) + I^-(aq) \rightarrow AgI(s)$

35.   a. $(NH_4)_2SO_4(aq) + Ba(NO_3)_2(aq) \rightarrow 2\ NH_4NO_3(aq) + BaSO_4(s)$

             $Ba^{2+}(aq) + SO_4^{2-}(aq) \rightarrow BaSO_4(s)$

       b. $Pb(NO_3)_2(aq) + 2\ NaCl(aq) \rightarrow PbCl_2(s) + 2\ NaNO_3(aq)$

             $Pb^{2+}(aq) + 2\ Cl^-(aq) \rightarrow PbCl_2(s)$

       c.   Potassium phosphate and sodium nitrate are both soluble in water. No reaction occurs.

       d.   No reaction occurs since all possible products are soluble.

       e. $CuCl_2(aq) + 2\ NaOH(aq) \rightarrow Cu(OH)_2(s) + 2\ NaCl(aq)$

             $Cu^{2+}(aq) + 2\ OH^-(aq) \rightarrow Cu(OH)_2(s)$

37.   Three possibilities are:

       Addition of K_2SO_4 solution to give a white ppt. of $PbSO_4$. Addition of NaCl solution to give a white ppt. of $PbCl_2$. Addition of K_2CrO_4 solution to give a bright yellow ppt. of $PbCrO_4$.

39. $2\ AgNO_3(aq) + Na_2CrO_4(aq) \rightarrow Ag_2CrO_4(s) + 2\ NaNO_3(aq)$

       $$0.0750\ L \times \frac{0.100\ mol\ AgNO_3}{L} \times \frac{1\ mol\ Na_2CrO_4}{2\ mol\ AgNO_3} \times \frac{161.98\ g\ Na_2CrO_4}{mol\ Na_2CrO_4} = 0.607\ g\ Na_2CrO_4$$

41. $Al(NO_3)_3(aq) + 3\ KOH(aq) \rightarrow Al(OH)_3(s) + 3\ KNO_3(aq)$

       $$0.0500\ L \times \frac{0.200\ mol\ Al(NO_3)_3}{L} = 0.0100\ mol\ Al(NO_3)_3$$

       $$0.2000\ L \times \frac{0.100\ mol\ KOH}{L} = 0.0200\ mol\ KOH$$

       From the balanced equation, 3 mol of KOH are required to react with 1 mol of $Al(NO_3)_3$ (3:1 mol ratio). The actual KOH to $Al(NO_3)_3$ mol ratio present is $0.0200/0.0100 = 2$ (2:1). Since the actual mol ratio present is less than the required mol ratio, KOH is the limiting reagent.

       $$0.0200\ mol\ KOH \times \frac{1\ mol\ Al(OH)_3}{3\ mol\ KOH} \times \frac{78.00\ g\ Al(OH)_3}{mol\ Al(OH)_3} = 0.520\ g\ Al(OH)_3$$

43. $2\ AgNO_3(aq) + CaCl_2(aq) \rightarrow 2\ AgCl(s) + Ca(NO_3)_2(aq)$

       $$mol\ AgNO_3 = 0.1000\ L \times \frac{0.20\ mol\ AgNO_3}{L} = 0.020\ mol\ AgNO_3$$

       $$mol\ CaCl_2 = 0.1000\ L \times \frac{0.15\ mol\ CaCl_2}{L} = 0.015\ mol\ CaCl_2$$

The required mol $AgNO_3$ to mol $CaCl_2$ ratio is 2:1 (from the balanced equation). The actual mol ratio present is 0.020/0.015 = 1.3 (1.3:1). Therefore, $AgNO_3$ is the limiting reagent.

$$\text{mass AgCl} = 0.020 \text{ mol AgNO}_3 \times \frac{1 \text{ mol AgCl}}{1 \text{ mol AgNO}_3} \times \frac{143.4 \text{ g AgCl}}{\text{mol AgCl}} = 2.9 \text{ g AgCl}$$

The net ionic equation is: $Ag^+(aq) + Cl^-(aq) \rightarrow AgCl(s)$. The ions remaining in solution are the unreacted Cl^- ions and the spectator ions, NO_3^- and Ca^{2+} (all Ag^+ is used up in forming AgCl). The mol of each ion present initially (before reaction) can be easily determined from the mol of each reactant. 0.020 mol $AgNO_3$ dissolves to form 0.020 mol Ag^+ and 0.020 mol NO_3^-. 0.015 mol $CaCl_2$ dissolves to form 0.015 mol Ca^{2+} and 2(0.015) = 0.030 mol Cl^-.

mol unreacted Cl^- = 0.030 mol Cl^- initially - 0.020 mol Cl^- reacted = 0.010 mol Cl^- unreacted

$$M_{Cl^-} = \frac{0.010 \text{ mol Cl}^-}{\text{total volume}} = \frac{0.010 \text{ mol Cl}^-}{0.1000 \text{ L} + 0.1000 \text{ L}} = 0.050 \; M \; Cl^-$$

The molarity of the spectator ions are:

$$M_{NO_3^-} = \frac{0.020 \text{ mol NO}_3^-}{0.2000 \text{ L}} = 0.10 \; M \; NO_3^-; \quad M_{Ca^{2+}} = \frac{0.015 \text{ mol Ca}^{2+}}{0.2000 \text{ L}} = 0.075 \; M \; Ca^{2+}$$

Acid-Base Reactions

45.   All the bases in this problem are ionic compounds containing OH^-. The acids are either strong or weak electrolytes. The best way to determine if an acid is a strong or weak electrolyte is to memorize all the strong electrolytes (strong acids). Any other acid you encounter that is not a strong acid will be a weak electrolyte (a weak acid) and the formula should be left unaltered in the complete ionic and net ionic equations. The strong acids to recognize are HCl, HBr, HI, HNO_3, $HClO_4$ and H_2SO_4. For the answers below, the order of the equations are molecular, complete ionic and net ionic.

a. $2 \text{ HClO}_4(aq) + \text{Mg(OH}_2)(s) \rightarrow 2 \text{ H}_2\text{O}(l) + \text{Mg(ClO}_4)_2(aq)$

$2 \text{ H}^+(aq) + 2 \text{ ClO}_4^-(aq) + \text{Mg(OH)}_2(s) \rightarrow 2 \text{ H}_2\text{O}(l) + \text{Mg}^{2+}(aq) + 2 \text{ ClO}_4^-(aq)$

$2 \text{ H}^+(aq) + \text{Mg(OH)}_2(s) \rightarrow 2 \text{ H}_2\text{O}(l) + \text{Mg}^{2+}(aq)$

b. $\text{HCN}(aq) + \text{NaOH}(aq) \rightarrow \text{H}_2\text{O}(l) + \text{NaCN}(aq)$

$\text{HCN}(aq) + \text{Na}^+(aq) + \text{OH}^-(aq) \rightarrow \text{H}_2\text{O}(l) + \text{Na}^+(aq) + \text{CN}^-(aq)$

$\text{HCN}(aq) + \text{OH}^-(aq) \rightarrow \text{H}_2\text{O}(l) + \text{CN}^-(aq)$

c. $\text{HCl}(aq) + \text{NaOH}(aq) \rightarrow \text{H}_2\text{O}(l) + \text{NaCl}(aq)$

$\text{H}^+(aq) + \text{Cl}^-(aq) + \text{Na}^+(aq) + \text{OH}^-(aq) \rightarrow \text{H}_2\text{O}(l) + \text{Na}^+(aq) + \text{Cl}^-(aq)$

$\text{H}^+(aq) + \text{OH}^-(aq) \rightarrow \text{H}_2\text{O}(l)$

47.   All the acids in this problem are strong electrolytes. The acids to recognize as strong electrolytes are HCl, HBr, HI, HNO_3, $HClO_4$ and H_2SO_4.

a. $KOH(aq) + HNO_3(aq) \rightarrow H_2O(l) + KNO_3(aq)$

$K^+(aq) + OH^-(aq) + H^+(aq) + NO_3^-(aq) \rightarrow H_2O(l) + K^+(aq) + NO_3^-(aq)$

$OH^-(aq) + H^+(aq) \rightarrow H_2O(l)$

b. $Ba(OH)_2(aq) + 2\ HCl(aq) \rightarrow 2\ H_2O(l) + BaCl_2(aq)$

$Ba^{2+}(aq) + 2\ OH^-(aq) + 2\ H^+(aq) + 2\ Cl^-(aq) \rightarrow 2\ H_2O(l) + Ba^{2+}(aq) + 2\ Cl^-(aq)$

$2\ OH^-(aq) + 2\ H^+(aq) \rightarrow 2\ H_2O(l)$ or $OH^-(aq) + H^+(aq) \rightarrow H_2O(l)$

c. $3\ HClO_4(aq) + Fe(OH)_3(s) \rightarrow 3\ H_2O(l) + Fe(ClO_4)_3(aq)$

$3\ H^+(aq) + 3\ ClO_4^-(aq) + Fe(OH)_3(s) \rightarrow 3\ H_2O(l) + Fe^{3+}(aq) + 3\ ClO_4^-(aq)$

$3\ H^+(aq) + Fe(OH)_3(s) \rightarrow 3\ H_2O(l) + Fe^{3+}(aq)$

49.   If we begin with 50.00 mL of 0.200 M NaOH, then:

$$50.00 \times 10^{-3}\ L \times \frac{0.200\ mol}{L} = 1.00 \times 10^{-2}\ mol\ NaOH\ is\ to\ be\ neutralized.$$

a. $NaOH(aq) + HCl(aq) \rightarrow NaCl(aq) + H_2O(l)$

$$1.00 \times 10^{-2}\ mol\ NaOH \times \frac{1\ mol\ HCl}{mol\ NaOH} \times \frac{1\ L}{0.100\ mol} = 0.100\ L\ or\ 100.\ mL$$

b. $HNO_3(aq) + NaOH(aq) \rightarrow H_2O(l) + NaNO_3(aq)$

$$1.00 \times 10^{-2}\ mol\ NaOH \times \frac{1\ mol\ HNO_3}{mol\ NaOH} \times \frac{1\ L}{0.150\ mol\ HNO_3} = 6.67 \times 10^{-2}\ L\ or\ 66.7\ mL$$

c. $HC_2H_3O_2(aq) + NaOH(aq) \rightarrow H_2O(l) + NaC_2H_3O_2(aq)$

$$1.00 \times 10^{-2}\ mol\ NaOH \times \frac{1\ mol\ HC_2H_3O_2}{mol\ NaOH} \times \frac{1\ L}{0.200\ mol\ HC_2H_3O_2} = 5.00 \times 10^{-2}\ L\ or\ 50.0\ mL$$

51. $HNO_3(aq) + NaOH(aq) \rightarrow NaNO_3(aq) + H_2O(l)$

$$15.0\ g\ NaOH \times \frac{1\ mol\ NaOH}{40.00\ g} = 0.375\ mol\ NaOH$$

$$0.1500\ L \times \frac{0.250\ mol\ HNO_3}{L} = 0.0375\ mol\ HNO_3$$

We have added more moles of NaOH than mol of HNO_3 present. Since NaOH and HNO_3 react in a 1:1 mol ratio, NaOH is in excess and the solution will be basic. The ions present after reaction will be the excess OH^- ions and the spectator ions, Na^+ and NO_3^-. The moles of ions present initially are:

$$\text{mol NaOH} = \text{mol Na}^+ = \text{mol OH}^- = 0.375 \text{ mol}$$

$$\text{mol HNO}_3 = \text{mol H}^+ = \text{mol NO}_3^- = 0.0375 \text{ mol}$$

The net ionic reaction occurring is: $H^+(aq) + OH^-(aq) \rightarrow H_2O(l)$

The mol of excess OH^- remaining after reaction will be the initial mol of OH^- minus the amount of OH^- neutralized by reaction with H^+:

$$\text{mol excess OH}^- = 0.375 \text{ mol} - 0.0375 \text{ mol} = 0.338 \text{ mol OH}^- \text{ excess}$$

The concentration of ions present is:

$$M_{OH^-} = \frac{\text{mol OH}^- \text{ excess}}{\text{volume}} = \frac{0.338 \text{ mol OH}^-}{0.1500 \text{ L}} = 2.25 \; M \text{ OH}^-$$

$$M_{NO_3^-} = \frac{0.0375 \text{ mol NO}_3^-}{0.1500 \text{ L}} = 0.250 \; M \text{ NO}_3^-; \quad M_{Na^+} = \frac{0.375 \text{ mol}}{0.1500 \text{ L}} = 2.50 \; M \text{ Na}^+$$

53. $HCl(aq) + NaOH(aq) \rightarrow H_2O(l) + NaCl(aq)$

$$24.16 \times 10^{-3} \text{ L NaOH} \times \frac{0.106 \text{ mol NaOH}}{\text{L NaOH}} \times \frac{1 \text{ mol HCl}}{\text{mol NaOH}} = 2.56 \times 10^{-3} \text{ mol HCl}$$

$$\text{Molarity of HCl} = \frac{2.56 \times 10^{-3} \text{ mol}}{25.00 \times 10^{-3} \text{ L}} = 0.102 \; M \text{ HCl}$$

55.   Since KHP is a monoprotic acid, the reaction is: $NaOH(aq) + KHP(aq) \rightarrow H_2O(l) + NaKP(aq)$

$$\text{Mass KHP} = 0.02046 \text{ L NaOH} \times \frac{0.1000 \text{ mol NaOH}}{\text{L NaOH}} \times \frac{1 \text{ mol KHP}}{\text{mol NaOH}} \times \frac{204.22 \text{ g KHP}}{\text{mol KHP}} = 0.4178 \text{ g KHP}$$

Oxidation-Reduction Reactions

57.   Apply rules in Table 4.2.

a. $KMnO_4$ is composed of K^+ and MnO_4^- ions. Assign oxygen a value of -2, which gives manganese a +7 oxidation state since the sum of oxidation states for all atoms in MnO_4^- must equal the -1 charge on MnO_4^-. K, +1; O, -2; Mn, +7.

b. Assign O a -2 oxidation state, which gives nickel a +4 oxidation state. Ni, +4; O, -2.

c. $K_4Fe(CN)_6$ is composed of K^+ cations and $Fe(CN)_6^{4-}$ anions. $Fe(CN)_6^{4-}$ is composed of iron and CN^- anions. For an overall anion charge of -4, iron must have a +2 oxidation state.

d. $(NH_4)_2HPO_4$ is made of NH_4^+ cations and HPO_4^{2-} anions. Assign +1 as the oxidation state of H and -2 as the oxidation state of O. In NH_4^+, $x + 4(+1) = +1$, $x = -3 =$ oxidation state of N. In HPO_4^{2-}, $+1 + y + 4(-2) = -2$, $y = +5 =$ oxidation state of P.

e. O, -2; P, +3   f. O, -2; Fe, + 8/3

g. O, -2; F, -1; Xe, +6   h. F, -1; S, +4

i. O, -2; C, +2   j. Na, +1; O, -2; C, +3

59. a. HBr: H, +1; Br, -1

b. HOBr: H, +1; O, -2; For Br, $+1 + 1(-2) + x = 0$, $x = +1$

c. Br_2: Br, 0

d. $HBrO_4$: H, +1; O, -2; For Br, $+1 + 4(-2) + x = 0$, $x = +7$

e. BrF_3: F, -1; For Br, $x + 3(-1) = 0$, $x = +3$

61. To determine if the reaction is an oxidation-reduction reaction, assign oxidation numbers. If the oxidation numbers change for some elements, the reaction is a redox reaction. If the oxidation numbers do not change, the reaction is not a redox reaction. In redox reactions, the species oxidized (called the reducing agent) shows an increase in oxidation numbers and the species reduced (called the oxidizing agent) shows a decrease in oxidation numbers.

	Redox?	Oxidizing Agent	Reducing Agent	Substance Oxidized	Substance Reduced
a.	Yes	O_2	CH_4	CH_4 (C)	O_2 (O)
b.	Yes	HCl	Zn	Zn	HCl (H)
c.	No	-	-	-	-
d.	Yes	O_3	NO	NO (N)	O_3 (O)
e.	Yes	H_2O_2	H_2O_2	H_2O_2 (O)	H_2O_2 (O)
f.	Yes	CuCl	CuCl	CuCl (Cu)	CuCl (Cu)

In c, no oxidation numbers change from reactants to products.

63. Use the method of half-reactions described in Section 4.10 of the text to balance these redox reactions. The first step always is to separate the reaction into the two half-reactions, then balance each half-reaction separately.

a. $Zn \rightarrow Zn^{2+} + 2\ e^-$ $2e^- + 2\ HCl \rightarrow H_2 + 2\ Cl^-$

Adding the two balanced half-reactions, $Zn(s) + 2\ HCl(aq) \rightarrow H_2(g) + Zn^{2+}(aq) + 2\ Cl^-(aq)$

b. $3\ I^- \rightarrow I_3^- + 2e^-$ $ClO^- \rightarrow Cl^-$

$2e^- + 2H^+ + ClO^- \rightarrow Cl^- + H_2O$

Adding the two balanced half-reactions so electrons cancel:

$$3\ I^-(aq) + 2\ H^+(aq) + ClO^-(aq) \rightarrow I_3^-(aq) + Cl^-(aq) + H_2O(l)$$

c. $As_2O_3 \rightarrow H_3AsO_4$ $NO_3^- \rightarrow NO + 2\ H_2O$

$As_2O_3 \rightarrow 2\ H_3AsO_4$ $4\ H^+ + NO_3^- \rightarrow NO + 2\ H_2O$

Left 3 - O; Right 8 - O $(3\ e^- + 4\ H^+ + NO_3^- \rightarrow NO + 2\ H_2O) \times 4$

Right hand side has 5 extra O.

Balance the oxygen atoms first using H_2O, then balance H using H^+, and finally balance charge using electrons.

$$(5\ H_2O + As_2O_3 \rightarrow 2\ H_3AsO_4 + 4\ H^+ + 4\ e^-) \times 3$$

Common factor is a transfer of 12 e^-. Add half-reactions so electrons cancel.

$$12\ e^- + 16\ H^+ + 4\ NO_3^- \rightarrow 4\ NO + 8\ H_2O$$
$$15\ H_2O + 3\ As_2O_3 \rightarrow 6\ H_3AsO_4 + 12\ H^+ + 12\ e^-$$

___

$$7\ H_2O(l) + 4\ H^+(aq) + 3\ As_2O_3(s) + 4\ NO_3^-(aq) \rightarrow 4\ NO(g) + 6\ H_3AsO_4(aq)$$

d. $(2\ Br^- \rightarrow Br_2 + 2\ e^-) \times 5$ $MnO_4^- \rightarrow Mn^{2+} + 4\ H_2O$

$(5\ e^- + 8\ H^+ + MnO_4^- \rightarrow Mn^{2+} + 4\ H_2O) \times 2$

Common factor is a transfer of 10 e^-.

$$10\ Br^- \rightarrow 5\ Br_2 + 10\ e^-$$
$$10\ e^- + 16\ H^+ + 2\ MnO_4^- \rightarrow 2\ Mn^{2+} + 8\ H_2O$$

___

$$16\ H^+(aq) + 2\ MnO_4^-(aq) + 10\ Br^-(aq) \rightarrow 5\ Br_2(l) + 2\ Mn^{2+}(aq) + 8\ H_2O(l)$$

e. $CH_3OH \rightarrow CH_2O$ $Cr_2O_7^{2-} \rightarrow Cr^{3+}$

$(CH_3OH \rightarrow CH_2O + 2\ H^+ + 2\ e^-) \times 3$ $14\ H^+ + Cr_2O_7^{2-} \rightarrow 2\ Cr^{3+} + 7\ H_2O$

$6\ e^- + 14\ H^+ + Cr_2O_7^{2-} \rightarrow 2\ Cr^{3+} + 7\ H_2O$

Common factor is a transfer of 6 e^-.

$$3\ CH_3OH \rightarrow 3\ CH_2O + 6\ H^+ + 6\ e^-$$
$$6\ e^- + 14\ H^+ + Cr_2O_7^{2-} \rightarrow 2\ Cr^{3+} + 7\ H_2O$$

___

$$8\ H^+(aq) + 3\ CH_3OH(aq) + Cr_2O_7^{2-}(aq) \rightarrow 2\ Cr^{3+}(aq) + 3\ CH_2O(aq) + 7\ H_2O(l)$$

65.   Use the same method as with acidic solutions. After the final balanced equation, convert H^+ to OH^- as described in section 4.10 of the text. The extra step involves converting H^+ into H_2O by adding equal moles of OH^- to each side of the reaction. This converts the reaction to a basic solution while keeping it balanced.

a. $Al \rightarrow Al(OH)_4^-$ $MnO_4^- \rightarrow MnO_2$

$4 H_2O + Al \rightarrow Al(OH)_4^- + 4 H^+$ $3 e^- + 4 H^+ + MnO_4^- \rightarrow MnO_2 + 2 H_2O$

$4 H_2O + Al \rightarrow Al(OH)_4^- + 4 H^+ + 3 e^-$

$$4 H_2O + Al \rightarrow Al(OH)_4^- + 4 H^+ + 3 e^-$$
$$3 e^- + 4 H^+ + MnO_4^- \rightarrow MnO_2 + 2 H_2O$$
$$\overline{}$$

$$2 H_2O(l) + Al(s) + MnO_4^-(aq) \rightarrow Al(OH)_4^-(aq) + MnO_2(s)$$

Since H^+ doesn't appear in the final balanced reaction, we are done.

b. $Cl_2 \rightarrow Cl^-$ $Cl_2 \rightarrow OCl^-$

$2 e^- + Cl_2 \rightarrow 2 Cl^-$ $2 H_2O + Cl_2 \rightarrow 2 OCl^- + 4 H^+ + 2 e^-$

$$2 e^- + Cl_2 \rightarrow 2 Cl^-$$
$$2 H_2O + Cl_2 \rightarrow 2 OCl^- + 4 H^+ + 2 e^-$$
$$\overline{}$$

$$2 H_2O + 2 Cl_2 \rightarrow 2 Cl^- + 2 OCl^- + 4 H^+$$

Now convert to a basic solution. Add 4 OH^- to both sides of the equation. The 4 OH^- will react with the 4 H^+ on the product side to give 4 H_2O. After this step, cancel identical species on both sides (2 H_2O). Applying these steps gives: $4 OH^- + 2 Cl_2 \rightarrow 2 Cl^- + 2 OCl^- + 2 H_2O$, which can be further simplified to:

$$2 OH^-(aq) + Cl_2(g) \rightarrow Cl^-(aq) + OCl^-(aq) + H_2O(l)$$

c. $NO_2^- \rightarrow NH_3$ $Al \rightarrow AlO_2^-$

$6 e^- + 7 H^+ + NO_2^- \rightarrow NH_3 + 2 H_2O$ $(2 H_2O + Al \rightarrow AlO_2^- + 4 H^+ + 3 e^-) \times 2$

Common factor is a transfer of 6 e^-.

$$6e^- + 7 H^+ + NO_2^- \rightarrow NH_3 + 2 H_2O$$
$$4 H_2O + 2 Al \rightarrow 2 AlO_2^- + 8 H^+ + 6 e^-$$
$$\overline{}$$

$$OH^- + 2 H_2O + NO_2^- + 2 Al \rightarrow NH_3 + 2 AlO_2^- + H^+ + OH^-$$

Reducing gives: $OH^-(aq) + H_2O(l) + NO_2^-(aq) + 2 Al(s) \rightarrow NH_3(g) + 2 AlO_2^-(aq)$

67. $NaCl + H_2SO_4 + MnO_2 \rightarrow Na_2SO_4 + MnCl_2 + Cl_2 + H_2O$

We could balance this reaction by the half-reaction method or by inspection. Let's try inspection. To balance Cl^-, we need 4 NaCl:

$4\ NaCl + H_2SO_4 + MnO_2 \rightarrow Na_2SO_4 + MnCl_2 + Cl_2 + H_2O$

Balance the Na^+ and SO_4^{2-} ions next:

$4\ NaCl + 2\ H_2SO_4 + MnO_2 \rightarrow 2\ Na_2SO_4 + MnCl_2 + Cl_2 + H_2O$

On the left side: 4-H and 10-O; On the right side: 8-O not counting H_2O

We need 2 H_2O on the right side to balance H and O:

$4\ NaCl(aq) + 2\ H_2SO_4(aq) + MnO_2(s) \rightarrow 2\ Na_2SO_4(aq) + MnCl_2(aq) + Cl_2(g) + 2\ H_2O(l)$

Additional Exercises

69. $0.100\ g\ Ca \times \dfrac{1\ mol\ Ca}{40.08\ g\ Ca} \times \dfrac{1\ mol\ Ca(OH)_2}{mol\ Ca} \times \dfrac{2\ mol\ OH^-}{mol\ Ca(OH)_2} = 4.99 \times 10^{-3}\ mol\ OH^-$

Molarity $= \dfrac{4.99 \times 10^{-3}\ mol}{450. \times 10^{-3}\ L} = 1.11 \times 10^{-2}\ M\ OH^-$

71.   There are other possible correct choices for the following answers. We have listed only three possible reactants in each case.

a. $AgNO_3$, $Pb(NO_3)_2$, and $Hg_2(NO_3)_2$ would form precipitates with the Cl^- ion.
$Ag^+(aq) + Cl^-(aq) \rightarrow AgCl(s);\ \ Pb^{2+}(aq) + 2\ Cl^-(aq) \rightarrow PbCl_2(s);$
$Hg_2^{2+}(aq) + 2\ Cl^-(aq) \rightarrow Hg_2Cl_2(s)$

b. Na_2SO_4, Na_2CO_3, and Na_3PO_4 would form precipitates with the Ca^{2+} ion.
$Ca^{2+}(aq) + SO_4^{2-}(aq) \rightarrow CaSO_4(s);\ \ Ca^{2+} + CO_3^{2-}(aq) \rightarrow CaCO_3(s)$
$3\ Ca^{2+}(aq) + 2\ PO_4^{3-}(aq) \rightarrow Ca_3(PO_4)_2(s)$

c. $NaOH$, Na_2S, and Na_2CO_3 would form precipitates with the Fe^{3+} ion.
$Fe^{3+}(aq) + 3\ OH^-(aq) \rightarrow Fe(OH)_3(s);\ \ 2\ Fe^{3+}(aq) + 3\ S^{2-}(aq) \rightarrow Fe_2S_3(s);$
$2\ Fe^{3+}(aq) + 3\ CO_3^{2-}(aq) \rightarrow Fe_2(CO_3)_3(s)$

d. $BaCl_2$, $Pb(NO_3)_2$, and $Ca(NO_3)_2$ would form precipitates with the SO_4^{2-} ion.
$Ba^{2+}(aq) + SO_4^{2-}(aq) \rightarrow BaSO_4(s);\ \ Pb^{2+}(aq) + SO_4^{2-}(aq) \rightarrow PbSO_4(s);$
$Ca^{2+}(aq) + SO_4^{2-}(aq) \rightarrow CaSO_4(s)$

e. Na_2SO_4, NaCl, and NaI would form precipitates with the Hg_2^{2+} ion.
$Hg_2^{2+}(aq) + SO_4^{2-}(aq) \rightarrow Hg_2SO_4(s)$; $Hg_2^{2+}(aq) + 2\ Cl^-(aq) \rightarrow Hg_2Cl_2(s)$;
$Hg_2^{2+}(aq) + 2\ I^-(aq) \rightarrow Hg_2I_2(s)$

f. NaBr, Na_2CrO_4, and Na_3PO_4 would form precipitates with the Ag^+ ion.
$Ag^+(aq) + Br^-(aq) \rightarrow AgBr(s)$; $2\ Ag^+(aq) + CrO_4^{2-}(aq) \rightarrow Ag_2CrO_4(s)$;
$3\ Ag^+(aq) + PO_4^{3-}(aq) \rightarrow Ag_3PO_4(s)$

73. $1.00\ L \times \dfrac{0.200\ mol\ Na_2S_2O_3}{L} \times \dfrac{1\ mol\ AgBr}{2\ mol\ Na_2S_2O_3} \times \dfrac{187.8\ g\ AgBr}{mol\ AgBr} = 18.8\ g\ AgBr$

75.   All the sulfur in $BaSO_4$ came from the saccharin. The conversion from $BaSO_4$ to saccharin utilizes the molar masses of each.

$0.5032\ g\ BaSO_4 \times \dfrac{32.07\ g\ S}{233.4\ g\ BaSO_4} \times \dfrac{183.19\ g\ saccharin}{32.07\ g\ S} = 0.3949\ g\ saccharin$

$\dfrac{Avg.\ mass}{Tablet} = \dfrac{0.3949\ g}{10\ tablets} = \dfrac{3.949 \times 10^{-2}\ g}{tablet} = \dfrac{39.49\ mg}{tablet}$

Avg. mass % $= \dfrac{0.3949\ g\ saccharin}{0.5894\ g} \times 100 = 67.00\%$ saccharin by mass

77. $Cr(NO_3)_3(aq) + 3\ NaOH(aq) \rightarrow Cr(OH)_3(s) + 3\ NaNO_3(aq)$

mol NaOH used
to form precipitate $= 2.06\ g\ Cr(OH)_3 \times \dfrac{1\ mol\ Cr(OH)_3}{103.02\ g} \times \dfrac{3\ mol\ NaOH}{1\ mol\ Cr(OH)_3} = 6.00 \times 10^{-2}\ mol\ NaOH$

$NaOH(aq) + HCl(aq) \rightarrow NaCl(aq) + H_2O(l)$

mol NaOH used
to react with HCl $= 0.1000\ L \times \dfrac{0.400\ mol\ HCl}{L} \times \dfrac{1\ mol\ NaOH}{mol\ HCl} = 4.00 \times 10^{-2}\ mol\ NaOH$

$M_{NaOH} = \dfrac{mol\ NaOH}{volume} = \dfrac{6.00 \times 10^{-2}\ mol + 4.00 \times 10^{-2}\ mol}{0.0500\ L} = 2.00\ M\ NaOH$

79. $HC_2H_3O_2(aq) + NaOH(aq) \rightarrow H_2O(l) + NaC_2H_3O_2(aq)$

a. $16.58 \times 10^{-3}\ L\ soln \times \dfrac{0.5062\ mol\ NaOH}{L\ soln} \times \dfrac{1\ mol\ acetic\ acid}{mol\ NaOH} = 8.393 \times 10^{-3}\ mol\ acetic\ acid$

Concentration of acetic acid $= \dfrac{8.393 \times 10^{-3}\ mol}{0.01000\ L} = 0.8393\ M$

b.   If we have 1.000 L of solution: total mass = 1000. mL $\times \dfrac{1.006 \text{ g}}{\text{mL}}$ = 1006 g

Mass of $HC_2H_3O_2$ = 0.8393 mol $\times \dfrac{60.05 \text{ g}}{\text{mol}}$ = 50.40 g

Mass % acetic acid = $\dfrac{50.40 \text{ g}}{1006 \text{ g}} \times 100$ = 5.010%

81.   Let HA = unknown acid; $HA(aq) + NaOH(aq) \rightarrow NaA(aq) + H_2O(l)$

mol HA present = 0.0250 L $\times \dfrac{0.500 \text{ mol NaOH}}{\text{L}} \times \dfrac{1 \text{ mol HA}}{1 \text{ mol NaOH}}$ = 0.0125 mol HA

$\dfrac{x \text{ g HA}}{\text{mol HA}} = \dfrac{2.20 \text{ g HA}}{0.0125 \text{ mol HA}}$, x = molar mass of HA = 176 g/mol

Empirical formula weight ≈ 3(12) + 4(1) + 3(16) = 88 g/mol

Since 176/88 = 2.0, the molecular formula is $(C_3H_4O_3)_2 = C_6H_8O_6$.

83. $Mn + HNO_3 \rightarrow Mn^{2+} + NO_2$

$Mn \rightarrow Mn^{2+} + 2 \text{ e}^-$ $\qquad\qquad\qquad$ $HNO_3 \rightarrow NO_2$
$\qquad\qquad\qquad\qquad\qquad\qquad\qquad$ $HNO_3 \rightarrow NO_2 + H_2O$
$\qquad\qquad\qquad\qquad\qquad\qquad$ $(\text{e}^- + H^+ + HNO_3 \rightarrow NO_2 + H_2O) \times 2$

$Mn \rightarrow Mn^{2+} + 2 \text{ e}^-$
$\underline{2 \text{ e}^- + 2 H^+ + 2 HNO_3 \rightarrow 2 NO_2 + 2 H_2O}$

$2 H^+(aq) + Mn(s) + 2 HNO_3(aq) \rightarrow Mn^{2+}(aq) + 2 NO_2(g) + 2 H_2O(l)$

$Mn^{2+} + IO_4^- \rightarrow MnO_4^- + IO_3^-$

$(4 H_2O + Mn^{2+} \rightarrow MnO_4^- + 8 H^+ + 5 \text{ e}^-) \times 2$ $\qquad\qquad$ $(2 \text{ e}^- + 2 H^+ + IO_4^- \rightarrow IO_3^- + H_2O) \times 5$

$8 H_2O + 2 Mn^{2+} \rightarrow 2 MnO_4^- + 16 H^+ + 10 \text{ e}^-$
$\underline{10 \text{ e}^- + 10 H^+ + 5 IO_4^- \rightarrow 5 IO_3^- + 5 H_2O}$

$3 H_2O(l) + 2 Mn^{2+}(aq) + 5 IO_4^-(aq) \rightarrow 2 MnO_4^-(aq) + 5 IO_3^-(aq) + 6 H^+(aq)$

Challenge Problems

85.   a.   0.308 g AgCl $\times \dfrac{35.45 \text{ g Cl}}{143.4 \text{ g AgCl}}$ = 0.0761 g Cl; %Cl = $\dfrac{0.0761 \text{ g}}{0.256 \text{ g}} \times 100$ = 29.7% Cl

Cobalt(III) oxide, Co_2O_3: 2(58.93) + 3(16.00) = 165.86 g/mol

0.145 g $Co_2O_3 \times \dfrac{117.86 \text{ g Co}}{165.86 \text{ g } Co_2O_3}$ = 0.103 g Co; %Co = $\dfrac{0.103 \text{ g}}{0.416 \text{ g}} \times 100$ = 24.8% Co

The remainder, 100.0 - (29.7 + 24.8) = 45.5%, is water.

Assuming 100.0 g of compound:

$$45.5 \text{ g H}_2\text{O} \times \frac{2.016 \text{ g H}}{18.02 \text{ g H}_2\text{O}} = 5.09 \text{ g H}; \quad \%\text{H} = \frac{5.09 \text{ g H}}{100.0 \text{ g compound}} \times 100 = 5.09\% \text{ H}$$

$$45.5 \text{ g H}_2\text{O} \times \frac{16.00 \text{ g O}}{18.02 \text{ g H}_2\text{O}} = 40.4 \text{ g O}; \quad \%\text{O} = \frac{40.4 \text{ g O}}{100.0 \text{ g compound}} \times 100 = 40.4\% \text{ O}$$

The mass percent composition is 24.8% Co, 29.7% Cl, 5.09% H and 40.4% O.

b. Out of 100.0 g of compound, there are:

$$24.8 \text{ g Co} \times \frac{1 \text{ mol}}{58.93 \text{ g Co}} = 0.421 \text{ mol Co}; \quad 29.7 \text{ g Cl} \times \frac{1 \text{ mol}}{35.45 \text{ g Cl}} = 0.838 \text{ mol Cl}$$

$$5.09 \text{ g H} \times \frac{1 \text{ mol}}{1.008 \text{ g H}} = 5.05 \text{ mol H}; \quad 40.4 \text{ g O} \times \frac{1 \text{ mol}}{16.00 \text{ g O}} = 2.53 \text{ mol O}$$

Dividing all results by 0.421, we get $CoCl_2 \cdot 6H_2O$.

c. $CoCl_2 \cdot 6H_2O(aq) + 2 \text{ AgNO}_3(aq) \rightarrow 2 \text{ AgCl}(s) + Co(NO_3)_2(aq) + 6 \text{ H}_2O(l)$

$CoCl_2 \cdot 6H_2O(aq) + 2 \text{ NaOH}(aq) \rightarrow Co(OH)_2(s) + 2 \text{ NaCl}(aq) + 6 \text{ H}_2O(l)$

$Co(OH)_2 \rightarrow Co_2O_3$ This is an oxidation-reduction reaction. Thus, we also need to include an oxidizing agent. The obvious choice is O_2.

$4 \text{ Co(OH)}_2(s) + O_2(g) \rightarrow 2 \text{ Co}_2O_3(s) + 4 \text{ H}_2O(l)$

87. a. $2 \text{ AgNO}_3(aq) + K_2CrO_4(aq) \rightarrow Ag_2CrO_4(s) + 2 \text{ KNO}_3(aq)$

Molar mass: 169.9 g/mol   194.20 g/mol   331.8 g/mol

The molar mass of Ag_2CrO_4 is 331.8 g/mol, so one mol of precipitate was formed.

We have equal masses of $AgNO_3$ and K_2CrO_4. Since the molar mass of $AgNO_3$ is less than that of K_2CrO_4, then we have more mol of $AgNO_3$ present. However, we will not have twice the mol of $AgNO_3$ present as compared to K_2CrO_4 as required by the balanced reaction; this is because the molar mass of $AgNO_3$ is no where near one-half the molar mass of K_2CrO_4. Therefore, $AgNO_3$ is limiting.

$$\text{mass AgNO}_3 = 1.000 \text{ mol Ag}_2\text{CrO}_4 \times \frac{2 \text{ mol AgNO}_3}{\text{mol Ag}_2\text{CrO}_4} \times \frac{169.9 \text{ g}}{\text{mol AgNO}_3} = 339.8 \text{ g AgNO}_3$$

Since equal masses of reactants are present, then 339.8 g K_2CrO_4 were present initially.

$$M_{K^+} = \frac{\text{mol K}^+}{\text{total volume}} = \frac{339.8 \text{ g K}_2\text{CrO}_4 \times \dfrac{1 \text{ mol K}_2\text{CrO}_4}{194.20 \text{ g}} \times \dfrac{2 \text{ mol K}^+}{\text{mol K}_2\text{CrO}_4}}{0.5000 \text{ L}} = 7.000 \, M \, K^+$$

b. mol CrO_4^{2-} present initially $= 339.8 \text{ g K}_2\text{CrO}_4 \times \dfrac{1 \text{ mol K}_2\text{CrO}_4}{194.20 \text{ g}} \times \dfrac{1 \text{ mol CrO}_4^{2-}}{\text{mol K}_2\text{CrO}_4} = 1.750 \text{ mol CrO}_4^{2-}$

mol CrO_4^{2-} in precipitate $= 1.000 \text{ mol Ag}_2\text{CrO}_4 \times \dfrac{1 \text{ mol CrO}_4^{2-}}{1 \text{ mol Ag}_2\text{CrO}_4} = 1.000 \text{ mol CrO}_4^{2-}$

$$M_{CrO_4^{2-}} = \frac{\text{excess mol CrO}_4^{2-}}{\text{total volume}} = \frac{1.750 \text{ mol} - 1.000 \text{ mol}}{0.5000 \text{ L} + 0.5000 \text{ L}} = \frac{0.750 \text{ mol}}{1.0000 \text{ L}} = 0.750 \, M$$

89. $0.298 \text{ g BaSO}_4 \times \dfrac{96.07 \text{ g SO}_4^{2-}}{233.4 \text{ g BaSO}_4} = 0.123 \text{ g SO}_4^{2-}$; % sulfate $= \dfrac{0.123 \text{ g SO}_4^{2-}}{0.205 \text{ g}} = 60.0\%$

Assume we have 100.0 g of the mixture of Na_2SO_4 and K_2SO_4. There are:

$$60.0 \text{ g SO}_4^{2-} \times \frac{1 \text{ mol}}{96.07 \text{ g}} = 0.625 \text{ mol SO}_4^{2-}$$

There must be $2 \times 0.625 = 1.25$ mol of +1 cations to balance the -2 charge of SO_4^{2-}.

Let x = number of moles of K^+ and y = number of moles of Na^+; then $x + y = 1.25$.

The total mass of Na^+ and K^+ must be 40.0 g in the assumed 100.0 g of mixture. Setting up an equation:

$$x \text{ mol K}^+ \times \frac{39.10 \text{ g}}{\text{mol}} + y \text{ mol Na}^+ \times \frac{22.99 \text{ g}}{\text{mol}} = 40.0 \text{ g}$$

So, we have two equations with two unknowns: $x + y = 1.25$ and $39.10 \, x + 22.99 \, y = 40.0$

Since $x = 1.25 - y$, then $39.10(1.25 - y) + 22.99 \, y = 40.0$

$48.9 - 39.10 \, y + 22.99 \, y = 40.0$, $-16.11 \, y = -8.9$

$y = 0.55$ mol Na^+ and $x = 1.25 - 0.55 = 0.70$ mol K^+

Therefore:

$$0.70 \text{ mol K}^+ \times \frac{1 \text{ mol K}_2\text{SO}_4}{2 \text{ mol K}^+} = 0.35 \text{ mol K}_2\text{SO}_4; \ 0.35 \text{ mol K}_2\text{SO}_4 \times \frac{174.27 \text{ g}}{\text{mol}} = 61 \text{ g K}_2\text{SO}_4$$

Since we assumed 100.0 g, the mixture is 61% K_2SO_4 and 39% Na_2SO_4.

91. $2\ H_3PO_4(aq) + 3\ Ba(OH)_2(aq) \rightarrow 6\ H_2O(l) + Ba_3(PO_4)_2(s)$

$$0.01420\ L \times \frac{0.141\ mol\ H_3PO_4}{L} \times \frac{3\ mol\ Ba(OH)_2}{2\ mol\ H_3PO_4} \times \frac{1\ L\ Ba(OH)_2}{0.0521\ mol\ Ba(OH)_2} = 0.0576\ L$$

$$= 57.6\ mL\ Ba(OH)_2$$

93.   a. $MgO(s) + 2\ HCl(aq) \rightarrow MgCl_2(aq) + H_2O(l)$

$Mg(OH)_2(s) + 2\ HCl(aq) \rightarrow MgCl_2(aq) + 2\ H_2O(l)$

$Al(OH)_3(s) + 3\ HCl(aq) \rightarrow AlCl_3(aq) + 3\ H_2O(l)$

b.   Let's calculate the number of moles of HCl neutralized per gram of substance. We can get these directly from the balanced equations and the molar masses of the substances.

$$\frac{2\ mol\ HCl}{mol\ MgO} \times \frac{1\ mol\ MgO}{40.31\ g\ MgO} = \frac{4.962 \times 10^{-2}\ mol\ HCl}{g\ MgO}$$

$$\frac{2\ mol\ HCl}{mol\ Mg(OH)_2} \times \frac{1\ mol\ Mg(OH)_2}{58.33\ g\ Mg(OH)_2} = \frac{3.429 \times 10^{-2}\ mol\ HCl}{g\ Mg(OH)_2}$$

$$\frac{3\ mol\ HCl}{mol\ Al(OH)_3} \times \frac{1\ mol\ Al(OH)_3}{78.00\ g\ Al(OH)_3} = \frac{3.846 \times 10^{-2}\ mol\ HCl}{g\ Al(OH)_3}$$

Therefore, one gram of magnesium oxide would neutralize the most 0.10 M HCl.

95. $$mol\ C_6H_8O_7 = 0.250\ g\ C_6H_8O_7 \times \frac{1\ mol\ C_6H_8O_7}{192.12\ g\ C_6H_8O_7} = 1.30 \times 10^{-3}\ mol\ C_6H_8O_7$$

Let H_xA represent citric acid where x is the number of acidic hydrogens. The balanced neutralization reaction is:

$H_xA(aq) + x\ OH^-(aq) \rightarrow x\ H_2O(l) + A^{x-}(aq)$

$$mol\ OH^-\ reacted = 0.0372\ L \times \frac{0.105\ mol\ OH^-}{L} = 3.91 \times 10^{-3}\ mol\ OH^-$$

$$x = \frac{mol\ OH^-}{mol\ citric\ acid} = \frac{3.91 \times 10^{-3}\ mol}{1.30 \times 10^{-3}\ mol} = 3.01$$

Therefore, the general acid formula for citric acid is H_3A, meaning that citric acid has three acidic hydrogens per citric acid molecule (citric acid is a triprotic acid).

97.   The amount of KHP used = $0.4016 \text{ g} \times \dfrac{1 \text{ mol}}{204.22 \text{ g}} = 1.967 \times 10^{-3}$ mol KHP

Since one mole of NaOH reacts completely with one mole of KHP, the NaOH solution contains 1.967×10^{-3} mol NaOH.

Molarity of NaOH = $\dfrac{1.967 \times 10^{-3} \text{ mol}}{25.06 \times 10^{-3} \text{ L}} = \dfrac{7.849 \times 10^{-2} \text{ mol}}{\text{L}}$

Maximum molarity = $\dfrac{1.967 \times 10^{-3} \text{ mol}}{25.01 \times 10^{-3} \text{ L}} = \dfrac{7.865 \times 10^{-2} \text{ mol}}{\text{L}}$

Minimum molarity = $\dfrac{1.967 \times 10^{-3} \text{ mol}}{25.11 \times 10^{-3} \text{ L}} = \dfrac{7.834 \times 10^{-2} \text{ mol}}{\text{L}}$

We can express this as $0.07849 \pm 0.00016 \; M$. An alternative is to express the molarity as $0.0785 \pm 0.0002 \; M$. The second way shows the actual number of significant figures in the molarity. The advantage of the first method is that it shows that we made all of our individual measurements to four significant figures.

CHAPTER FIVE

GASES

Questions

17. $PV = nRT$ = constant at constant n and T. At two sets of conditions, P_1V_1 = constant = P_2V_2.

$P_1V_1 = P_2V_2$ (Boyle's law).

$\dfrac{V}{T} = \dfrac{nR}{P}$ = constant at constant n and P. At two sets of conditions, $\dfrac{V_1}{T_1}$ = constant = $\dfrac{V_2}{T_2}$.

$\dfrac{V_1}{T_1} = \dfrac{V_2}{T_2}$ (Charles's law)

19.   The kinetic molecular theory assumes that gas particles do not exert forces on each other and that gas particles are volumeless. Real gas particles do exert attractive forces for each other, and real gas particles do have volumes. A gas behaves most ideally at low pressures and high temperatures. The effect of attractive forces is minimized at high temperatures since the gas particles are moving very rapidly. At low pressure, the container volume is relatively large (P and V are inversely related) so the volume of the container taken up by the gas particles is negligible.

21.   Method 1: molar mass = $\dfrac{dRT}{P}$

Determine the density of a gas at a measurable temperature and pressure, then use the above equation to determine the molar mass.

Method 2: $\dfrac{\text{effusion rate for gas 1}}{\text{effusion rate for gas 2}} = \sqrt{\dfrac{(\text{molar mass})_2}{(\text{molar mass})_1}}$

Determine the effusion rate of the unknown gas relative to some known gas; then use Graham's law of effusion (the above equation) to determine the molar mass.

23.   Rigid container (constant volume): As reactants are converted to products, the mol of gas particles present decrease by one-half. As n decreases, the pressure will decrease (by one-half). Density is the mass per unit volume. Mass is conserved in a chemical reaction, so the density of the gas will not change since mass and volume do not change.

Flexible container (constant pressure): Pressure is constant since the container changes volume in order to keep a constant pressure. As the mol of gas particles decrease by a factor of 2, the volume of the container will decrease (by one-half). We have the same mass of gas in a smaller volume, so the gas density will increase (is doubled).

Exercises

Pressure

25.   a. $4.8 \text{ atm} \times \dfrac{760 \text{ mm Hg}}{\text{atm}} = 3.6 \times 10^3 \text{ mm Hg}$; b. $3.6 \times 10^3 \text{ mm Hg} \times \dfrac{1 \text{ torr}}{\text{mm Hg}} = 3.6 \times 10^3 \text{ torr}$

      c. $4.8 \text{ atm} \times \dfrac{1.013 \times 10^5 \text{ Pa}}{\text{atm}} = 4.9 \times 10^5 \text{ Pa}$;   d. $4.8 \text{ atm} \times \dfrac{14.7 \text{ psi}}{\text{atm}} = 71 \text{ psi}$

27. $6.5 \text{ cm} \times \dfrac{10 \text{ mm}}{\text{cm}} = 65 \text{ mm Hg or 65 torr}$; $65 \text{ torr} \times \dfrac{1 \text{ atm}}{760 \text{ torr}} = 8.6 \times 10^{-2} \text{ atm}$

      $8.6 \times 10^{-2} \text{ atm} = \dfrac{1.013 \times 10^5 \text{ Pa}}{\text{atm}} = 8.7 \times 10^3 \text{ Pa}$

29.   If the levels of Hg in each arm of the manometer are equal, the pressure in the flask is equal to atmospheric pressure. When they are unequal, the difference in height in mm will be equal to the difference in pressure in mm Hg between the flask and the atmosphere. Which level is higher will tell us whether the pressure in the flask is less than or greater than atmospheric.

      a. $P_{flask} < P_{atm}$; $P_{flask} = 760. - 118 = 642 \text{ mm Hg} = 642 \text{ torr}$; $642 \text{ torr} \times \dfrac{1 \text{ atm}}{760 \text{ torr}} = 0.845 \text{ atm}$

        $0.845 \text{ atm} \times \dfrac{1.013 \times 10^5 \text{ Pa}}{\text{atm}} = 8.56 \times 10^4 \text{ Pa}$

      b. $P_{flask} > P_{atm}$; $P_{flask} = 760. \text{ torr} + 215 \text{ torr} = 975 \text{ torr}$; $975 \text{ torr} \times \dfrac{1 \text{ atm}}{760 \text{ torr}} = 1.28 \text{ atm}$

        $1.28 \text{ atm} \times \dfrac{1.013 \times 10^5 \text{ Pa}}{\text{atm}} = 1.30 \times 10^5 \text{ Pa}$

      c. $P_{flask} = 635 - 118 = 517 \text{ torr}$; $P_{flask} = 635 + 215 = 850. \text{ torr}$

Gas Laws

31.   From Boyle's law, $P_1V_1 = P_2V_2$ at constant n and T.

      $P_2 = \dfrac{P_1V_1}{V_2} = \dfrac{5.20 \text{ atm} \times 0.400 \text{ L}}{2.14 \text{ L}} = 0.972 \text{ atm}$

As expected, as the volume increased, the pressure decreased.

33.   From Avogadro's law, $V_1/n_1 = V_2/n_2$ at constant T and P.

$$V_2 = \frac{V_1 n_2}{n_1} = \frac{11.2 \text{ L} \times 2.00 \text{ mol}}{0.500 \text{ mol}} = 44.8 \text{ L}$$

As expected, as the mol of gas present increases, volume increases.

35.   a. $PV = nRT$, $V = \dfrac{nRT}{P} = \dfrac{2.00 \text{ mol} \times \dfrac{0.08206 \text{ L atm}}{\text{mol K}} \times (155 + 273) \text{ K}}{5.00 \text{ atm}} = 14.0 \text{ L}$

        b. $PV = nRT$, $n = \dfrac{PV}{RT} = \dfrac{0.300 \text{ atm} \times 2.00 \text{ L}}{\dfrac{0.08206 \text{ L atm}}{\text{mol K}} \times 155 \text{ K}} = 4.72 \times 10^{-2} \text{ mol}$

        c. $PV = nRT$, $T = \dfrac{PV}{nR} = \dfrac{4.47 \text{ atm} \times 25.0 \text{ L}}{2.01 \text{ mol} \times \dfrac{0.08206 \text{ L atm}}{\text{mol K}}} = 678 \text{ K} = 405°\text{C}$

        d. $PV = nRT$, $P = \dfrac{nRT}{V} = \dfrac{10.5 \text{ mol} \times \dfrac{0.08206 \text{ L atm}}{\text{mol K}} \times (273 + 75) \text{ K}}{2.25 \text{ L}} = 133 \text{ atm}$

37. $n = \dfrac{PV}{RT} = \dfrac{135 \text{ atm} \times 200.0 \text{ L}}{0.08206 \dfrac{\text{L atm}}{\text{mol K}} \times (273 + 24) \text{ K}} = 1.11 \times 10^3 \text{ mol}$

        For He: $1.11 \times 10^3 \text{ mol} \times \dfrac{4.003 \text{ g He}}{\text{mol}} = 4.44 \times 10^3 \text{ g He}$

        For H_2: $1.11 \times 10^3 \text{ mol} \times \dfrac{2.016 \text{ g He}}{\text{mol}} = 2.24 \times 10^3 \text{ g H}_2$

39.   a. $PV = nRT$; $175 \text{ g Ar} \times \dfrac{1 \text{ mol Ar}}{39.95 \text{ g Ar}} = 4.38 \text{ mol Ar}$

            $T = \dfrac{PV}{nR} = \dfrac{10.0 \text{ atm} \times 2.50 \text{ L}}{4.38 \text{ mol} \times \dfrac{0.08206 \text{ L atm}}{\text{mol K}}} = 69.6 \text{ K}$

        b. $PV = nRT$, $P = \dfrac{nRT}{V} = \dfrac{4.38 \text{ mol} \times \dfrac{0.08206 \text{ L atm}}{\text{mol K}} \times 225 \text{ K}}{2.50 \text{ L}} = 32.3 \text{ atm}$

41.   At constant n and T, $PV = nRT = $ constant, $P_1V_1 = P_2V_2$; At sea level, $P = 1.00$ atm $= 760.$ mm Hg.

$$V_2 = \frac{P_1V_1}{P_2} = \frac{760. \text{ mm Hg} \times 2.0 \text{ L}}{500. \text{ mm Hg}} = 3.0 \text{ L}$$

The balloon will burst at this pressure since the volume must expand beyond the 2.5 L limit of the balloon.

Note: To solve this problem, we did not have to convert the pressure units into atm; the units of mm Hg canceled each other. In general, only convert units if you have to. Whenever the gas constant R is not used to solve a problem, pressure and volume units must only be consistent, and not necessarily in units of atm and L. The exception is temperature as T must <u>always</u> be converted to the Kelvin scale.

43. $PV = nRT$, V and n constant, so $\dfrac{P}{T} = \dfrac{nR}{V} = $ constant and $\dfrac{P_1}{T_1} = \dfrac{P_2}{T_2}$.

$$P_2 = \frac{P_1T_2}{T_1} = 13.7 \text{ MPa} \times \frac{(273 + 450.) \text{ K}}{(273 + 23) \text{ K}} = 33.5 \text{ MPa}$$

45. $PV = nRT$, n constant; $\dfrac{PV}{T} = nR = $ constant, $\dfrac{P_1V_1}{T_1} = \dfrac{P_2V_2}{T_2}$

$$P_2 = \frac{P_1V_1T_2}{V_2T_1} = 710. \text{ torr} \times \frac{5.0 \times 10^2 \text{ mL}}{25 \text{ mL}} \times \frac{(273 + 820.) \text{ K}}{(273 + 30.) \text{ K}} = 5.1 \times 10^4 \text{ torr}$$

47. $PV = nRT$, n is constant. $\dfrac{PV}{T} = nR = $ constant, $\dfrac{P_1V_1}{T_1} = \dfrac{P_2V_2}{T_2}$, $V_2 = \dfrac{V_1P_1T_2}{P_2T_1}$

$$V_2 = 1.00 \text{ L} \times \frac{760. \text{ torr}}{220. \text{ torr}} \times \frac{(273-31) \text{ K}}{(273 + 23) \text{ K}} = 2.82 \text{ L}; \quad \Delta V = 2.82 - 1.00 = 1.82 \text{ L}$$

Gas Density, Molar Mass, and Reaction Stoichiometry

49.   STP: $T = 273$ K and $P = 1.00$ atm; $n = \dfrac{PV}{RT} = \dfrac{1.00 \text{ atm} \times 1.5 \text{ L}}{\dfrac{0.08206 \text{ L atm}}{\text{mol K}} \times 273 \text{ K}} = 6.7 \times 10^{-2}$ mol He

Or we can use the fact that at STP, 1 mol of an ideal gas occupies 22.42 L.

$$1.5 \text{ L} \times \frac{1 \text{ mol He}}{22.42 \text{ L}} = 6.7 \times 10^{-2} \text{ mol He}; \quad 6.7 \times 10^{-2} \text{ mol He} \times \frac{4.003 \text{ g He}}{\text{mol He}} = 0.27 \text{ g He}$$

51. $C_6H_{12}O_6(s) + 6\,O_2(g) \rightarrow 6\,CO_2(g) + 6\,H_2O(g)$

$$5.00\text{ g }C_6H_{12}O_6 \times \frac{1\text{ mol }C_6H_{12}O_6}{180.16\text{ g}} \times \frac{6\text{ mol }O_2}{\text{mol }C_6H_{12}O_6} = 0.167\text{ mol }O_2$$

$$V = \frac{nRT}{P} = \frac{0.167\text{ mol} \times 0.08206\,\dfrac{L\text{ atm}}{\text{mol K}} \times 301\text{ K}}{0.976\text{ atm}} = 4.23\text{ L }O_2$$

Since T and P are constant, the volume of each gas will be directly proportional to the mol of gas present. The balanced equation says that equal mol of CO_2 and H_2O will be produced as mol of O_2 reacted. So the volumes of CO_2 and H_2O produced will equal the volume of O_2 reacted.

$$V_{CO_2} = V_{H_2O} = V_{O_2} = 4.23\text{ L}$$

53. $n_{H_2} = \dfrac{PV}{RT} = \dfrac{1.0\text{ atm} \times \left[4800\text{ m}^3 \times \left(\dfrac{100\text{ cm}}{\text{m}} \right)^3 \times \dfrac{1\text{ L}}{1000\text{ cm}^3} \right]}{\dfrac{0.08206\text{ L atm}}{\text{mol K}} \times 273\text{ K}} = 2.1 \times 10^5\text{ mol}$

2.1×10^5 mol H_2 are in the balloon. This is 80.% of the total amount of H_2 that had to be generated:

   $0.80\text{ (total mol }H_2) = 2.1 \times 10^5,\ \ \text{total mol }H_2 = 2.6 \times 10^5\text{ mol }H_2$

$2.6 \times 10^5\text{ mol }H_2 \times \dfrac{1\text{ mol Fe}}{\text{mol }H_2} \times \dfrac{55.85\text{ g Fe}}{\text{mol Fe}} = 1.5 \times 10^7\text{ g Fe}$

$2.6 \times 10^5\text{ mol }H_2 \times \dfrac{1\text{ mol }H_2SO_4}{\text{mol }H_2} \times \dfrac{98.09\text{ g }H_2SO_4}{\text{mol }H_2SO_4} \times \dfrac{100\text{ g reagent}}{98\text{ g }H_2SO_4} = 2.6 \times 10^7\text{ g of 98\% sulfuric acid}$

55. $CH_3OH + 3/2\,O_2 \rightarrow CO_2 + 2\,H_2O$ or $2\,CH_3OH(l) + 3\,O_2(g) \rightarrow 2\,CO_2(g) + 4\,H_2O(g)$

$50.0\text{ mL} \times \dfrac{0.850\text{ g}}{\text{mL}} \times \dfrac{1\text{ mol}}{32.04\text{ g}} = 1.33\text{ mol }CH_3OH(l)\text{ available}$

$$n_{O_2} = \frac{PV}{RT} = \frac{2.00\text{ atm} \times 22.8\text{ L}}{\dfrac{0.08206\text{ L atm}}{\text{mol K}} \times 300.\text{ K}} = 1.85\text{ mol }O_2\text{ available}$$

$1.33\text{ mol }CH_3OH \times \dfrac{3\text{ mol }O_2}{2\text{ mol }CH_3OH} = 2.00\text{ mol }O_2$

2.00 mol O_2 are required to react completely with all of the CH_3OH available. We only have 1.85 mol O_2, so O_2 is limiting.

$$1.85 \text{ mol } O_2 \times \frac{4 \text{ mol } H_2O}{3 \text{ mol } O_2} = 2.47 \text{ mol } H_2O$$

57.   a. $CH_4(g) + NH_3(g) + O_2(g) \rightarrow HCN(g) + H_2O(g)$; Balancing H first, then O, gives:

$$CH_4 + NH_3 + \frac{3}{2}O_2 \rightarrow HCN + 3 \text{ } H_2O \text{ or } 2 \text{ } CH_4(g) + 2 \text{ } NH_3(g) + 3 \text{ } O_2(g) \rightarrow 2 \text{ } HCN(g) + 6 \text{ } H_2O(g)$$

   b. $PV = nRT$, T and P constant; $\dfrac{V_1}{n_1} = \dfrac{V_2}{n_2}$, $\dfrac{V_1}{V_2} = \dfrac{n_1}{n_2}$

Since the volumes are all measured at constant T and P, the volumes of gas present are directly proportional to the mol of gas present (Avogadro's law). Because Avogadro's law applies, the balanced reaction gives mol relationships as well as volume relationships. Therefore, 2 L of CH_4, 2 L of NH_3 and 3 L of O_2 are required by the balanced equation for the production of 2 L of HCN. The actual volume ratio is 20.0 L CH_4:20.0 L NH_3:20.0 L O_2 (or 1:1:1). The volume of O_2 required to react with all of the CH_4 and NH_3 present is 20.0 L $\times$(3/2) = 30.0 L. Since only 20.0 L of O_2 are present, O_2 is the limiting reagent. The volume of HCN produced is:

$$20.0 \text{ L } O_2 \times \frac{2 \text{ L HCN}}{3 \text{ L } O_2} = 13.3 \text{ L HCN}$$

59.   One of the equations developed in the text to determine molar mass is:

$$\text{molar mass} = \frac{dRT}{P} \text{ where d = density in units of g/L}$$

$$\text{molar mass} = \frac{1.65 \text{ g/L} \times \dfrac{0.08206 \text{ L atm}}{\text{mol K}} \times (273 + 27) \text{ K}}{734 \text{ torr} \times \dfrac{1 \text{ atm}}{760 \text{ torr}}} = 42.1 \text{ g/mol}$$

The empirical formula mass of CH_2 = 12.01 + 2(1.008) = 14.03 g/mol.

$$\frac{42.1}{14.03} = 3.00; \text{ Molecular formula} = C_3H_6$$

61. $P \times$ (molar mass) = dRT, d = density = $\dfrac{P \times (\text{molar mass})}{RT}$

For $SiCl_4$, molar mass = M = 28.09 + 4(35.45) = 169.89 g/mol

$$d = \frac{(758 \text{ torr} \times \dfrac{1 \text{ atm}}{760 \text{ torr}}) \times \dfrac{169.89 \text{ g}}{\text{mol}}}{\dfrac{0.08206 \text{ L atm}}{\text{mol K}} \times 358 \text{ K}} = 5.77 \text{ g/L for } SiCl_4$$

For SiHCl$_3$, molar mass = M = 28.09 + 1.008 + 3(35.45) = 135.45 g/mol

$$d = \frac{PM}{RT} = \frac{(758 \text{ torr} \times \dfrac{1 \text{ atm}}{760 \text{ torr}}) \times \dfrac{135.45 \text{ g}}{\text{mol}}}{\dfrac{0.08206 \text{ L atm}}{\text{mol K}} \times 358 \text{ K}} = 4.60 \text{ g/L for SiHCl}_3$$

Partial Pressure

63. $$P_{CO_2} = \frac{nRT}{V} = \frac{\left(7.8 \text{ g} \times \dfrac{1 \text{ mol}}{44.01 \text{ g}}\right) \times \dfrac{0.08206 \text{ L atm}}{\text{mol K}} \times 300. \text{ K}}{4.0 \text{ L}} = 1.1 \text{ atm}$$

With air present, the partial pressure of CO$_2$ will still be 1.1 atm. The total pressure will be the sum of the partial pressures, $P_{total} = P_{CO_2} + P_{air}$.

$$P_{total} = 1.1 \text{ atm} + \left(740 \text{ torr} \times \frac{1 \text{ atm}}{760 \text{ torr}}\right) = 1.1 + 0.97 = 2.1 \text{ atm}$$

65.   Use the relationship $P_1V_1 = P_2V_2$ for each gas, since T and n for each gas is constant.

For H$_2$: $P_2 = \dfrac{P_1V_1}{V_2} = 475 \text{ torr} \times \dfrac{2.00 \text{ L}}{3.00 \text{ L}} = 317 \text{ torr}$

For N$_2$: $P_2 = 0.200 \text{ atm} \times \dfrac{1.00 \text{ L}}{3.00 \text{ L}} = 0.0667 \text{ atm};\ 0.0667 \text{ atm} \times \dfrac{760 \text{ torr}}{\text{atm}} = 50.7 \text{ torr}$

$P_{total} = P_{H_2} + P_{N_2} = 317 + 50.7 = 368 \text{ torr}$

67.   a.   mol fraction CH$_4$ = $\chi_{CH_4} = \dfrac{P_{CH_4}}{P_{total}} = \dfrac{0.175 \text{ atm}}{0.175 \text{ atm} + 0.250 \text{ atm}} = 0.412;\ \chi_{O_2} = 1.000 - 0.412 = 0.588$

b.   PV = nRT, $n_{total} = \dfrac{P_{total} \times V}{RT} = \dfrac{0.425 \text{ atm} \times 10.5 \text{ L}}{\dfrac{0.08206 \text{ L atm}}{\text{mol K}} \times 338 \text{ K}} = 0.161 \text{ mol}$

c. $\chi_{CH_4} = \dfrac{n_{CH_4}}{n_{total}}$, $n_{CH_4} = \chi_{CH_4} \times n_{total} = 0.412 \times 0.161 \text{ mol} = 6.63 \times 10^{-2} \text{ mol CH}_4$

$6.63 \times 10^{-2} \text{ mol CH}_4 \times \dfrac{16.04 \text{ g CH}_4}{\text{mol CH}_4} = 1.06 \text{ g CH}_4$

$n_{O_2} = 0.588 \times 0.161 \text{ mol} = 9.47 \times 10^{-2} \text{ mol O}_2;\ 9.47 \times 10^{-2} \text{ mol O}_2 \times \dfrac{32.00 \text{ g O}_2}{\text{mol O}_2} = 3.03 \text{ g O}_2$

69. $P_{TOT} = P_{H_2} + P_{H_2O}$, 1.032 atm $= P_{H_2} + 32$ torr $\times \dfrac{1\ atm}{760\ torr}$, $P_{H_2} = 1.032 - 0.042 = 0.990$ atm

$$n_{H_2} = \dfrac{P_{H_2}V}{RT} = \dfrac{0.990\ atm \times 0.240\ L}{0.08206\ \dfrac{L\ atm}{mol\ K} \times 303\ K} = 9.56 \times 10^{-3}\ mol\ H_2$$

$$9.56 \times 10^{-3}\ mol\ H_2 \times \dfrac{1\ mol\ Zn}{mol\ H_2} \times \dfrac{65.38\ g\ Zn}{mol\ Zn} = 0.625\ g\ Zn$$

71. $2\ NaClO_3(s) \rightarrow 2\ NaCl(s) + 3\ O_2(g)$

$P_{total} = P_{O_2} + P_{H_2O}$, $P_{O_2} = P_{total} - P_{H_2O} = 734$ torr $- 19.8$ torr $= 714$ torr

$$n_{O_2} = \dfrac{P_{O_2} \times V}{RT} = \dfrac{\left(714\ torr \times \dfrac{1\ atm}{760\ torr}\right) \times 0.0572\ L}{\dfrac{0.08206\ L\ atm}{mol\ K} \times (273 + 22)\ K} = 2.22 \times 10^{-3}\ mol\ O_2$$

Mass $NaClO_3$ decomposed $= 2.22 \times 10^{-3}\ mol\ O_2 \times \dfrac{2\ mol\ NaClO_3}{3\ mol\ O_2} \times \dfrac{106.44\ g\ NaClO_3}{mol\ NaClO_3} = 0.158$ g $NaClO_3$

Mass % $NaClO_3 = \dfrac{0.158\ g}{0.8765\ g} \times 100 = 18.0\%$

Kinetic Molecular Theory and Real Gases

73. $(KE)_{avg} = (3/2)\ RT$; At 273 K: $(KE)_{avg} = \dfrac{3}{2} \times \dfrac{8.3145\ J}{mol\ K} \times 273\ K = 3.40 \times 10^3$ J/mol

At 546 K: $(KE)_{avg} = \dfrac{3}{2} \times \dfrac{8.3145\ J}{mol\ K} \times 546\ K = 6.81 \times 10^3$ J/mol

75. $u_{rms} = \left(\dfrac{3RT}{M}\right)^{1/2}$, where $R = \dfrac{8.3145\ J}{mol\ K}$ and M = molar mass in kg $= 1.604 \times 10^{-2}$ kg/mol for CH_4

For CH_4 at 273 K: $u_{rms} = \left(\dfrac{\dfrac{3 \times 8.3145\ J}{mol\ K} \times 273\ K}{1.604 \times 10^{-2}\ kg/mol}\right)^{1/2} = 652$ m/s

Similarly u_{rms} for CH_4 at 546 K is 921 m/s.

For N_2 at 273 K: $u_{rms} = \left(\dfrac{\dfrac{3 \times 8.3145\ J}{mol\ K} \times 273\ K}{2.802 \times 10^{-2}\ kg/mol}\right)^{1/2} = 493$ m/s

Similarly for N_2 at 546 K, $u_{rms} = 697$ m/s.

77. $KE_{ave} = (3/2)\,RT$ and $KE = (1/2)\,mv^2$; As the temperature increases, the average kinetic energy of the gas sample will increase. The average kinetic energy increases because the increased temperature results in an increase in the average velocity of the gas molecules.

79.   a.   They will all have the same average kinetic energy since they are all at the same temperature.

       b.   Flask C; H_2 has the smallest molar mass. At constant T, the lightest molecules are the fastest (on the average). This must be true in order for the average kinetic energies to be constant.

81.   Graham's law of effusion:

$$\frac{Rate_1}{Rate_2} = \left(\frac{M_2}{M_1}\right)^{1/2} \text{ where M = molar mass;} \quad \frac{31.50}{30.50} = \left(\frac{32.00}{M}\right)^{1/2} = 1.033$$

$$\frac{32.00}{M} = 1.067, \text{ so M = 29.99 g/mol; Of the choices, the gas would be NO, nitrogen monoxide.}$$

83. $$\frac{Rate_1}{Rate_2} = \left(\frac{M_2}{M_1}\right)^{1/2}, \quad \frac{Rate\,(^{12}C\,^{17}O)}{Rate\,(^{12}C\,^{18}O)} = \left(\frac{30.0}{29.0}\right)^{1/2} = 1.02; \quad \frac{Rate\,(^{12}C\,^{16}O)}{Rate\,(^{12}C\,^{18}O)} = \left(\frac{30.0}{28.0}\right)^{1/2} = 1.04$$

The relative rates of effusion of $^{12}C^{16}O$: $^{12}C^{17}O$: $^{12}C^{18}O$ are 1.04, 1.02, 1.00.

Advantage: CO_2 isn't as toxic as CO.

Major disadvantages of using CO_2 instead of CO:

       1.   Can get a mixture of oxygen isotopes in CO_2.

       2.   Some species, e.g., $^{12}C^{16}O^{18}O$ and $^{12}C^{17}O_2$, would effuse (gaseously diffuse) at about the same rate since the masses are about equal. Thus, some species cannot be separated from each other.

85.   a. $$P = \frac{nRT}{V} = \frac{0.5000 \text{ mol} \times \dfrac{0.08206 \text{ L atm}}{\text{mol K}} \times (25.0 + 273.2) \text{ K}}{1.0000 \text{ L}} = 12.24 \text{ atm}$$

       b. $$\left[P + a\left(\frac{n}{V}\right)^2\right] \times (V - nb) = nRT; \text{ For } N_2\text{: } a = 1.39 \text{ atm L}^2/\text{mol}^2 \text{ and } b = 0.0391 \text{ L/mol}$$

$$\left[P + 1.39\left(\frac{0.5000}{1.0000}\right)^2 \text{atm}\right] \times (1.0000 \text{ L} - 0.5000 \times 0.0391 \text{ L}) = 12.24 \text{ L atm}$$

$$(P + 0.348 \text{ atm}) \times (0.9805 \text{ L}) = 12.24 \text{ L atm}$$

$$P = \frac{12.24 \text{ L atm}}{0.9805 \text{ L}} - 0.348 \text{ atm} = 12.48 - 0.348 = 12.13 \text{ atm}$$

c.   The ideal gas law is high by 0.11 atm or $\frac{0.11}{12.13} \times 100 = 0.91\%$.

Atmospheric Chemistry

87. $\chi_{NO} = 5 \times 10^{-7}$ from Table 5.4. $P_{NO} = \chi_{NO} \times P_{total} = 5 \times 10^{-7} \times 1.0 \text{ atm} = 5 \times 10^{-7} \text{ atm}$

$$PV = nRT, \quad \frac{n}{V} = \frac{P}{RT} = \frac{5 \times 10^{-7} \text{ atm}}{\dfrac{0.08206 \text{ L atm}}{\text{mol K}} \times 273 \text{ K}} = 2 \times 10^{-8} \text{ mol NO/L}$$

$$\frac{2 \times 10^{-8} \text{ mol}}{L} \times \frac{1 \text{ L}}{1000 \text{ cm}^3} \times \frac{6.022 \times 10^{23} \text{ molecules}}{\text{mol}} = 1 \times 10^{13} \text{ molecules NO/cm}^3$$

89.   At 100. km, $T \approx -75 °C$ and $P \approx 10^{-4.5} \approx 3 \times 10^{-5}$ atm.

$$PV = nRT, \quad \frac{PV}{T} = nR = \text{constant}, \quad \frac{P_1 V_1}{T_1} = \frac{P_2 V_2}{T_2}$$

$$V_2 = \frac{V_1 P_1 T_2}{P_2 T_1} = \frac{10.0 \text{ L} \times (3 \times 10^{-5} \text{ atm}) \times 273 \text{ K}}{1.0 \text{ atm} \times 198 \text{ K}} = 4 \times 10^{-4} \text{ L} = 0.4 \text{ mL}$$

91. $N_2(g) + O_2(g) \rightarrow 2 \text{ NO}(g)$, automobile combustion or formed by lightning

$2 \text{ NO}(g) + O_2(g) \rightarrow 2 \text{ NO}_2(g)$, reaction with atmospheric O_2

$2 \text{ NO}_2(g) + H_2O(l) \rightarrow HNO_3(aq) + HNO_2(aq)$, reaction with atmospheric H_2O

$S(s) + O_2(g) \rightarrow SO_2(g)$, combustion of coal

$2 \text{ SO}_2(g) + O_2(g) \rightarrow 2SO_3(g)$, reaction with atmospheric O_2

$H_2O(l) + SO_3(g) \rightarrow H_2SO_4(aq)$, reaction with atmospheric H_2O

Additional Exercises

93.   a. $PV = nRT$ b. $PV = nRT$ c. $PV = nRT$

   $PV = \text{Constant}$ $P = \left(\dfrac{nR}{V}\right) \times T = \text{Const} \times T$ $T = \left(\dfrac{P}{nR}\right) \times V = \text{Const} \times V$

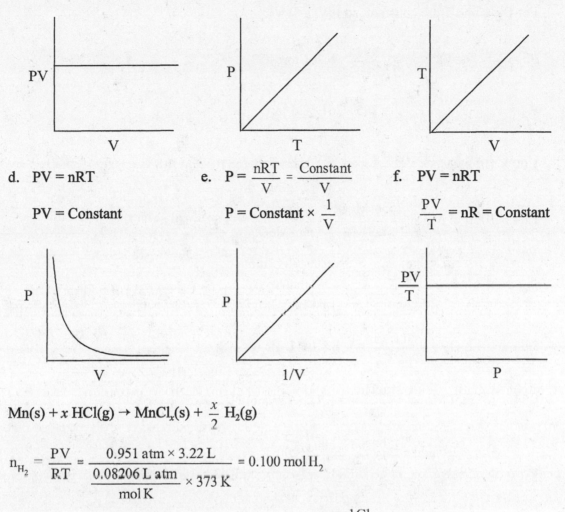

d. $PV = nRT$

$PV = Constant$

e. $P = \dfrac{nRT}{V} = \dfrac{Constant}{V}$

$P = Constant \times \dfrac{1}{V}$

f. $PV = nRT$

$\dfrac{PV}{T} = nR = Constant$

95. $Mn(s) + x\,HCl(g) \rightarrow MnCl_x(s) + \dfrac{x}{2}\,H_2(g)$

$$n_{H_2} = \frac{PV}{RT} = \frac{0.951\ \text{atm} \times 3.22\ \text{L}}{\dfrac{0.08206\ \text{L atm}}{\text{mol K}} \times 373\ \text{K}} = 0.100\ \text{mol}\ H_2$$

mol Cl in compound = mol HCl = $0.100\ \text{mol}\ H_2 \times \dfrac{x\ \text{mol Cl}}{\dfrac{x}{2}\ \text{mol}\ H_2} = 0.200\ \text{mol Cl}$

$$\frac{\text{mol Cl}}{\text{mol Mn}} = \frac{0.200\ \text{mol Cl}}{2.747\ \text{g Mn} \times \dfrac{1\ \text{mol Mn}}{54.94\ \text{g Mn}}} = \frac{0.200\ \text{mol Cl}}{0.05000\ \text{mol Mn}} = 4.00$$

The formula of compound is $MnCl_4$.

97.   We will apply Boyle's law to solve. $PV = nRT = $ contstant, $P_1V_1 = P_2V_2$

Let condition (1) correspond to He from the tank that can be used to fill balloons. We must leave 1.0 atm of He in the tank, so $P_1 = 200.$ atm - 1.00 = 199 atm and $V_1 = 15.0$ L. Condition (2) will correspond to the filled balloons with $P_2 = 1.00$ atm and $V_2 = N(2.00\ \text{L})$ where N is the number of filled balloons, each at a volume of 2.00 L.

199 atm × 15.0 L = 1.00 atm × N(2.00 L), N = 1492.5; We can't fill 0.5 of a balloon, so N = 1492 balloons or to 3 significant figures, 1490 balloons.

99.   For O_2, n and T are constant, so $P_1V_1 = P_2V_2$.

$$P_1 = \frac{P_2V_2}{V_1} = 785 \text{ torr} \times \frac{1.94 \text{ L}}{2.00 \text{ L}} = 761 \text{ torr} = P_{O_2}$$

$$P_{tot} = P_{O_2} + P_{H_2O}, \quad P_{H_2O} = 785 - 761 = 24 \text{ torr}$$

101. $$1.00 \times 10^3 \text{ kg Mo} \times \frac{1000 \text{ g}}{\text{kg}} \times \frac{1 \text{ mol Mo}}{95.94 \text{ g Mo}} = 1.04 \times 10^4 \text{ mol Mo}$$

$$1.04 \times 10^4 \text{ mol Mo} \times \frac{1 \text{ mol MoO}_3}{\text{mol Mo}} \times \frac{7/2 \text{ mol O}_2}{\text{mol MoO}_3} = 3.64 \times 10^4 \text{ mol O}_2$$

$$V_{O_2} = \frac{n_{O_2}RT}{P} = \frac{3.64 \times 10^4 \text{ mol} \times \dfrac{0.08206 \text{ L atm}}{\text{mol K}} \times 290. \text{ K}}{1.00 \text{ atm}} = 8.66 \times 10^5 \text{ L of O}_2$$

$$8.66 \times 10^5 \text{ L O}_2 \times \frac{100 \text{ L air}}{21 \text{ L O}_2} = 4.1 \times 10^6 \text{ L air}$$

$$1.04 \times 10^4 \text{ mol Mo} \times \frac{3 \text{ mol H}_2}{\text{mol Mo}} = 3.12 \times 10^4 \text{ mol H}_2$$

$$V_{H_2} = \frac{3.12 \times 10^4 \text{ mol} \times \dfrac{0.08206 \text{ L atm}}{\text{mol K}} \times 290. \text{ K}}{1.00 \text{ atm}} = 7.42 \times 10^5 \text{ L of H}_2$$

103. $$750. \text{ mL juice} \times \frac{12 \text{ mL C}_2\text{H}_5\text{OH}}{100 \text{ mL juice}} = 90. \text{ mL C}_2\text{H}_5\text{OH present}$$

$$90. \text{ mL C}_2\text{H}_5\text{OH} \times \frac{0.79 \text{ g C}_2\text{H}_5\text{OH}}{\text{mL C}_2\text{H}_5\text{OH}} \times \frac{1 \text{ mol C}_2\text{H}_5\text{OH}}{46.07 \text{ g C}_2\text{H}_5\text{OH}} \times \frac{2 \text{ mol CO}_2}{2 \text{ mol C}_2\text{H}_5\text{OH}} = 1.5 \text{ mol CO}_2$$

The CO_2 will occupy (825 - 750. =) 75 mL not occupied by the liquid (headspace).

$$P_{CO_2} = \frac{n_{CO_2} \times RT}{V} = \frac{1.5 \text{ mol} \times \dfrac{0.08206 \text{ L atm}}{\text{mol K}} \times 298 \text{ K}}{75 \times 10^{-3} \text{ L}} = 490 \text{ atm}$$

Actually, enough CO_2 will dissolve in the wine to lower the pressure of CO_2 to a much more reasonable value.

105.   If Be^{3+}, the formula is $Be(C_5H_7O_2)_3$ and $M \approx 13.5 + 15(12) + 21(1) + 6(16) = 311$ g/mol.

If Be^{2+}, the formula is $Be(C_5H_7O_2)_2$ and $M \approx 9.0 + 10(12) + 14(1) + 4(16) = 207$ g/mol.

Data Set I (M = dRT/P and d = mass/V):

$$M = \frac{mass \times RT}{PV} = \frac{0.2022 \text{ g} \times \dfrac{0.08206 \text{ L atm}}{\text{mol K}} \times 286 \text{ K}}{(765.2 \text{ torr} \times \dfrac{1 \text{ atm}}{760 \text{ torr}}) \times 22.6 \times 10^{-3} \text{ L}} = 209 \text{ g/mol}$$

Data Set II:

$$M = \frac{mass \times RT}{PV} = \frac{0.2224 \text{ g} \times \dfrac{0.08206 \text{ L atm}}{\text{mol K}} \times 290. \text{ K}}{(764.6 \text{ torr} \times \dfrac{1 \text{ atm}}{760 \text{ torr}}) \times 26.0 \times 10^{-3} \text{ L}} = 202 \text{ g/mol}$$

These results are close to the expected value of 207 g/mol for $Be(C_5H_7O_2)_2$. Thus, we conclude from these data that beryllium is a divalent element with an atomic mass of 9.0 amu.

107. $P_{total} = P_{N_2} + P_{H_2O}$, $P_{N_2} = 726$ torr - 23.8 torr = 702 torr $\times \dfrac{1 \text{ atm}}{760 \text{ torr}} = 0.924$ atm

$$PV = nRT, \quad n_{N_2} = \frac{P_{N_2} \times V}{RT} = \frac{0.924 \text{ atm} \times 31.8 \times 10^{-3} \text{ L}}{\dfrac{0.08206 \text{ L atm}}{\text{mol K}} \times 298 \text{ K}} = 1.20 \times 10^{-3} \text{ mol N}_2$$

Mass of N in compound = 1.20×10^{-3} mol $\times \dfrac{28.02 \text{ g N}_2}{\text{mol}} = 3.36 \times 10^{-2}$ g

$\% N = \dfrac{3.36 \times 10^{-2} \text{ g}}{0.253 \text{ g}} \times 100 = 13.3\% \text{ N}$

109.   At constant T, the lighter the gas molecules, the faster the average velocity. Therefore, the pressure will increase initially because the lighter H_2 molecules will effuse into container A faster than air will escape. However, the pressures will eventually equalize once the gases have had time to mix thoroughly.

111.   The values of *a* are: H_2, $\dfrac{0.244 \text{ L}^2 \text{ atm}}{\text{mol}^2}$; CO_2, 3.59; N_2, 1.39; CH_4, 2.25

Since *a* is a measure of interparticle attractions, the attractions are greatest for CO_2.

Challenge Problems

113. $BaO(s) + CO_2(g) \rightarrow BaCO_3(s);\ CaO(s) + CO_2(g) \rightarrow CaCO_3(s)$

$$n_i = \frac{P_i V}{RT} = \text{initial moles of } CO_2 = \frac{\dfrac{750.}{760} \text{ atm} \times 1.50 \text{ L}}{\dfrac{0.08206 \text{ L atm}}{\text{mol K}} \times 303.2 \text{ K}} = 0.0595 \text{ mol } CO_2$$

$$n_f = \frac{P_f V}{RT} = \text{final moles of } CO_2 = \frac{\dfrac{230.}{760} \text{ atm} \times 1.50 \text{ L}}{\dfrac{0.08206 \text{ L atm}}{\text{mol K}} \times 303.2 \text{ K}} = 0.0182 \text{ mol } CO_2$$

$0.0595 - 0.0182 = 0.0413 \text{ mol } CO_2$ reacted.

Since each metal reacts 1:1 with CO_2, the mixture contains 0.0413 mol of BaO and CaO. The molar masses of BaO and CaO are 153.3 g/mol and 56.08 g/mol, respectively.

Let x = g BaO and y = g CaO, so:

$$x + y = 5.14 \text{ g and } \frac{x}{153.3} + \frac{y}{56.08} = 0.0413 \text{ mol}$$

Solving by simultaneous equations:

$$
\begin{array}{r}
x + 2.734\,y = 6.33 \\
\underline{-x -y = -5.14} \\
1.734\,y = 1.19
\end{array}
$$

$y = 0.686$ g CaO and $5.14 - y = x = 4.45$ g BaO

$\% \text{ BaO} = \dfrac{4.45 \text{ g BaO}}{5.14 \text{ g}} \times 100 = 86.6\% \text{ BaO};\ \% \text{ CaO} = 100.0 - 86.6 = 13.4\% \text{ CaO}$

115.   a.   The reaction is: $CH_4(g) + 2\,O_2(g) \rightarrow CO_2(g) + 2\,H_2O(g)$

$PV = nRT,\ \dfrac{PV}{n} = RT = \text{constant},\ \dfrac{P_{CH_4} V_{CH_4}}{n_{CH_4}} = \dfrac{P_{air} V_{air}}{n_{air}}$

The balanced equation requires 2 mol O_2 for every mol of CH_4 that reacts. For three times as much oxygen, we would need 6 mol O_2 per mol of CH_4 reacted ($n_{O_2} = 6\,n_{CH_4}$). Air is 21% mol percent O_2, so $n_{O_2} = 0.21\,n_{air}$. Therefore, the mol of air we would need to delivery the excess O_2 are:

$n_{O_2} = 0.21\,n_{air} = 6\,n_{CH_4},\ n_{air} = 29\,n_{CH_4},\ \dfrac{n_{air}}{n_{CH_4}} = 29$

In one minute:

$$V_{air} = V_{CH_4} \times \frac{n_{air}}{n_{CH_4}} \times \frac{P_{CH_4}}{P_{air}} = 200.\ L \times 29 \times \frac{1.50\ atm}{1.00\ atm} = 8.7 \times 10^3\ L\ air/min$$

b. If x moles of CH_4 were reacted, then $6\ x$ mol O_2 were added, producing $0.950\ x$ mol CO_2 and $0.050\ x$ mol of CO. In addition, $2\ x$ mol H_2O must be produced to balance the hydrogens.

$$CH_4(g) + 2\ O_2(g) \rightarrow CO_2(g) + 2\ H_2O(g);\ \ CH_4(g) + 3/2\ O_2(g) \rightarrow CO(g) + 2\ H_2O(g)$$

Amount O_2 reacted:

$$0.950\ x\ mol\ CO_2 \times \frac{2\ mol\ O_2}{mol\ CO_2} = 1.90\ x\ mol\ O_2$$

$$0.050\ x\ mol\ CO \times \frac{1.5\ mol\ O_2}{mol\ CO} = 0.075\ x\ mol\ O_2$$

Amount of O_2 left in reaction mixture = $6.00\ x - 1.90\ x - 0.075\ x = 4.03\ x$ mol O_2

$$\text{Amount of }N_2\ = 6.00\ x\ mol\ O_2 \times \frac{79\ mol\ N_2}{21\ mol\ O_2} = 22.6\ x \approx 23\ x\ mol\ N_2$$

The reaction mixture contains:

$$0.950\ x\ mol\ CO_2 + 0.050\ x\ mol\ CO + 4.03\ x\ mol\ O_2 + 2.00\ x\ mol\ H_2O$$

$$+ 23\ x\ mol\ N_2 = 30.\ x\ total\ mol\ of\ gas$$

$$\chi_{CO} = \frac{0.050\ x}{30.\ x} = 0.0017;\ \ \chi_{CO_2} = \frac{0.950\ x}{30.\ x} = 0.032;\ \ \chi_{O_2} = \frac{4.03\ x}{30.\ x} = 0.13;$$

$$\chi_{H_2O} = \frac{2.00\ x}{30.\ x} = 0.067;\ \ \ \chi_{N_2} = \frac{23\ x}{30.\ x} = 0.77$$

117.   a. Volume of hot air: $V = \dfrac{4}{3}\pi r^3 = \dfrac{4}{3}\pi(2.50\ m)^3 = 65.4\ m^3$

(Note: radius = diameter/2 = 5.00/2 = 2.50 m)

$$65.4\ m^3 \times \left(\frac{10\ dm}{m}\right)^3 \times \frac{1\ L}{dm^3} = 6.54 \times 10^4\ L$$

$$n = \frac{PV}{RT} = \frac{\left(745\ torr \times \dfrac{1\ atm}{760\ torr}\right) \times 6.54 \times 10^4\ L}{\dfrac{0.08206\ L\ atm}{mol\ K} \times (273 + 65)\ K} = 2.31 \times 10^3\ mol\ air$$

Mass of hot air $= 2.31 \times 10^3$ mol $\times \dfrac{29.0\text{ g}}{\text{mol}} = 6.70 \times 10^4$ g

Mass of air displaced:

$$n = \frac{PV}{RT} = \frac{\dfrac{745}{760}\text{ atm} \times 6.54 \times 10^4\text{ L}}{\dfrac{0.08206\text{ L atm}}{\text{mol K}} \times (273 + 21)\text{ K}} = 2.66 \times 10^3\text{ mol air}$$

Mass $= 2.66 \times 10^3$ mol $\times \dfrac{29.0\text{ g}}{\text{mol}} = 7.71 \times 10^4$ g of air displaced

Lift $= 7.71 \times 10^4$ g $- 6.70 \times 10^4$ g $= 1.01 \times 10^4$ g

b. Mass of air displaced is the same, 7.71×10^4 g. Moles of He in balloon will be the same as moles of air displaced, 2.66×10^3 mol, since P, V and T are the same.

Mass of He $= 2.66 \times 10^3$ mol $\times \dfrac{4.003\text{ g}}{\text{mol}} = 1.06 \times 10^4$ g

Lift $= 7.71 \times 10^4$ g $- 1.06 \times 10^4$ g $= 6.65 \times 10^4$ g

c. Mass of hot air:

$$n = \frac{PV}{RT} = \frac{\dfrac{630.}{760}\text{ atm} \times 6.54 \times 10^4\text{ L}}{\dfrac{0.08206\text{ L atm}}{\text{mol K}} \times 338\text{ K}} = 1.95 \times 10^3\text{ mol air}$$

1.95×10^3 mol $\times \dfrac{29.0\text{ g}}{\text{mol}} = 5.66 \times 10^4$ g of hot air

Mass of air displaced:

$$n = \frac{PV}{RT} = \frac{\dfrac{630.}{760}\text{ atm} \times 6.54 \times 10^4\text{ L}}{\dfrac{0.08206\text{ L atm}}{\text{mol K}} \times 294\text{ K}} = 2.25 \times 10^3\text{ mol air}$$

2.25×10^3 mol $\times \dfrac{29.0\text{ g}}{\text{mol}} = 6.53 \times 10^4$ g of air displaced

Lift $= 6.53 \times 10^4$ g $- 5.66 \times 10^4$ g $= 8.7 \times 10^3$ g

119.   a. If we have 1.0×10^6 L of air, then there are 3.0×10^2 L of CO.

$P_{CO} = \chi_{CO} \times P_{total}$; $\chi_{CO} = \dfrac{V_{CO}}{V_{total}}$ since $V \propto n$; $P_{CO} = \dfrac{3.0 \times 10^2\text{ L}}{1.0 \times 10^6\text{ L}} \times 628\text{ torr} = 0.19\text{ torr}$

b. $n_{CO} = \dfrac{P_{CO} \times V}{RT}$;   Assuming 1.0 cm^3 of air = 1.0 mL = 1.0 × 10^{-3} L:

$$n_{CO} = \dfrac{\dfrac{0.19}{760} \text{ atm} \times 1.0 \times 10^{-3} \text{ L}}{\dfrac{0.08206 \text{ L atm}}{\text{mol K}} \times 273 \text{ K}} = 1.1 \times 10^{-8} \text{ mol CO}$$

$1.1 \times 10^{-8} \text{ mol} \times \dfrac{6.022 \times 10^{23} \text{ molecules}}{\text{mol}} = 6.6 \times 10^{15}$ molecules CO in the 1.0 cm^3 of air

CHAPTER SIX

THERMOCHEMISTRY

Questions

9.   A coffee-cup calorimeter is at constant (atmospheric) pressure. The heat released or gained at constant pressure is ΔH. A bomb calorimeter is at constant volume. The heat released or gained at constant volume is ΔE.

11. The specific heat capacities are: 0.89 J/g•°C (Al) and 0.45 J/g•°C (Fe)
     Al would be the better choice. It has a higher heat capacity and a lower density than Fe. Using Al, the same amount of heat could be dissipated by a smaller mass, keeping the mass of the amplifier down.

13. A state function is a function whose change depends only on the initial and final states and not on how one got from the initial to the final state. If H and E were not state functions, the law of conservation of energy (first law) would not be true.

15. In order to compare values of ΔH to each other, a common reference (or zero) point must be chosen. The definition of ΔH_f° establishes the pure elements in their standard states as that common reference point.

Exercises

Potential and Kinetic Energy

17. $KE = \frac{1}{2}mv^2$; Convert mass and velocity to SI units. $1 \text{ J} = \dfrac{1 \text{ kg m}^2}{\text{s}^2}$

$$\text{Mass} = 5.25 \text{ oz} \times \frac{1 \text{ lb}}{16 \text{ oz}} \times \frac{1 \text{ kg}}{2.205 \text{ lb}} = 0.149 \text{ kg}$$

$$\text{Velocity} = \frac{1.0 \times 10^2 \text{ mi}}{\text{hr}} \times \frac{1 \text{ hr}}{60 \text{ min}} \times \frac{1 \text{ min}}{60 \text{ s}} \times \frac{1760 \text{ yd}}{\text{mi}} \times \frac{1 \text{ m}}{1.094 \text{ yd}} = \frac{45 \text{ m}}{\text{s}}$$

$$KE = \frac{1}{2}mv^2 = \frac{1}{2} \times 0.149 \text{ kg} \times \left(\frac{45 \text{ m}}{\text{s}}\right)^2 = 150 \text{ J}$$

19. $KE = \frac{1}{2}mv^2 = \frac{1}{2} \times 2.0 \text{ kg} \times \left(\frac{1.0 \text{ m}}{\text{s}}\right)^2 = 1.0 \text{ J}; \quad KE = \frac{1}{2}mv^2 = \frac{1}{2} \times 1.0 \text{ kg} \times \left(\frac{2.0 \text{ m}}{\text{s}}\right)^2 = 2.0 \text{ J}$

The 1.0 kg object with a velocity of 2.0 m/s has the greater kinetic energy.

Heat and Work

21.   a. $\Delta E = q + w = 51 \text{ kJ} + (-15 \text{ kJ}) = 36 \text{ kJ}$

b. $\Delta E = 100. \text{ kJ} + (-65 \text{ kJ}) = 35 \text{ kJ}$ c. $\Delta E = -65 + (-20.) = -85 \text{ kJ}$

d. When the system delivers work to the surroundings, $w < 0$. This is the case in all these examples, a, b and c.

23. $\Delta E = q + w = -125 + 104 = -21 \text{ kJ}$

25. $w = -P\Delta V = -P \times (V_f - V_i) = -2.0 \text{ atm} \times (5.0 \times 10^{-3} \text{ L} - 5.0 \text{ L}) = -2.0 \text{ atm} \times (-5.0 \text{ L}) = 10. \text{ L atm}$

We can also calculate the work in Joules.

$1 \text{ atm} = 1.013 \times 10^5 \text{ Pa} = 1.013 \times 10^5 \frac{\text{kg}}{\text{m s}^2}; \quad 1 \text{ L} = 1000 \text{ cm}^3 = 1 \times 10^{-3} \text{ m}^3$

$1 \text{ L atm} = 1 \times 10^{-3} \text{ m}^3 \times 1.013 \times 10^5 \frac{\text{kg}}{\text{m s}^2} = 101.3 \frac{\text{kg m}^2}{\text{s}^2} = 101.3 \text{ J}$

$w = 10. \text{ L atm} \times \frac{101.3 \text{ J}}{\text{L atm}} = 1013 \text{ J} = 1.0 \times 10^3 \text{ J}$

27. $q = \text{molar heat capacity} \times \text{mol} \times \Delta T = \frac{20.8 \text{ J}}{^\circ\text{C mol}} \times 39.1 \text{ mol} \times (38.0 - 0.0) \,^\circ\text{C} = 30{,}900 \text{ J} = 30.9 \text{ kJ}$

$w = -P\Delta V = -1.00 \text{ atm} \times (998 \text{ L} - 876 \text{ L}) = -122 \text{ L atm} \times \frac{101.3 \text{ J}}{\text{L atm}} = -12{,}400 \text{ J} = -12.4 \text{ kJ}$

$\Delta E = q + w = 30.9 \text{ kJ} + (-12.4 \text{ kJ}) = 18.5 \text{ kJ}$

Properties of Enthalpy

29.   This is an endothermic reaction so heat must be absorbed in order to convert reactants into products. The high temperature environment of internal combustion engines provides the heat.

31.   a. Heat is absorbed from the water (it gets colder) as KBr dissolves, so this is an endothermic process.

b. Heat is released as CH_4 is burned, so this is an exothermic process.

c. Heat is released to the water (it gets hot) as H_2SO_4 is added, so this is an exothermic process.

d. Heat must be added (absorbed) to boil water, so this is an endothermic process.

33. $4\ Fe(s) + 3\ O_2(g) \rightarrow 2\ Fe_2O_3(s)$ $\Delta H = -1652$ kJ; Note that 1652 kJ of heat are released when 4 mol Fe react with 3 mol O_2 to produce 2 mol Fe_2O_3.

a. 4.00 mol Fe $\times \dfrac{-1652\ kJ}{4\ mol\ Fe} = -1650$ kJ heat released

b. 1.00 ml $Fe_2O_3 \times \dfrac{-1652\ kJ}{2\ mol\ Fe_2O_3} = -826$ kJ heat released

c. 1.00 g Fe $\times \dfrac{1\ mol\ Fe}{55.85\ g} \times \dfrac{-1652\ kJ}{4\ mol\ Fe} = -7.39$ kJ heat released

d. 10.0 g Fe $\times \dfrac{1\ mol\ Fe}{55.85\ g} = 0.179$ mol Fe; 2.00 g $O_2 \times \dfrac{1\ mol\ O_2}{32.00\ g} = 0.0625$ mol O_2

0.179 mol Fe/0.0625 mol O_2 = 2.86; The balanced equation requires a 4 mol Fe/3 mol O_2 = 1.33 mol ratio. O_2 is limiting since the actual mol Fe/mol O_2 ratio is less than the required mol ratio.

0.0625 mol $O_2 \times \dfrac{-1652\ kJ}{3\ mol\ O_2} = -34.4$ kJ heat released

35.   From Sample Exercise 6.3, $q = 1.3 \times 10^8$ J. Since the heat transfer process is only 60.% efficient, the total energy required is: 1.3×10^8 J $\times \dfrac{100.\ J}{60.\ J} = 2.2 \times 10^8$ J

mass $C_3H_8 = 2.2 \times 10^8$ J $\times \dfrac{1\ mol\ C_3H_8}{2221 \times 10^3\ J} \times \dfrac{44.09\ g\ C_3H_8}{mol\ C_3H_8} = 4.4 \times 10^3$ g C_3H_8

Calorimetry and Heat Capacity

37.   Specific heat capacity is defined as the amount of heat necessary to raise the temperature of one gram of substance by one degree Celsius. Therefore, $H_2O(l)$ with the largest heat capacity value requires the largest amount of heat for this process. The amount of heat for $H_2O(l)$ is:

$$energy = s \times m \times \Delta T = \frac{4.18\ J}{g\ °C} \times 25.0\ g \times (37.0°C - 15.0°C) = 2.30 \times 10^3\ J$$

The largest temperature change when a certain amount of energy is added to a certain mass of substance will occur for the substance with the smallest specific heat capacity. This is Hg(l), and the temperature change for this process is:

$$\Delta T = \frac{\text{energy}}{s \times m} = \frac{10.7 \text{ kJ} \times \dfrac{1000 \text{ J}}{\text{kJ}}}{\dfrac{0.14 \text{ J}}{g \, °C} \times 550. \text{ g}} = 140 °C$$

39.   The units for specific heat capacity (s) are J/g•°C. $s = \dfrac{78.2 \text{ J}}{45.6 \text{ g} \times 13.3 °C} = \dfrac{0.129 \text{ J}}{g \, °C}$

Molar heat capacity $= \dfrac{0.129 \text{ J}}{g \, °C} \times \dfrac{207.2 \text{ g}}{\text{mol Pb}} = \dfrac{26.7 \text{ J}}{\text{mol} \, °C}$

41. | Heat loss by hot water | = | Heat gain by cooler water |

The magnitude of heat loss and heat gain are equal in calorimetry problems. The only difference is the sign (positive or negative). To avoid sign errors, keep all quantities positive and, if necessary, deduce the correct signs at the end of the problem. Water has a specific heat capacity = s = 4.18 J/°C•g = 4.18 J/K•g (ΔT in °C = ΔT in K).

Heat loss by hot water $= s \times m \times \Delta T = \dfrac{4.18 \text{ J}}{g \, K} \times 50.0 \text{ g} \times (330. \text{ K} - T_f)$

Heat gain by cooler water $= \dfrac{4.18 \text{ J}}{g \, K} \times 30.0 \text{ g} \times (T_f - 280. \text{ K})$;   Heat loss = Heat gain, so:

$\dfrac{209 \text{ J}}{K} \times (330. \text{ K} - T_f) = \dfrac{125 \text{ J}}{K} \times (T_f - 280. \text{ K})$, $6.90 \times 10^4 - 209 \, T_f = 125 \, T_f - 3.50 \times 10^4$

334 $T_f = 1.040 \times 10^5$, $T_f = 311$ K

Note that the final temperature is closer to the temperature of the more massive hot water, which is as it should be.

43.   Heat gained by water = heat loss by metal = $s \times m \times \Delta T$ where s = specific heat capacity.

Heat gain $= \dfrac{4.18 \text{ J}}{g \, °C} \times 150.0 \text{ g} \times (18.3 °C - 15.0 °C) = 2100 \text{ J}$

A common error in calorimetry problems is sign errors. Keeping all quantities positive helps eliminate sign errors.

heat loss = 2100 J = $s \times 150.0 \text{ g} \times (75.0 °C - 18.3 °C)$, $s = \dfrac{2100 \text{ J}}{150.0 \text{ g} \times 56.7 °C} = 0.25 \text{ J/g•°C}$

45. 50.0×10^{-3} L $\times$ 0.100 mol/L = 5.00×10^{-3} mol of both $AgNO_3$ and HCl are reacted. Thus, 5.00×10^{-3} mol of AgCl will be produced since there is a 1:1 mol ratio between reactants.

Heat lost by chemicals = Heat gained by solution

Heat gain $= \dfrac{4.18 \text{ J}}{g \, °C} \times 100.0 \text{ g} \times (23.40 - 22.60) °C = 330 \text{ J}$

Heat loss = 330 J; This is the heat evolved (exothermic reaction) when 5.00×10^{-3} mol of AgCl is produced. So q = -330 J and ΔH (heat per mol AgCl formed) is negative with a value of:

$$\Delta H = \frac{-330 \text{ J}}{5.00 \times 10^{-3} \text{ mol}} \times \frac{1 \text{ kJ}}{1000 \text{ J}} = \text{-66 kJ/mol}$$

Note: Sign errors are common with calorimetry problems. However, the correct sign for ΔH can easily be determined from the ΔT data, i.e., if ΔT of the solution increases, then the reaction is exothermic since heat was released, and if ΔT of the solution decreases, then the reaction is endothermic since the reaction absorbed heat from the water. For calorimetry problems, keep all quantities positive until the end of the calculation, then decide the sign for ΔH. This will help eliminate sign errors.

47.   Since ΔH is exothermic, the temperature of the solution will increase as $CaCl_2(s)$ dissolves. Keeping all quantities positive:

$$\text{Heat loss as } CaCl_2 \text{ dissolves} = 11.0 \text{ g } CaCl_2 \times \frac{1 \text{ mol } CaCl_2}{110.98 \text{ g } CaCl_2} \times \frac{81.5 \text{ kJ}}{\text{mol } CaCl_2} = 8.08 \text{ kJ}$$

$$\text{Heat gained by solution} = 8.08 \times 10^3 \text{ J} = \frac{4.18 \text{ J}}{\text{g } °C} \times (125 + 11.0) \text{ g} \times (T_f - 25.0°C)$$

$$T_f - 25.0°C = \frac{8.08 \times 10^3}{4.18 \times 136} = 14.2°C, \quad T_f = 14.2°C + 25.0°C = 39.2°C$$

49.   First, we need to get the heat capacity of the calorimeter from the combustion of benzoic acid.

Heat lost by combustion = Heat gained by calorimeter

$$\text{Heat loss} = 0.1584 \text{ g} \times \frac{26.42 \text{ kJ}}{\text{g}} = 4.185 \text{ kJ}$$

$$\text{Heat gain} = 4.185 \text{ kJ} = C_{cal} \times \Delta T, \quad C_{cal} = \frac{4.185 \text{ kJ}}{2.54°C} = 1.65 \text{ kJ/}°C$$

Now we can calculate the heat of combustion of vanillin. Heat loss = Heat gain

$$\text{Heat gain by calorimeter} = \frac{1.65 \text{ kJ}}{°C} \times 3.25°C = 5.36 \text{ kJ}$$

Heat loss = 5.36 kJ, which is the heat evolved by the combustion of the vanillin.

$$\Delta E_{comb} = \frac{-5.36 \text{ kJ}}{0.2130 \text{ g}} = \text{-25.2 kJ/g}; \quad \Delta E_{comb} = \frac{-25.2 \text{ kJ}}{\text{g}} \times \frac{152.14 \text{ g}}{\text{mol}} = \text{-3830 kJ/mol}$$

Hess's Law

51.   Information given:

$$C(s) + O_2(g) \rightarrow CO_2(g) \qquad\qquad \Delta H = -393.7 \text{ kJ}$$
$$CO(g) + 1/2\ O_2(g) \rightarrow CO_2(g) \qquad \Delta H = -283.3 \text{ kJ}$$

Using Hess's Law:

$$2\ C(s) + 2\ O_2(g) \rightarrow 2\ CO_2(g) \qquad\qquad \Delta H_1 = 2(-393.7 \text{ kJ})$$
$$2\ CO_2(g) \rightarrow 2\ CO(g) + O_2(g) \qquad\qquad \Delta H_2 = -2(-283.3 \text{ kJ})$$

$$\overline{2\ C(s) + O_2(g) \rightarrow 2\ CO(g) \qquad\qquad \Delta H = \Delta H_1 + \Delta H_2 = -220.8 \text{ kJ}}$$

Note: The enthalpy change for a reaction that is reversed is the negative quantity of the enthalpy change for the original reaction. If the coefficients in a balanced reaction are multiplied by an integer, the value of ΔH is multiplied by the same integer.

53. $2\ NO_2 \rightarrow N_2 + 2\ O_2 \qquad\qquad \Delta H = -(67.7 \text{ kJ})$
       $N_2 + 2\ O_2 \rightarrow N_2O_4 \qquad\qquad \Delta H = 9.7 \text{ kJ}$

$$\overline{2\ NO_2(g) \rightarrow N_2O_4(g) \qquad\qquad \Delta H = -58.0 \text{ kJ}}$$

55. $NO + O_3 \rightarrow NO_2 + O_2 \qquad\qquad \Delta H = -199 \text{ kJ}$
       $3/2\ O_2 \rightarrow O_3 \qquad\qquad\qquad \Delta H = -1/2(-427 \text{ kJ})$
       $O \rightarrow 1/2\ O_2 \qquad\qquad\qquad \Delta H = -1/2(495 \text{ kJ})$

$$\overline{NO(g) + O(g) \rightarrow NO_2(g) \qquad\qquad \Delta H = -233 \text{ kJ}}$$

57. $4\ HNO_3 \rightarrow 2\ N_2O_5 + 2\ H_2O \qquad\qquad \Delta H = -2(-76.6 \text{ kJ})$
       $2\ N_2 + 6\ O_2 + 2\ H_2 \rightarrow 4\ HNO_3 \qquad\qquad \Delta H = 4(-174.1 \text{ kJ})$
       $2\ H_2O \rightarrow 2\ H_2 + O_2 \qquad\qquad \Delta H = -2(-285.8 \text{ kJ})$

$$\overline{2\ N_2(g) + 5\ O_2(g) \rightarrow 2\ N_2O_5(g) \qquad\qquad \Delta H = 28.4 \text{ kJ}}$$

Standard Enthalpies of Formation

59.   The change in enthalpy that accompanies the formation of one mole of a compound from its elements, with all substances in their standard states, is the standard enthalpy of formation for a compound. The reactions that refer to ΔH_f° are:

$$Na(s) + 1/2\ Cl_2(g) \rightarrow NaCl(s); \quad H_2(g) + 1/2\ O_2(g) \rightarrow H_2O(l)$$

$$6\ C(\text{graphite, } s) + 6\ H_2(g) + 3\ O_2(g) \rightarrow C_6H_{12}O_6(s); \quad Pb(s) + S(\text{rhombic, } s) + 2\ O_2(g)$$
$$\rightarrow PbSO_4(s)$$

61.   In general: $\Delta H^\circ = \Sigma n_p \Delta H^\circ_{f,\,products} - \Sigma n_r \Delta H^\circ_{f,\,reactants}$ and all elements in their standard state have $\Delta H^\circ_f = 0$ by definition.

a.   The balanced equation is: $2\,NH_3(g) + 3\,O_2(g) + 2\,CH_4(g) \rightarrow 2\,HCN(g) + 6\,H_2O(g)$

$\Delta H^\circ = [2\;mol\;HCN \times \Delta H^\circ_{f,\,HCN} + 6\;mol\;H_2O(g) \times \Delta H^\circ_{f,\,H_2O}]$

$- [2\;mol\;NH_3 \times \Delta H^\circ_{f,\,NH_3} + 2\;mol\;CH_4 \times \Delta H^\circ_{f,\,CH_4}]$

$\Delta H^\circ = [2(135.1) + 6(-242)] - [2(-46) + 2(-75)] = -940.\;kJ$

b. $Ca_3(PO_4)_2(s) + 3\,H_2SO_4(l) \rightarrow 3\,CaSO_4(s) + 2\,H_3PO_4(l)$

$\Delta H^\circ = \left[3\;mol\;CaSO_4\left(\dfrac{-1433\;kJ}{mol}\right) + 2\;mol\;H_3PO_4(l)\left(\dfrac{-1267\;kJ}{mol}\right)\right]$

$- \left[1\;mol\;Ca_3(PO_4)_2\left(\dfrac{-4126\;kJ}{mol}\right) + 3\;mol\;H_2SO_4(l)\left(\dfrac{-814\;kJ}{mol}\right)\right]$

$\Delta H^\circ = -6833\;kJ - (-6568\;kJ) = -265\;kJ$

c. $NH_3(g) + HCl(g) \rightarrow NH_4Cl(s)$

$\Delta H^\circ = [1\;mol\;NH_4Cl \times \Delta H^\circ_{f,\,NH_4Cl}] - [1\;mol\;NH_3 \times \Delta H^\circ_{f,\,NH_3} + 1\;mol\;HCl \times \Delta H^\circ_{f,\,HCl}]$

$\Delta H^\circ = \left[1\;mol\left(\dfrac{-314\;kJ}{mol}\right)\right] - \left[1\;mol\left(\dfrac{-46\;kJ}{mol}\right) + 1\;mol\left(\dfrac{-92\;kJ}{mol}\right)\right]$

$\Delta H^\circ = -314\;kJ + 138\;kJ = -176\;kJ$

63.   a. $4\,NH_3(g) + 5\,O_2(g) \rightarrow 4\,NO(g) + 6\,H_2O(g)$; $\Delta H^\circ = \Sigma n_p \Delta H^\circ_{f,\,products} - \Sigma n_r \Delta H^\circ_{f,\,reactants}$

$\Delta H^\circ = \left[4\;mol\left(\dfrac{90.\;kJ}{mol}\right) + 6\;mol\left(\dfrac{-242\;kJ}{mol}\right)\right] - \left[4\;mol\left(\dfrac{-46\;kJ}{mol}\right)\right] = -908\;kJ$

$2\,NO(g) + O_2(g) \rightarrow 2\,NO_2(g)$

$\Delta H^\circ = \left[2\;mol\left(\dfrac{34\;kJ}{mol}\right)\right] - \left[2\;mol\left(\dfrac{90.\;kJ}{mol}\right)\right] = -112\;kJ$

$$3\ NO_2(g) + H_2O(l) \rightarrow 2\ HNO_3(aq) + NO(g)$$

$$\Delta H^\circ = \left[2\ mol\left(\frac{-207\ kJ}{mol}\right) + 1\ mol\left(\frac{90.\ kJ}{mol}\right)\right]$$

$$- \left[3\ mol\left(\frac{34\ kJ}{mol}\right) + 1\ mol\left(\frac{-286\ kJ}{mol}\right)\right] = -140.\ kJ$$

Note: All ΔH_f° values are assumed $\pm\ 1$ kJ.

b. $12\ NH_3(g) + 15\ O_2(g) \rightarrow 12\ NO(g) + 18\ H_2O(g)$
$12\ NO(g) + 6\ O_2(g) \rightarrow 12\ NO_2(g)$
$12\ NO_2(g) + 4\ H_2O(l) \rightarrow 8\ HNO_3(aq) + 4\ NO(g)$
$4\ H_2O(g) \rightarrow 4\ H_2O(l)$

$$\overline{12\ NH_3(g) + 21\ O_2(g) \rightarrow 8\ HNO_3(aq) + 4\ NO(g) + 14\ H_2O(g)}$$

The overall reaction is exothermic since each step is exothermic.

65. $3\ Al(s) + 3\ NH_4ClO_4(s) \rightarrow Al_2O_3(s) + AlCl_3(s) + 3\ NO(g) + 6\ H_2O(g)$

$$\Delta H^\circ - \left[6\ mol\left(\frac{-242\ kJ}{mol}\right) + 3\ mol\left(\frac{90.\ kJ}{mol}\right) + 1\ mol\left(\frac{-704\ kJ}{mol}\right) + 1\ mol\left(\frac{-1676\ kJ}{mol}\right)\right]$$

$$- \left[3\ mol\left(\frac{-295\ kJ}{mol}\right)\right] = -2677\ kJ$$

67. $2\ ClF_3(g) + 2\ NH_3(g) \rightarrow N_2(g) + 6\ HF(g) + Cl_2(g)$ $\Delta H^\circ = -1196$ kJ

$$\Delta H^\circ = [6\ \Delta H_{f,\ HF}^\circ] - [2\ \Delta H_{f\ ClF_3}^\circ + 2\ \Delta H_{f,\ NH_3}^\circ]$$

$$-1196\ kJ = 6\ mol\left(\frac{-271\ kJ}{mol}\right) - 2\ \Delta H_{f\ ClF_3}^\circ - 2\ mol\left(\frac{-46\ kJ}{mol}\right)$$

$$-1196\ kJ = -1626\ kJ - 2\ \Delta H_{f\ ClF_3}^\circ + 92\ kJ,\ \Delta H_{f\ ClF_3}^\circ = \frac{(-1626 + 92 + 1196)\ kJ}{2\ mol} = \frac{-169\ kJ}{mol}$$

Energy Consumption and Sources

69. $C_2H_5OH(l) + 3\ O_2(g) \rightarrow 2\ CO_2(g) + 3\ H_2O(l)$

$\Delta H^\circ = [2\ (-393.5\ kJ) + 3(-286\ kJ)] - (-278\ kJ) = -1367$ kJ/mol ethanol

$$\frac{-1367\ kJ}{mol} \times \frac{1\ mol}{46.07\ g} = -29.67\ kJ/g$$

71. $C_3H_8(g) + 5 O_2(g) \rightarrow 3 CO_2(g) + 4 H_2O(l)$

$\Delta H° = [3(-393.5 \text{ kJ}) + 4(-286 \text{ kJ})] - [-104 \text{ kJ}] = -2221 \text{ kJ/mol } C_3H_8$

$$\frac{-2221 \text{ kJ}}{\text{mol}} \times \frac{1 \text{ mol}}{44.09 \text{ g}} = \frac{-50.37 \text{ kJ}}{\text{g}} \text{ vs. -47.7 kJ/g for octane (Sample Exercise 6.11)}$$

The fuel values are very close. An advantage of propane is that it burns more cleanly. The boiling point of propane is -42°C. Thus, it is more difficult to store propane and there are extra safety hazards associated with using high pressure compressed gas tanks.

73.   The molar volume of a gas at STP is 22.42 L (from Chapter 5).

$$4.19 \times 10^6 \text{ kJ} \times \frac{1 \text{ mol CH}_4}{891 \text{ kJ}} \times \frac{22.42 \text{ L CH}_4}{\text{mol CH}_4} = 1.05 \times 10^5 \text{ L CH}_4$$

Additional Exercises

75.   a. $2 SO_2(g) + O_2(g) \rightarrow 2 SO_3(g)$ $(w = -P\Delta V)$; Since the volume of the piston apparatus decreased as reactants were converted to products, w is positive (w > 0).

b. $COCl_2(g) \rightarrow CO(g) + Cl_2(g)$; Since the volume increased, w is negative (w < 0).

c. $N_2(g) + O_2(g) \rightarrow 2 NO(g)$; Since the volume did not change, no PV work is done (w = 0).

In order to predict the sign of w for a reaction, compare the coefficients of all the product gases in the balanced equation to the coefficients of all the reactant gases. When a balanced reaction has more mol of product gases than mol of reactant gases (as in b), the reaction will expand in volume (ΔV positive), and the system does work on the surroundings. When a balanced reaction has a decrease in the mol of gas from reactants to products (as in a), the reaction will contract in volume (ΔV negative), and the surroundings will do compression work on the system. When there is no change in the mol of gas from reactants to products (as in c), $\Delta V = 0$ and w = 0.

77. $\Delta E_{overall} = \Delta E_{step 1} + \Delta E_{step 2}$; This is a cyclic process which means that the overall initial state and final state are the same. Since ΔE is a state function, $\Delta E_{overall} = 0$ and $\Delta E_{step 1} = -\Delta E_{step 2}$.

$\Delta E_{step 1} = q + w = 45 \text{ J} + (-10. \text{ J}) = 35 \text{ J}$

$\Delta E_{step 2} = -\Delta E_{step 1} = -35 \text{ J} = q + w, \quad -35 \text{ J} = -60 \text{ J} + w, \quad w = 25 \text{ J}$

79.   Heat loss by hot water = heat gain by cold water; Keeping all quantities positive to avoid sign errors:

$$\frac{4.18 \text{ J}}{\text{g} \,^{\circ}\text{C}} \times m_{hot} \times (55.0\,^{\circ}\text{C} - 37.0\,^{\circ}\text{C}) = \frac{4.18 \text{ J}}{\text{g} \,^{\circ}\text{C}} \times 90.0 \text{ g} \times (37.0\,^{\circ}\text{C} - 22.0\,^{\circ}\text{C})$$

$$m_{hot} = \frac{90.0 \text{ g} \times 15.0\,^{\circ}\text{C}}{18.0\,^{\circ}\text{C}} = 75.0 \text{ g hot water needed}$$

81. $HCl(aq) + NaOH(aq) \rightarrow H_2O(l) + NaCl(aq)$ $\Delta H = -56$ kJ

$$0.2000 \text{ L} \times \frac{0.400 \text{ mol HCl}}{\text{L}} = 8.00 \times 10^{-2} \text{ mol HCl}$$

$$0.1500 \text{ L} \times \frac{0.500 \text{ mol NaOH}}{\text{L}} = 7.50 \times 10^{-2} \text{ mol NaOH}$$

Since the balanced reaction requires a 1:1 mol ratio between HCl and NaOH, and since fewer mol of NaOH are actually present as compared to HCl, then NaOH is the limiting reagent.

Heat released = 7.50×10^{-2} mol NaOH $\times \dfrac{-56 \text{ kJ}}{\text{mol NaOH}} = -4.2$ kJ heat released

83.   Heat released = 1.056 g × 26.42 kJ/g = 27.90 kJ = Heat gain by water and calorimeter

Heat gain = 27.90 kJ = $\dfrac{4.18 \text{ kJ}}{\text{kg} \,^{\circ}\text{C}} \times 0.987 \text{ kg} \times \Delta T + \dfrac{6.66 \text{ kJ}}{^{\circ}\text{C}} \times \Delta T$

27.90 = (4.13 + 6.66) ΔT = 10.79 ΔT, $\Delta T = 2.586\,^{\circ}\text{C}$

$2.586\,^{\circ}\text{C} = T_f - 23.32\,^{\circ}\text{C}$, $T_f = 25.91\,^{\circ}\text{C}$

85.   a. ΔH° = 3 mol (227 kJ/mol) - 1 mol (49 kJ/mol) = 632 kJ

       b.   Since 3 $C_2H_2(g)$ is higher in energy than $C_6H_6(l)$, acetylene will release more energy per gram when burned in air.

87.   a. $C_2H_4(g) + O_3(g) \rightarrow CH_3CHO(g) + O_2(g)$, ΔH° = -166 kJ - [143 kJ + 52 kJ] = -361 kJ

       b. $O_3(g) + NO(g) \rightarrow NO_2(g) + O_2(g)$, ΔH° = 34 kJ - [90. kJ + 143 kJ] = -199 kJ

       c. $SO_3(g) + H_2O(l) \rightarrow H_2SO_4(aq)$, ΔH° = -909 kJ - [-396 kJ + (-286 kJ)] = -227 kJ

       d. $2 NO(g) + O_2(g) \rightarrow 2 NO_2(g)$, ΔH° = 2(34) kJ - 2(90.) kJ = -112 kJ

Challenge Problems

89.   a. $C_{12}H_{22}O_{11}(s) + 12\ O_2(g) \rightarrow 12\ CO_2(g) + 11\ H_2O(l)$

        b.   A bomb calorimeter is at constant volume, so heat released = $q_v = \Delta E$:

$$\Delta E = \frac{-24.00\ \text{kJ}}{1.46\ \text{g}} \times \frac{342.30\ \text{g}}{\text{mol}} = -5630\ \text{kJ/mol}\ C_{12}H_{22}O_{11}$$

        c.   Since $PV = nRT$, then $P\Delta V = RT\Delta n$ where Δn = mol gaseous products - mol gaseous reactants.

$$\Delta H = \Delta E + P\Delta V = \Delta E + RT\Delta n$$

For this reaction, $\Delta n = 12 - 12 = 0$, so $\Delta H = \Delta E = -5630$ kJ/mol.

91.   Energy used in 8.0 hours = 40. kWh = $\dfrac{40.\ \text{kJ h}}{\text{s}} \times \dfrac{3600\ \text{s}}{\text{h}} = 1.4 \times 10^5$ kJ

Energy from the sun in 8.0 hours = $\dfrac{1.0\ \text{kJ}}{\text{s m}^2} \times \dfrac{60\ \text{s}}{\text{min}} \times \dfrac{60\ \text{min}}{\text{h}} \times 8.0\ \text{h} = 2.9 \times 10^4$ kJ/m^2

Only 13% of the sunlight is converted into electricity:

$$0.13 \times (2.9 \times 10^4\ \text{kJ/m}^2) \times \text{Area} = 1.4 \times 10^5\ \text{kJ}, \quad \text{Area} = 37\ \text{m}^2$$

93. $400\ \text{kcal} \times \dfrac{4.18\ \text{kJ}}{\text{kcal}} = 1.67 \times 10^3\ \text{kJ} \approx 2 \times 10^3\ \text{kJ}$

$$PE = mgz = \left(180\ \text{lb} \times \frac{1\ \text{kg}}{2.205\ \text{lb}} \right) \times \frac{9.80\ \text{m}}{\text{s}^2} \times \left(8\ \text{in} \times \frac{2.54\ \text{cm}}{\text{in}} \times \frac{1\ \text{m}}{100\ \text{cm}} \right) = 160\ \text{J} \approx 200\ \text{J}$$

200 J of energy are needed to climb one step. The total number of steps to climb are:

$$2 \times 10^6\ \text{J} \times \frac{1\ \text{step}}{200\ \text{J}} = 1 \times 10^4\ \text{steps}$$

CHAPTER SEVEN

ATOMIC STRUCTURE AND PERIODICITY

Questions

15. Planck's discovery that heated bodies give off only certain frequencies of light and Einstein's study of the photoelectric effect.

17. Only very small particles with a tiny mass exhibit wave and particle properties, e.g., an electron. Some evidence supporting the wave properties of matter are:

   1) Electrons can be diffracted like light.

   2) The electron microscope uses electrons in a fashion similar to the way in which light is used in a light microscope.

19. n: Gives the energy (it completely specifies the energy only for the H-atom or ions with one electron) and the relative size of the orbitals.

   ℓ: Gives the type (shape) of orbital.

   m_ℓ: Gives information about the direction in which the orbital is pointing.

21. A nodal surface in an atomic orbital is a surface in which the probability of finding an electron is zero.

23. No, the spin is a convenient model. Since we cannot locate or "see" the electron, we cannot see if it is spinning.

25. The outermost electrons are the valence electrons. When atoms interact with each other, it will be the outermost electrons that are involved in these interactions. In addition, how tightly the nucleus holds these outermost electrons determines atomic size, ionization energy and other properties of atoms.

   Elements in the same group have similar valence electron configurations and, as a result, have similar chemical properties.

27. Across a period, the positive charge from the nucleus increases as protons are added. The number of electrons also increase, but these outer electrons do not completely shield the increasing nuclear charge from each other. The general result is that the outer electrons are more strongly bound as one goes across a period which results in larger ionization energies (and smaller size).

Aluminum is out of order because the electrons in the filled 3s orbital shield some of the nuclear charge from the 3p electron. Hence, the 3p electron is less tightly bound than a 3s electron resulting in a lower ionization energy for aluminum as compared to magnesium. The ionization energy of sulfur is lower than phosphorus because of the extra electron-electron repulsions in the doubly occupied sulfur 3p orbital. These added repulsions, which are not present in phosphorus, make it slightly easier to remove an electron from sulfur as compared to phosphorus.

29.   Ionization energy: $P(g) \rightarrow P^+(g) + e^-$; electron affinity: $P(g) + e^- \rightarrow P^-(g)$

31.   The valence electrons are strongly attracted to the nucleus for elements with large ionization energies. One would expect these species to readily accept another electron and have very exothermic electron affinities. The noble gases are an exception; they have a large IE but have an endothermic EA. Noble gases have a stable arrangement of electrons. Adding an electron disrupts this stable arrangement, resulting in unfavorable electron affinities.

33.   For hydrogen and one-electron ions (hydrogenlike ions), all atomic orbitals with the same n value have the same energy. For polyatomic atoms/ions, the energy of the atomic orbitals also depends on ℓ. Since there are more nondegenerate energy levels for polyatomic atoms/ions as compared to hydrogen, there are many more possible electronic transitions resulting in more complicated line spectra.

35.   Yes, the maximum number of unpaired electrons in any configuration corresponds to a minimum in electron-electron repulsions.

37.   Ionization energy is for removal of the electron from the atom in the gas phase. The work function is for the removal of an electron from the solid.

$M(g) \rightarrow M^+(g) + e^-$ ionization energy; $M(s) \rightarrow M^+(s) + e^-$ work function

Exercises

Light and Matter

39. 1×10^{-9} m = 1 nm; $\nu = \dfrac{c}{\lambda} = \dfrac{2.998 \times 10^8 \text{ m/s}}{780. \times 10^{-9} \text{ m}} = 3.84 \times 10^{14} \text{ s}^{-1}$

41. $\nu = \dfrac{c}{\lambda} = \dfrac{3.00 \times 10^8 \text{ m/s}}{1.0 \times 10^{-2} \text{ m}} = 3.0 \times 10^{10} \text{ s}^{-1}$

$E = h\nu = 6.63 \times 10^{-34} \text{ J s} \times 3.0 \times 10^{10} \text{ s}^{-1} = 2.0 \times 10^{-23} \text{ J/photon}$

$\dfrac{2.0 \times 10^{-23} \text{ J}}{\text{photon}} \times \dfrac{6.02 \times 10^{23} \text{ photons}}{\text{mol}} = 12 \text{ J/mol}$

43.   The wavelength is the distance between consecutive wave peaks. Wave a shows 4 wavelengths and wave b shows 8 wavelengths.

Wave a: $\lambda = \dfrac{1.6 \times 10^{-3} \text{ m}}{4} = 4.0 \times 10^{-4} \text{ m}$

Wave b: $\lambda = \dfrac{1.6 \times 10^{-3} \text{ m}}{8} = 2.0 \times 10^{-4} \text{ m}$

Wave a has the longer wavelength. Since frequency and photon energy are both inversely proportional to wavelength, then wave b will have the higher frequency and larger photon energy since it has the shorter wavelength.

$\nu = \dfrac{c}{\lambda} = \dfrac{3.00 \times 10^8 \text{ m/s}}{2.0 \times 10^{-4} \text{ m}} = 1.5 \times 10^{12} \text{ s}^{-1}$

$E = \dfrac{hc}{\lambda} = \dfrac{6.63 \times 10^{-34} \text{ J s} \times 3.00 \times 10^8 \text{ m/s}}{2.0 \times 10^{-4} \text{ m}} = 9.9 \times 10^{-22} \text{ J}$

Since both waves are examples of electromagnetic radiation, then both waves travel at the same speed, c, the speed of light. From Figure 7.2 of the text, both of these waves represent infrared electromagnetic radiation.

45.   The energy needed to remove a single electron is:

$\dfrac{279.7 \text{ kJ}}{\text{mol}} \times \dfrac{1 \text{ mol}}{6.0221 \times 10^{23}} = 4.645 \times 10^{-22} \text{ kJ} = 4.645 \times 10^{-19} \text{ J}$

$E = \dfrac{hc}{\lambda}, \quad \lambda = \dfrac{hc}{E} = \dfrac{6.6261 \times 10^{-34} \text{ J s} \times 2.9979 \times 10^8 \text{ m/s}}{4.645 \times 10^{-19} \text{ J}} = 4.277 \times 10^{-7} \text{ m} = 427.7 \text{ nm}$

47.   a.   90.% of speed of light = $0.90 \times 3.00 \times 10^8$ m/s = 2.7×10^8 m/s

$\lambda = \dfrac{h}{mv}, \quad \lambda = \dfrac{6.63 \times 10^{-34} \text{ J s}}{1.67 \times 10^{-27} \text{ kg} \times 2.7 \times 10^8 \text{ m/s}} = 1.5 \times 10^{-15} \text{ m} = 1.5 \times 10^{-6} \text{ nm}$

Note: For units to come out, the mass must be in kg since $1 \text{ J} = \dfrac{1 \text{ kg m}^2}{\text{s}^2}$.

b. $\lambda = \dfrac{h}{mv} = \dfrac{6.63 \times 10^{-34} \text{ J s}}{0.15 \text{ kg} \times 10.0 \text{ m/s}} = 4.4 \times 10^{-34} \text{ m} = 4.4 \times 10^{-25} \text{ nm}$

This number is so small that it is essentially zero. We cannot detect a wavelength this small. The meaning of this number is that we do not have to consider the wave properties of large objects.

49. $\lambda = \dfrac{h}{mv} = \dfrac{6.63 \times 10^{-34} \text{ J s}}{9.11 \times 10^{-31} \text{ kg} \times (1.0 \times 10^{-3} \times 3.00 \times 10^8 \text{ m/s})} = 2.4 \times 10^{-9} \text{ m} = 2.4 \text{ nm}$

Hydrogen Atom: The Bohr Model

51.   For the H atom (Z = 1): $E_n = -2.178 \times 10^{-18}$ J/n^2; For a spectral transition, $\Delta E = E_f - E_i$:

$$\Delta E = -2.178 \times 10^{-18} \text{ J} \left(\frac{1}{n_f^2} - \frac{1}{n_i^2} \right)$$

where n_i and n_f are the levels of the initial and final states, respectively. A positive value of ΔE always corresponds to an absorption of light, and a negative value of ΔE always corresponds to an emission of light.

a. $\Delta E = -2.178 \times 10^{-18} \text{ J} \left(\frac{1}{2^2} - \frac{1}{3^2} \right) = -2.178 \times 10^{-18} \text{ J} \left(\frac{1}{4} - \frac{1}{9} \right)$

$\Delta E = -2.178 \times 10^{-18} \text{ J} \times (0.2500 - 0.1111) = -3.025 \times 10^{-19}$ J

The photon of light must have precisely this energy (3.025×10^{-19} J).

$| \Delta E | = E_{photon} = h\nu = \dfrac{hc}{\lambda}$ or $\lambda = \dfrac{hc}{|\Delta E|} = \dfrac{6.6261 \times 10^{-34} \text{ J s} \times 2.9979 \times 10^8 \text{ m/s}}{3.025 \times 10^{-19} \text{ J}}$

$$= 6.567 \times 10^{-7} \text{ m} = 656.7 \text{ nm}$$

From Figure 7.2, this is visible electromagnetic radiation (red light).

b. $\Delta E = -2.178 \times 10^{-18} \text{ J} \left(\frac{1}{2^2} - \frac{1}{4^2} \right) = -4.084 \times 10^{-19}$ J

$\lambda = \dfrac{hc}{|\Delta E|} = \dfrac{6.6261 \times 10^{-34} \text{ J s} \times 2.9979 \times 10^8 \text{ m/s}}{4.084 \times 10^{-19} \text{ J}} = 4.864 \times 10^{-7} \text{ m} = 486.4 \text{ nm}$

This is visible electromagnetic radiation (green-blue light).

c. $\Delta E = -2.178 \times 10^{-18} \text{ J} \left(\frac{1}{1^2} - \frac{1}{2^2} \right) = -1.634 \times 10^{-18}$ J

$\lambda = \dfrac{6.6261 \times 10^{-34} \text{ J s} \times 2.9979 \times 10^8 \text{ m/s}}{1.634 \times 10^{-18} \text{ J}} = 1.216 \times 10^{-7} \text{ m} = 121.6 \text{ nm}$

This is ultraviolet electromagnetic radiation.

53.

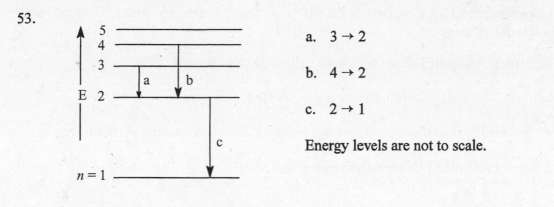

a. $3 \rightarrow 2$

b. $4 \rightarrow 2$

c. $2 \rightarrow 1$

Energy levels are not to scale.

55. $\Delta E = -2.178 \times 10^{-18}$ J $\left(\dfrac{1}{n_f^2} - \dfrac{1}{n_i^2} \right) = -2.178 \; 10^{-18}$ J $\left(\dfrac{1}{5^2} - \dfrac{1}{1^2} \right) = 2.091 \times 10^{-18}$ J $= E_{photon}$

$\lambda = \dfrac{hc}{E} = \dfrac{6.6261 \times 10^{-34} \text{ J s} \times 2.9979 \times 10^8 \text{ m/s}}{2.091 \times 10^{-18} \text{ J}} = 9.500 \times 10^{8}$ m $= 95.00$ nm

Since wavelength and energy are inversely related, visible light ($\lambda \approx 400$ - 700 nm) is not energetic enough to excite an electron in hydrogen from $n = 1$ to $n = 5$.

$\Delta E = -2.178 \times 10^{-18}$ J $\left(\dfrac{1}{6^2} - \dfrac{1}{2^2} \right) = 4.840 \times 10^{-19}$ J

$\lambda = \dfrac{hc}{E} = \dfrac{6.6261 \times 10^{-34} \text{ J s} \times 2.9979 \times 10^8 \text{ m/s}}{4.840 \times 10^{-19} \text{ J}} = 4.104 \times 10^{-7}$ m $= 410.4$ nm

Visible light with $\lambda = 410.4$ nm will excite an electron from the $n = 2$ to the $n = 6$ energy level.

57. $\Delta E = h\nu = 6.6261 \times 10^{-34}$ J s $\times 1.141 \times 10^{14}$ s$^{-1} = 7.560 \times 10^{-20}$ J

$\Delta E = -7.560 \times 10^{-20}$ J since light is emitted.

$\Delta E = E_4 - E_n$, -7.560×10^{-20} J $= -2.178 \times 10^{-18}$ J $\left(\dfrac{1}{4^2} - \dfrac{1}{n^2} \right)$

$3.471 \times 10^{-2} = 6.250 \times 10^{-2} - \dfrac{1}{n^2}$, $\dfrac{1}{n^2} = 2.779 \times 10^{-2}$, $n^2 = 35.98$, $n = 6$

Quantum Mechanics, Quantum Numbers, and Orbitals

59.   a. $\Delta(mv) = m\Delta v = 9.11 \times 10^{-31}$ kg $\times 0.100$ m/s $= \dfrac{9.11 \times 10^{-32} \text{ kg m}}{\text{s}}$

$\Delta(mv) \bullet \Delta x \geq \dfrac{h}{4\pi}$, $\Delta x = \dfrac{h}{4\pi\Delta(mv)} = \dfrac{6.626 \times 10^{-34} \text{ J s}}{4 \times 3.142 \times (9.11 \times 10^{-32} \text{ kg m/s})} = 5.79 \times 10^{-4}$ m

b. $\Delta x = \dfrac{h}{4\pi m\Delta v} = \dfrac{6.626 \times 10^{-34} \text{ J s}}{4 \times 3.142 \times 0.145 \text{ kg} \times 0.100 \text{ m/s}} = 3.64 \times 10^{-33}$ m

   c. The diameter of an H atom is roughly 1.0×10^{-8} cm. The uncertainty in position is much larger than the size of the atom.

   d. The uncertainty is insignificant compared to the size of a baseball.

61. $n = 1, 2, 3, ... ;$ $\ell = 0, 1, 2, ... (n-1);$ $m_\ell = -\ell \, ... \, -2, -1, 0, 1, 2, ... +\ell$

63.   b. ℓ must be smaller than n.   d. For $\ell = 0$, $m_\ell = 0$ is the only allowed value.

65. ψ^2 gives the probability of finding the electron at that point.

Polyelectronic Atoms

67.   5p: three orbitals; $3d_{z^2}$: one orbital; 4d: five orbitals

      $n = 5$: $\ell = 0$ (1 orbital), $\ell = 1$ (3 orbitals), $\ell = 2$ (5 orbitals), $\ell = 3$ (7 orbitals), $\ell = 4$ (9 orbitals)

          Total for $n = 5$ is 25 orbitals.

      $n = 4$: $\ell = 0$ (l), $\ell = 1$ (3), $\ell = 2$ (5), $\ell = 3$ (7); Total for $n = 4$ is 16 orbitals.

69.   a. $n = 4$: ℓ can be 0, 1, 2, or 3. Thus we have s (2 e⁻), p (6 e⁻), d (10 e⁻) and f (14 e⁻) orbitals present. Total number of electrons to fill these orbitals is 32.

    b. $n = 5$, $m_\ell = +1$: For $n = 5$, $\ell = 0, 1, 2, 3, 4$. For $\ell = 1, 2, 3, 4$, all can have $m_\ell = +1$. Four distinct orbitals, thus 8 electrons.

    c. $n = 5$, $m_s = +1/2$: For $n = 5$, $\ell = 0, 1, 2, 3, 4$. Number of orbitals = 1, 3, 5, 7, 9 for each value of ℓ, respectively. There are 25 orbitals with $n = 5$. They can hold 50 electrons and 25 of these electrons can have $m_s = +1/2$.

    d. $n = 3$, $\ell = 2$: These quantum numbers define a set of 3d orbitals. There are 5 degenerate 3d orbitals which can hold a total of 10 electrons.

    e. $n = 2$, $\ell = 1$: These define a set of 2p orbitals. There are 3 degenerate 2p orbitals which can hold a total of 6 electrons.

71.   Si: $1s^2 2s^2 2p^6 3s^2 3p^2$ or [Ne]$3s^2 3p^2$; Ga: $1s^2 2s^2 2p^6 3s^2 3p^6 4s^2 3d^{10} 4p^1$ or [Ar]$4s^2 3d^{10} 4p^1$

    As: [Ar]$4s^2 3d^{10} 4p^3$; Ge: [Ar]$4s^2 3d^{10} 4p^2$; Al: [Ne]$3s^2 3p^1$; Cd: [Kr]$5s^2 4d^{10}$

    S: [Ne]$3s^2 3p^4$; Se: [Ar]$4s^2 3d^{10} 4p^4$

73.   The following are complete electron configurations. Noble gas shorthand notation could also be used.

    Sc: $1s^2 2s^2 2p^6 3s^2 3p^6 4s^2 3d^1$; Fe: $1s^2 2s^2 2p^6 3s^2 3p^6 4s^2 3d^6$

P: $1s^2 2s^2 2p^6 3s^2 3p^3$;   Cs: $1s^2 2s^2 2p^6 3s^2 3p^6 4s^2 3d^{10} 4p^6 5s^2 4d^{10} 5p^6 6s^1$

Eu: $1s^2 2s^2 2p^6 3s^2 3p^6 4s^2 3d^{10} 4p^6 5s^2 4d^{10} 5p^6 6s^2 4f^6 5d^1$*

Pt: $1s^2 2s^2 2p^6 3s^2 3p^6 4s^2 3d^{10} 4p^6 5s^2 4d^{10} 5p^6 6s^2 4f^{14} 5d^8$*

Xe: $1s^2 2s^2 2p^6 3s^2 3p^6 4s^2 3d^{10} 4p^6 5s^2 4d^{10} 5p^6$;   Br: $1s^2 2s^2 2p^6 3s^2 3p^6 4s^2 3d^{10} 4p^5$

*Note: These electron configurations were predicted using only the periodic table.

Actual electron configurations are: Eu: $[Xe]6s^2 4f^7$ and Pt: $[Xe]6s^1 4f^{14} 5d^9$

75.   Exceptions: Cr, Cu, Nb, Mo, Tc, Ru, Rh, Pd, Ag, Pt, Au; Tc, Ru, Rh, Pd and Pt do not correspond to the supposed extra stability of half-filled and filled subshells.

77.   a. Both In and I have one unpaired 5p electron, but only the nonmetal I would be expected to form a covalent compound with the nonmetal F. One would predict an ionic compound to form between the metal In and the nonmetal F.

        I: $[Kr]5s^2 4d^{10} 5p^5$ ↑↓ ↑↓ ↑
                                      5p

       b. From the periodic table, this will be element 120.   Element 120: $[Rn]7s^2 5f^{14} 6d^{10} 7p^6 8s^2$

       c. Rn: $[Xe]6s^2 4f^{14} 5d^{10} 6p^6$; Note that the next discovered noble gas will also have 4f electrons (as well as 5f electrons).

       d. This is chromium, which is an exception to the predicted filling order. Cr has 6 unpaired electrons and the next most is 5 unpaired electrons for Mn.

        Cr: $[Ar]4s^1 3d^5$ ↑   ↑ ↑ ↑ ↑ ↑
                             4s   3d

79.   B : $1s^2 2s^2 2p^1$

	n	ℓ	m_ℓ	m_s
1s	1	0	0	+1/2
1s	1	0	0	-1/2
2s	2	0	0	+1/2
2s	2	0	0	-1/2
2p*	2	1	-1	+1/2

*This is only one of several possibilities for the 2p electron. The 2p electron in B could have m_ℓ = -1, 0 or +1, and m_s = +1/2 or -1/2, a total of six possibilities.

N : $1s^2 2s^2 2p^3$

	n	ℓ	m_ℓ	m_s
1s	1	0	0	+1/2
1s	1	0	0	-1/2
2s	2	0	0	+1/2
2s	2	0	0	-1/2
2p	2	1	-1	+1/2
2p	2	1	0	+1/2
2p	2	1	+1	+1/2

(Or all 2p electrons could have $m_s = -1/2$.)

81.   O: $1s^2 2s^2 2p_x^2 2p_y^2$ ($\uparrow\downarrow$ $\uparrow\downarrow$ __); There are no unpaired electrons in this oxygen atom. This configuration would be an excited state, and in going to the more stable ground state ($\uparrow\downarrow$ $\uparrow$ $\uparrow$), energy would be released.

83.   We get the number of unpaired electrons by examining the incompletely filled subshells. The paramagnetic substances have unpaired electrons, and the ones with no unpaired electrons are not paramagnetic (they are called diamagnetic).

Li: $1s^2 2s^1$ $\uparrow$; Paramagnetic with 1 unpaired electron.
      2s

N: $1s^2 2s^2 2p^3$ $\uparrow$ $\uparrow$ $\uparrow$; Paramagnetic with 3 unpaired electrons.
                2p

Ni: [Ar]$4s^2 3d^8$ $\uparrow\downarrow$ $\uparrow\downarrow$ $\uparrow\downarrow$ $\uparrow$ $\uparrow$; Paramagnetic with 2 unpaired electrons.
              3d

Te: [Kr]$5s^2 4d^{10} 5p^4$ $\uparrow\downarrow$ $\uparrow$ $\uparrow$; Paramagnetic with 2 unpaired electrons.
                    5p

Ba: [Xe]$6s^2$ $\uparrow\downarrow$; Not paramagnetic since no unpaired electrons.
          6s

Hg: [Xe]$6s^2 4f^{14} 5d^{10}$ $\uparrow\downarrow$ $\uparrow\downarrow$ $\uparrow\downarrow$ $\uparrow\downarrow$ $\uparrow\downarrow$; Not paramagnetic since no unpaired electrons.
            5d

The Periodic Table and Periodic Properties

85.   Size (radii) decreases left to right across the periodic table, and size increases from top to bottom of the periodic table.

a. Be < Mg < Ca   b. Xe < I < Te   c. Ge < Ga < In

87.   The ionization energy trend is the opposite of the radii trend; ionization energy (IE), in general, increases left to right across the periodic table and decreases from top to bottom of the periodic table.

a. Ca < Mg < Be   b. Te < I < Xe   c. In < Ga < Ge

89.   a.   Li   b.   P

       c. O^+. This ion has the fewest electrons as compared to the other oxygen species present. O^+ has
            the smallest amount of electron-electron repulsions, which makes it the smallest ion with the
            largest ionization energy.

       d.   From the radii trend, $Ar < Cl < S$ and $Kr > Ar$. Since variation in size down a family is greater
            than the variation across a period, we would predict Cl to be the smallest of the three.

       e.   Cu

91.   a.   106: $[Rn]7s^2 5f^{14} 6d^4$ b.   W   c. SgO_3 or Sg_2O_3 and SgO_4^{2-} or $Sg_2O_7^{2-}$
                                                               (similar to Cr; Sg = 106)

93.   As: $[Ar]4s^2 3d^{10} 4p^3$; Se: $[Ar]4s^2 3d^{10} 4p^4$; The general ionization energy trend predicts that Se
       should have a higher ionization energy than As. Se is an exception to the general ionization energy
       trend. There are extra electron-electron repulsions in Se because two electrons are in the same 4p
       orbital, resulting in a lower ionization energy for Se than predicted.

95.   a.   More favorable EA: C and Br; The electron affinity trend is very erratic. Both N and Ar have
            positive EA values (unfavorable) due to their electron configurations (see text for detailed
            explanation).

       b.   Higher IE: N and Ar (follows the IE trend)

       c.   Larger size: C and Br (follows the radii trend)

97.   The electron affinity trend is very erratic. In general, EA decreases down the periodic table, and the
       trend across the table is too erratic to be of much use.

       a.   Se < S; S is more exothermic.   b.   I < Br < F < Cl; Cl is most exothermic (F is an
                                                          exception).

99.   Electron-electron repulsions are much greater in O^- than in S^- because the electron goes into a
       smaller 2p orbital vs. the larger 3p orbital in sulfur. This results in a more favorable (more
       exothermic) EA for sulfur.

101.   a.   The electron affinity of Mg^{2+} is ΔH for $Mg^{2+}(g) + e^- \rightarrow Mg^+(g)$. This is just the reverse of the
            second ionization energy, or $EA(Mg^{2+}) = -IE_2(Mg) = -1445$ kJ/mol (Table 7.5).

       b.   EA of Al^+ is ΔH for $Al^+(g) + e^- \rightarrow Al(g)$. $EA(Al^+) = -IE_1(Al) = -580$ kJ/mol (Table 7.5)

Alkali Metals

103.   It should be potassium peroxide, K_2O_2, since K^+ ions are much more stable than K^{2+} ions. The
       second ionization energy of K is very large as compared to the first ionization energy.

105. $\nu = \dfrac{c}{\lambda} = \dfrac{2.9979 \times 10^8 \text{ m/s}}{455.5 \times 10^{-9} \text{ m}} = 6.582 \times 10^{14} \text{ s}^{-1}$

$E = h\nu = 6.6261 \times 10^{-34} \text{ J s} \times 6.582 \times 10^{14} \text{ s}^{-1} = 4.361 \times 10^{-19} \text{ J}$

107.   Yes; the ionization energy general trend is to decrease down a group, and the atomic radius trend is to increase down a group. The data in Table 7.8 confirm both of these general trends.

109.   a. $4 \text{ Li(s)} + O_2(g) \rightarrow 2 \text{ Li}_2O(s)$ b. $2 \text{ K(s)} + \text{S(s)} \rightarrow \text{K}_2\text{S(s)}$

Additional Exercises

111. $E = \dfrac{310. \text{ kJ}}{\text{mol}} \times \dfrac{1 \text{ mol}}{6.022 \times 10^{23}} = 5.15 \times 10^{-22} \text{ kJ} = 5.15 \times 10^{-19} \text{ J}$

$E = \dfrac{hc}{\lambda}, \quad \lambda = \dfrac{hc}{E} = \dfrac{6.626 \times 10^{-34} \text{ J s} \times 2.998 \times 10^8 \text{ m/s}}{5.15 \times 10^{-19}} = 3.86 \times 10^{-7} \text{ m} = 386 \text{ nm}$

113. $60 \times 10^6 \text{ km} \times \dfrac{1000 \text{ m}}{\text{km}} \times \dfrac{1 \text{ s}}{3.00 \times 10^8 \text{ m}} = 200 \text{ s}$ (about 3 minutes)

115. $\Delta E = -R_H \left(\dfrac{1}{n_f^2} - \dfrac{1}{n_i^2} \right) = -2.178 \times 10^{-18} \text{ J} \left(\dfrac{1}{2^2} - \dfrac{1}{6^2} \right) = -4.840 \times 10^{-19} \text{ J}$

$\lambda = \dfrac{hc}{|\Delta E|} = \dfrac{6.6261 \times 10^{-34} \text{ J s} \times 2.9979 \times 10^8 \text{ m/s}}{4.840 \times 10^{-19} \text{ J}} = 4.104 \times 10^{-7} \text{ m} \times \dfrac{100 \text{ cm}}{\text{m}} = 4.104 \times 10^{-5} \text{ cm}$

From the spectrum, $\lambda = 4.104 \times 10^{-5}$ cm is violet light, so the $n = 6$ to $n = 2$ visible spectrum line is violet.

117.   a.   True for H only.   b.   True for all atoms.   c.   True for all atoms.

119.   Cd: $1s^2 2s^2 2p^6 3s^2 3p^6 4s^2 3d^{10} 4p^6 5s^2 4d^{10}$

a. $\ell = 2$ are d orbitals. There are 20 electrons in d orbitals ($3d^{10}$ and $4d^{10}$).

b.   For $n = 4$, there are 4s, 4p, 4d and 4f orbitals. Cd only uses the 4s, 4p and 4d orbitals. Since all are filled with electrons, then Cd has $2 + 6 + 10 = 18$ electrons with $n = 4$.

c.   All p, d and f orbitals have one orbital which has $m_\ell = -1$. So each of the 2p, 3p, 4p, 3d and 4d orbitals (which are filled) have an orbital with $m_\ell = -1$; $5(2) = 10$ electrons with $m_\ell = -1$.

d. Each orbital used has one of the two paired electrons with $m_s = -1/2$. Therefore, one-half of the total electrons in Cd have $m_s = -1/2$; $(48/2) = 24$ electrons with $m_s = -1/2$.

121. The general ionization energy trend says that ionization energy increases going left to right across the periodic table. However, one of the exceptions to this trend occurs between groups 2A and 3A. Between these two groups, group 3A elements usually have a lower ionization energy than group 2A elements. Therefore, Al should have the lowest first ionization energy value, followed by Mg, with Si having the largest ionization energy. Looking at the values for the first ionization energy in the graph, the green plot is Al, the blue plot is Mg, and the red plot is Si.

Mg (the blue plot) is the element with the huge jump between I_2 and I_3. Mg has two valence electrons, so the third electron removed is an inner core electron. Inner core electrons are always much more difficult to remove compared to valence electrons since they are closer to the nucleus, on average, than the valence electrons.

123. Valence electrons are easier to remove than inner core electrons. The large difference in energy between I_2 and I_3 indicates that this element has two valence electrons. This element is most likely an alkaline earth metal since alkaline earth metal elements all have two valence electrons.

125. a.
$$Na(g) \rightarrow Na^+(g) + e^- \qquad\qquad I_1 = 495 \text{ kJ}$$
$$Cl(g) + e^- \rightarrow Cl^-(g) \qquad\qquad EA = -348.7 \text{ kJ}$$
$$\overline{Na(g) + Cl(g) \rightarrow Na^+(g) + Cl^-(g) \qquad\qquad \Delta H = 146 \text{ kJ}}$$

b.
$$Mg(g) \rightarrow Mg^+(g) + e^- \qquad\qquad I_1 = 735 \text{ kJ}$$
$$F(g) + e^- \rightarrow F^-(g) \qquad\qquad EA = -327.8 \text{ kJ}$$
$$\overline{Mg(g) + F(g) \rightarrow Mg^+(g) + F^-(g) \qquad\qquad \Delta H = 407 \text{ kJ}}$$

c.
$$Mg^+(g) \rightarrow Mg^{2+}(g) + e^- \qquad\qquad I_2 = 1445 \text{ kJ}$$
$$F(g) + e^- \rightarrow F^-(g) \qquad\qquad EA = -327.8 \text{ kJ}$$
$$\overline{Mg^+(g) + F(g) \rightarrow Mg^{2+}(g) + F^-(g) \qquad\qquad \Delta H = 1117 \text{ kJ}}$$

d. From parts b and c we get:
$$Mg(g) + F(g) \rightarrow Mg^+(g) + F^-(g) \qquad\qquad \Delta H = 407 \text{ kJ}$$
$$Mg^+(g) + F(g) \rightarrow Mg^{2+}(g) + F^-(g) \qquad\qquad \Delta H = 1117 \text{ kJ}$$
$$\overline{Mg(g) + 2\,F(g) \rightarrow Mg^{2+}(g) + 2\,F^-(g) \qquad\qquad \Delta H = 1524 \text{ kJ}}$$

Challenge Problems

127. $E_{photon} = \dfrac{hc}{\lambda} = \dfrac{6.6261 \times 10^{-34} \text{ J s} \times 2.9979 \times 10^8 \text{ m/s}}{253.4 \times 10^{-9} \text{ m}} = 7.839 \times 10^{-19} \text{ J}; \ \Delta E = -7.839 \times 10^{-19} \text{ J}$

The general energy equation for one-electron ions is $E_n = -2.178 \times 10^{-18} \text{ J} \ (Z^2)/n^2$ where Z = atomic number.

$$\Delta E = -2.178 \times 10^{-18} \text{ J} \ (Z)^2 \left(\dfrac{1}{n_f^2} - \dfrac{1}{n_i^2} \right), \ Z = 4 \text{ for Be}^{3+}$$

$$-7.839 \times 10^{-19} \text{ J} = -2.178 \times 10^{-18} \ (4)^2 \left(\dfrac{1}{n_f^2} - \dfrac{1}{5^2} \right)$$

$$\dfrac{7.839 \times 10^{-19}}{2.178 \times 10^{-18} \times 16} + \dfrac{1}{25} = \dfrac{1}{n_f^2}, \ \dfrac{1}{n_f^2} = 0.06249, \ n_f = 4$$

This emission line corresponds to the $n = 5 \rightarrow n = 4$ electronic transition.

129. For $r = a_o$ and $\theta = 0°$ (Z = 1 for H):

$$\psi_{2p_z} = \dfrac{1}{4(2\pi)^{1/2}} \left(\dfrac{1}{5.29 \times 10^{-11}} \right)^{3/2} (1) \ e^{-1/2} \cos 0 = 1.57 \times 10^{14}; \ \psi^2 = 2.46 \times 10^{28}$$

For $r = a_o$ and $\theta = 90°$, $\psi_{2p_z} = 0$ since $\cos 90° = 0$; $\psi^2 = 0$

There is no probability of finding an electron in the $2p_z$ orbital with $\theta = 0°$. As expected, the xy plane, which corresponds to $\theta = 0°$, is a node for the $2p_z$ atomic orbital.

131. a. 1st period: $p = 1, \ q = 1, \ r = 0, \ s = \pm 1/2$ (2 elements)

2nd period: $p = 2, \ q = 1, \ r = 0, \ s = \pm 1/2$ (2 elements)

3rd period: $p = 3, \ q = 1, \ r = 0, \ s = \pm 1/2$ (2 elements)

$p = 3, \ q = 3, \ r = -2, \ s = \pm 1/2$ (2 elements)

$p = 3, \ q = 3, \ r = 0, \ s = \pm 1/2$ (2 elements)

$p = 3, \ q = 3, \ r = +2, \ s = \pm 1/2$ (2 elements)

4th period: $p = 4$; q and r values are the same as with $p = 3$ (8 total elements)

1							2
3							4
5	6	7	8	9	10	11	12
13	14	15	16	17	18	19	20

b. Elements 2, 4, 12 and 20 all have filled shells and will be least reactive.

c. Draw similarities to the modern periodic table.

XY could be X^+Y^-, $X^{2+}Y^{2-}$ or $X^{3+}Y^{3-}$. Possible ions for each are:

   X^+ could be elements 1, 3, 5 or 13; Y^- could be 11 or 19.

   X^{2+} could be 6 or 14; Y^{2-} could be 10 or 18.

   X^{3+} could be 7 or 15; Y^{3-} could be 9 or 17.

Note: X^{4+} and Y^{4+} ions probably won't form.

XY_2 will be $X^{2+}(Y^-)_2$; See above for possible ions.

X_2Y will be $(X^+)_2Y^{2-}$; See above for possible ions.

XY_3 will be $X^{3+}(Y^-)_3$; See above for possible ions.

X_2Y_3 will be $(X^{3+})_2(Y^{2-})_3$; See above for possible ions.

d. $p = 4$, $q = 3$, $r = -2$, $s = \pm 1/2$ (2)

   $p = 4$, $q = 3$, $r = 0$, $s = \pm 1/2$ (2)

   $p = 4$, $q = 3$, $r = +2$, $s = \pm 1/2$ (2)

   A total of 6 electrons can have $p = 4$ and $q = 3$.

e. $p = 3$, $q = 0$, $r = 0$:   This is not allowed; q must be odd. Zero electrons can have these
                          quantum numbers.

f. $p = 6$, $q = 1$, $r = 0$, $s = \pm 1/2$ (2)

   $p = 6$, $q = 3$, $r = -2, 0, +2$; $s = \pm 1/2$ (6)

   $p = 6$, $q = 5$, $r = -4, -2, 0, +2, +4$; $s = \pm 1/2$ (10)

Eighteen electrons can have $p = 6$.

133. a. As we remove succeeding electrons, the electron being removed is closer to the nucleus, and there are fewer electrons left repelling it. The remaining electrons are more strongly attracted to the nucleus, and it takes more energy to remove these electrons.

  b. Al : $1s^22s^22p^63s^23p^1$; For I_4, we begin by removing an electron with $n = 2$. For I_3, we remove an electron with $n = 3$. In going from $n = 3$ to $n = 2$, there is a big jump in ionization energy because the $n = 2$ electrons (inner core electrons) are much closer to the nucleus on average than $n = 3$ electrons (valence electrons).   Since the $n = 2$ electrons are closer to the nucleus,

  they are held more tightly and require a much larger amount of energy to remove them compared to the $n = 3$ electrons.

  c. Al^{4+}; The electron affinity for Al^{4+} is ΔH for the reaction:

  $$Al^{4+}(g) + e^- \rightarrow Al^{3+}(g) \quad \Delta H = -I_4 = -11,600 \text{ kJ/mol}$$

  d. The greater the number of electrons, the greater the size.

  Size trend: $Al^{4+} < Al^{3+} < Al^{2+} < Al^+ < Al$

CHAPTER EIGHT

BONDING: GENERAL CONCEPTS

Questions

13.   a.   Electronegativity: The ability of an atom <u>in a molecule</u> to attract electrons to itself.

Electron affinity: The energy change for $M(g) + e^- \rightarrow M^-(g)$. EA deals with isolated atoms in the gas phase.

b.   Covalent bond: Sharing of electron pair(s); Polar covalent bond: Unequal sharing of electron pair(s).

c.   Ionic bond: Electrons are no longer shared, i.e., complete transfer of electron(s) from one atom to another to form ions.

15. H_2O and NH_3 have lone pair electrons on the central atoms. These lone pair electrons require more room than the bonding electrons, which tends to compress the angles between the bonding pairs. The bond angle for H_2O is the smallest since oxygen has two lone pairs on the central atom, and the bond angle is compressed more than in NH_3 where N has only one lone pair.

17.   The two general requirements for a polar molecular are:

1.   polar bonds

2.   a structure such that the bond dipoles of the polar bonds do not cancel.

19.   Of the compounds listed, P_2O_5 is the only compound containing only covalent bonds. $(NH_4)_2SO_4$, $Ca_3(PO_4)_2$, K_2O, and KCl are all compounds composed of ions so they exhibit ionic bonding. The ions in $(NH_4)_2SO_4$ are NH_4^+ and SO_4^{2-}. Covalent bonds exist between the N and H atoms in NH_4^+ and between the S and O atoms in SO_4^{2-}. Therefore, $(NH_4)_2SO_4$ contains both ionic and covalent bonds. The same is true for $Ca_3(PO_4)_2$. The bonding is ionic between the Ca^{2+} and PO_4^{3-} ions and covalent between the P and O atoms in PO_4^{3-}. Therefore, $(NH_4)_2SO_4$ and $Ca_3(PO_4)_2$ are the compounds with both ionic and covalent bonds.

Exercises

Chemical Bonds and Electronegativity

21.   The general trend for electronegativity is:
        1) increase as we go from left to right across a period and
        2) decrease as we go down a group

        Using these trends, the expected orders are:

        a. C < N < O   b. Se < S < Cl   c. Sn < Ge < Si   d. Tl < Ge < S

23.   The most polar bond will have the greatest difference in electronegativity between the two atoms.
        From positions in the periodic table, we would predict:

        a. Ge–F   b. P–Cl   c. S–F   d. Ti–Cl

25.   The general trends in electronegativity used in Exercises 8.21 and 8.23 are only rules of thumb. In
        this exercise, we use experimental values of electronegativities and can begin to see several
        exceptions. The order of EN from Figure 8.3 is:

        a. C (2.5) < N (3.0) < O (3.5)   same as predicted

        b. Se (2.4) < S (2.5) < Cl (3.0)   same

        c. Si = Ge = Sn (1.8)   different

        d. Tl (1.8) = Ge (1.8) < S (2.5)   different

        Most polar bonds using actual EN values:

        a. Si–F and Ge–F have equal polarity (Ge–F predicted).

        b. P–Cl (same as predicted)

        c. S–F (same as predicted)   d. Ti–Cl (same as predicted)

27.   Electronegativity values increase from left to right across the periodic table. The order of
        electronegativities for the atoms from smallest to largest electronegativity will be H = P < C < N <
        O < F. The most polar bond will be F–H since it will have the largest difference in
        electronegativities, and the least polar bond will be P–H since it will have the smallest difference in
        electronegativities (ΔEN = 0). The order of the bonds in decreasing polarity will be F–H > O–H >
        N–H > C–H > P–H.

Ions and Ionic Compounds

29. Rb^+: $[Ar]4s^23d^{10}4p^6$; Ba^{2+}: $[Kr]5s^24d^{10}5p^6$; Se^{2-}: $[Ar]4s^23d^{10}4p^6$

I^-: $[Kr]5s^24d^{10}5p^6$

31.   a. Sc^{3+} b. Te^{2-} c. Ce^{4+} and Ti^{4+} d. Ba^{2+}

All of these have the number of electrons of a noble gas.

33.   There are many possible ions with 10 electrons. Some are: N^{3-}, O^{2-}, F^-, Na^+, Mg^{2+} and Al^{3+}. In terms of size, the ion with the most protons will hold the electrons the tightest and will be the smallest. The largest ion will be the ion with the fewest protons. The size trend is:

$$Al^{3+} < Mg^{2+} < Na^+ < F^- < O^{2-} < N^{3-}$$
smallest   largest

35.   a. $Cu > Cu^+ > Cu^{2+}$ b. $Pt^{2+} > Pd^{2+} > Ni^{2+}$ c. $O^{2-} > O^- > O$

d. $La^{3+} > Eu^{3+} > Gd^{3+} > Yb^{3+}$ e. $Te^{2-} > I^- > Cs^+ > Ba^{2+} > La^{3+}$

For answer a, as electrons are removed from an atom, size decreases. Answers b and d follow the radii trend. For answer c, as electrons are added to an atom, size increases. Answer e follows the trend for an isoelectronic series, i.e., the smallest ion has the most protons.

37.   a. Li^+ and N^{3-} are the expected ions. The formula of the compound would be Li_3N (lithium nitride).

b. Ga^{3+} and O^{2-}; Ga_2O_3, gallium(III) oxide or gallium oxide

c. Rb^+ and Cl^-; $RbCl$, rubidium chloride   d. Ba^{2+} and S^{2-}; BaS, barium sulfide

39.   Lattice energy is proportional to Q_1Q_2/r where Q is the charge of the ions and r is the distance between the ions. In general, charge effects on lattice energy are much greater than size effects.

a. NaCl; Na^+ is smaller than K^+.   b. LiF; F^- is smaller than Cl^-.

c. MgO; O^{2-} has a greater charge than OH^-.   d. $Fe(OH)_3$; Fe^{3+} has a greater charge than Fe^{2+}.

e. Na_2O; O^{2-} has a greater charge than Cl^-.   f. MgO; The ions are smaller in MgO.

41.
$Na(s) \rightarrow Na(g)$	$\Delta H = 109$ kJ (sublimation)
$Na(g) \rightarrow Na^+(g) + e^-$	$\Delta H = 495$ kJ (ionization energy)
$1/2\ Cl_2(g) \rightarrow Cl(g)$	$\Delta H = 239/2$ kJ (bond energy)
$Cl(g) + e^- \rightarrow Cl^-(g)$	$\Delta H = -349$ kJ (electron affinity)
$Na^+(g) + Cl^-(g) \rightarrow NaCl(s)$	$\Delta H = -786$ kJ (lattice energy)

$Na(s) + 1/2\ Cl_2(g) \rightarrow NaCl(s)$ $\Delta H^\circ_f = -412$ kJ/mol

43.   From the data given, it takes less energy to produce $Mg^+(g) + O^-(g)$ than to produce $Mg^{2+}(g) + O^{2-}(g)$. However, the lattice energy for $Mg^{2+}O^{2-}$ will be much more exothermic than that for Mg^+O^- due to the greater charges in $Mg^{2+}O^{2-}$. The favorable lattice energy term dominates and $Mg^{2+}O^{2-}$ forms.

45. Ca^{2+} has a greater charge than Na^+, and Se^{2-} is smaller than Te^{2-}. The effect of charge on the lattice energy is greater than the effect of size. We expect the trend from most exothermic to least exothermic to be:

$$CaSe > CaTe > Na_2Se > Na_2Te$$
$$(-2862) \quad (-2721) \quad (-2130) \quad (-2095) \quad \text{This is what we observe.}$$

Bond Energies

47.   a.   H—H + Cl——Cl ⟶ 2 H—Cl

Bonds broken: Bonds formed:

   1 H – H (432 kJ/mol)   2 H – Cl (427 kJ/mol)
   1 Cl – Cl (239 kJ/mol)

$\Delta H = \Sigma D_{broken} - \Sigma D_{formed}$, $\Delta H = 432 \text{ kJ} + 239 \text{ kJ} - 2(427) \text{ kJ} = -183 \text{ kJ}$

   b.   N≡N + 3 H—H ⟶ 2 H—N—H
                                          |
                                          H

Bonds broken: Bonds formed:

   1 N ≡ N (941 kJ/mol)   6 N – H (391 kJ/mol)
   3 H – H (432 kJ/mol)

$\Delta H = 941 \text{ kJ} + 3(432) \text{ kJ} - 6(391) \text{ kJ} = -109 \text{ kJ}$

49.
```
     H                      H
     |                      |
 H—C—N≡C   ⟶   H—C—C≡N
     |                      |
     H                      H
```

Bonds broken: 1 C – N (305 kJ/mol)   Bonds formed: 1 C – C (347 kJ/mol)

$\Delta H = \Sigma D_{broken} - \Sigma D_{formed}$, $\Delta H = 305 - 347 = -42 \text{ kJ}$

Note: Sometimes some of the bonds remain the same between reactants and products. To save time, only break and form bonds that are involved in the reaction.

51.

Bonds broken:

    5 C – H (413 kJ/mol)
    1 C – C (347 kJ/mol)
    1 C – O (358 kJ/mol)
    1 O – H (467 kJ/mol)
    3 O = O (495 kJ/mol)

Bonds formed:

    2×2 C = O (799 kJ/mol)
    3×2 O – H (467 kJ/mol)

$\Delta H = 5(413 \text{ kJ}) + 347 \text{ kJ} + 358 \text{ kJ} + 467 \text{ kJ} + 3(495 \text{ kJ}) - [4(799 \text{ kJ}) + 6(467 \text{ kJ})]$

$$= -1276 \text{ kJ}$$

53.   H—H + O=O $\longrightarrow$ H—O—O—H $\Delta H = -153 \text{ kJ}$

Bonds broken:

    1 H – H (432 kJ/mol)
    1 O = O (495 kJ/mol)

Bonds formed:

    2 O – H (467 kJ/mol)
    1 O – O (D_{OO} kJ/mol)

$\Delta H = -153 \text{ kJ} = 432 \text{ kJ} + 495 \text{ kJ} - [2(467 \text{ kJ}) + D_{OO}]$, $D_{OO} = 146 \text{ kJ/mol}$

55.   a. $\Delta H^\circ = 2 \; \Delta H^\circ_{\text{f, HCl}} = 2 \text{ mol } (-92 \text{ kJ/mol}) = -184 \text{ kJ} \; (= -183 \text{ kJ from bond energies})$

    b. $\Delta H^\circ = 2 \; \Delta H^\circ_{\text{f, NH}_3} = 2 \text{ mol } (-46 \text{ kJ/mol}) = -92 \text{ kJ} \; (= -109 \text{ kJ from bond energies})$

Comparing the values for each reaction, bond energies seem to give a reasonably good estimate of the enthalpy change for a reaction. The estimate is especially good for gas phase reactions.

57.   a. Using SF_4 data: $SF_4(g) \rightarrow S(g) + 4 F(g)$

        $\Delta H^\circ = 4 D_{SF} = 278.8 + 4 (79.0) - (-775) = 1370. \text{ kJ}$

        $D_{SF} = \dfrac{1370. \text{ kJ}}{4 \text{ mol SF bonds}} = 342.5 \text{ kJ/mol}$

        Using SF_6 data: $SF_6(g) \rightarrow S(g) + 6 F(g)$

        $\Delta H^\circ = 6 D_{SF} = 278.8 + 6 (79.0) - (-1209) = 1962 \text{ kJ}$

        $D_{SF} = \dfrac{1962 \text{ kJ}}{6} = 327.0 \text{ kJ/mol}$

b. The $S - F$ bond energy in the table is 327 kJ/mol. The value in the table was based on the $S - F$ bond in SF_6.

c. $S(g)$ and $F(g)$ are not the most stable forms of the elements at $25\,^\circ C$. The most stable forms are $S_8(s)$ and $F_2(g)$; $\Delta H_f^\circ = 0$ for these two species.

59. $N_2 + 3\,H_2 \rightarrow 2\,NH_3$; $\Delta H = D_{N_2} + 3\,D_{H_2} - 6\,D_{NH}$; $\Delta H^\circ = 2(-46\ kJ) = -92\ kJ$

$-92\ kJ = 941\ kJ + 3(432\ kJ) - (6\ D_{N-H})$, $6\ D_{N-H} = 2329\ kJ$, $D_{N-H} = 388.2\ kJ/mol$

Table in text: 391 kJ/mol;   There is good agreement.

Lewis Structures and Resonance

61.   Drawing Lewis structures is mostly trial and error. However, the first two steps are always the same. These steps are 1) count the valence electrons available in the molecule, and 2) attach all atoms to each other with single bonds (called the skeletal structure). Unless noted otherwise, the atom listed first is assumed to be the atom in the middle (called the central atom) and all other atoms in the formula are attached to this atom. The most notable exceptions to the rule are formulas which begin with H, e.g., H_2O, H_2CO, etc. Hydrogen can never be a central atom since this would require H to have more than two electrons.

After counting valence electrons and drawing the skeletal structure, the rest is trial and error. We place the remaining electrons around the various atoms in an attempt to satisfy the octet rule (or duet rule for H). Keep in mind that practice makes perfect. After practicing you can (and will) become very adept at drawing Lewis structures.

a. HCN has $1 + 4 + 5 = 10$ valence electrons.

b. PH_3 has $5 + 3(1) = 8$ valence electrons.

```
H—C—N        H—C≡N:
```

Skeletal   Lewis
structure   structure

Skeletal structures uses 4 e⁻ ; 6 e⁻ remain

```
H—P—H        H—P̈—H
   |              |
   H              H
```

Skeletal   Lewis
structure   structure

Skeletal structure uses 6 e⁻; 2 e⁻ remain

c. $CHCl_3$ has $4 + 1 + 3(7) = 26$ valence electrons.

d. NH_4^+ has $5 + 4(1) - 1 = 8$ valence electrons.

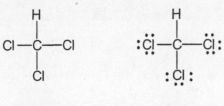

Note: Subtract valence electrons for positive charged ions.

Skeletal structure   Lewis structure   Lewis structure

Skeletal structure uses 8 e⁻; 18 e⁻ remain

e. H_2CO has $2(1) + 4 + 6 = 12$ valence electrons.

f. SeF_2 has $6 + 2(7) = 20$ valence electrons.

g. CO_2 has $4 + 2(6) = 16$ valence electrons.

h. O_2 has $2(6) = 12$ valence electrons.

i. HBr has $1 + 7 = 8$ valence electrons.

63.   In each case in this problem, the octet rule cannot be satisfied for the central atom. BeH_2 and BH_3 have too few electrons around the central atom, and all the others have too many electrons around the central atom. Always try to satisfy the octet rule for every atom, but when it is impossible, the central atom is the species which will disobey the octet rule.

PF_5, $5 + 5(7) = 40$ e⁻ BeH_2, $2 + 2(1) = 4$ e⁻

H – Be – H

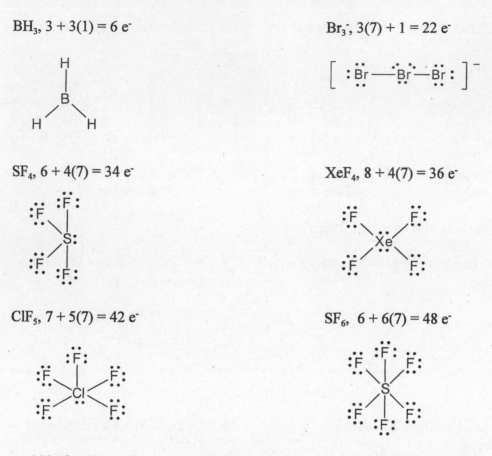

BH_3, $3 + 3(1) = 6$ e⁻

Br_3^-, $3(7) + 1 = 22$ e⁻

SF_4, $6 + 4(7) = 34$ e⁻

XeF_4, $8 + 4(7) = 36$ e⁻

ClF_5, $7 + 5(7) = 42$ e⁻

SF_6, $6 + 6(7) = 48$ e⁻

65.   a. NO_2^- has $5 + 2(6) + 1 = 18$ valence electrons. The skeletal structure is: O – N – O

To get an octet about the nitrogen and only use 18 e⁻, we must form a double bond to one of the oxygen atoms.

Since there is no reason to have the double bond to a particular oxygen atom, we can draw two resonance structures. Each Lewis structure uses the correct number of electrons and satisfies the octet rule, so each is a valid Lewis structure. Resonance structures occur when you have multiple bonds that can be in various positions. We say the actual structure is an average of these two resonance structures.

NO_3^- has $5 + 3(6) + 1 = 24$ valence electrons. We can draw three resonance structures for NO_3^-, with the double bond rotating among the three oxygen atoms.

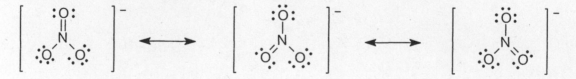

N$_2$O$_4$ has 2(5) + 4(6) = 34 valence electrons. We can draw four resonance structures for N$_2$O$_4$.

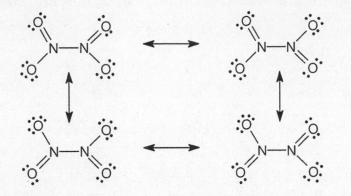

b.   OCN$^-$ has 6 + 4 + 5 + 1 = 16 valence electrons. We can draw three resonance structures for OCN$^-$.

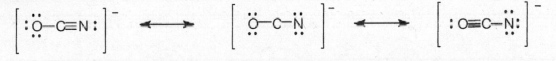

SCN$^-$ has 6 + 4 + 5 + 1 = 16 valence electrons. Three resonance structures can be drawn.

N$_3^-$ has 3(5) + 1 = 16 valence electrons. As with OCN$^-$ and SCN$^-$, three different resonance structures can be drawn.

67.   Benzene has 6(4) + 6(1) = 30 valence electrons. Two resonance structures can be drawn for benzene. The actual structure of benzene is an average of these two resonance structures, that is, all carbon-carbon bonds are equivalent with a bond length and bond strength somewhere between a single and a double bond.

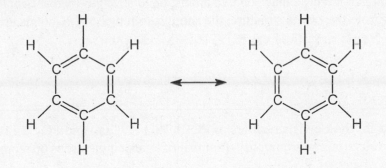

69.   We will use a hexagon to represent the six-member carbon ring, and we will omit the 4 hydrogen atoms and the three lone pairs of electrons on each chlorine. If no resonance existed, we could draw 4 different molecules:

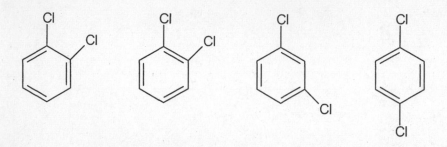

If the double bonds in the benzene ring exhibit resonance, then we can draw only three different dichlorobenzenes. The circle in the hexagon represents the delocalization of the three double bonds in the benzene ring (see Exercise 8.67).

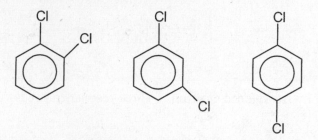

With resonance, all carbon-carbon bonds are equivalent. We can't distinguish between a single and double bond between adjacent carbons that have a chlorine attached. That only 3 isomers are observed provides evidence for the existence of resonance.

71. N_2 (10 e⁻):   :N≡N: Triple bond between N and N.

      N_2F_4 (38 e⁻):   :F̈—N̈—N̈—F̈:   Single bond between N and N.
                           | |
                          :F̈: :F̈:

      N_2F_2 (24 e⁻):   :F̈—N=N—F̈:   Double bond between N and N.

      As the number of bonds increase between two atoms, bond strength increases and bond length decreases. From the Lewis structure, the shortest to longest N-N bonds are: $N_2 < N_2F_2 < N_2F_4$.

Formal Charge

73.   See Exercise 8.62a for the Lewis structures of $POCl_3$, SO_4^{2-}, ClO_4^- and PO_4^{3-}. Formal charge = [number of valence electrons on free atom] - [number of lone pair electrons on atom - 1/2 (number of shared electrons of atom)].

a. $POCl_3$: P, FC = 5 - 1/2(8) = +1   b. SO_4^{2-}: S, FC = 6 - 1/2(8) = +2

c. ClO_4^-: Cl, FC = 7 - 1/2(8) = +3   d. PO_4^{3-}: P, FC = 5 - 1/2(8) = +1

e. SO_2Cl_2, 6 + 2(6) + 2(7) = 32 e⁻   f. XeO_4, 8 + 4(6) = 32 e⁻

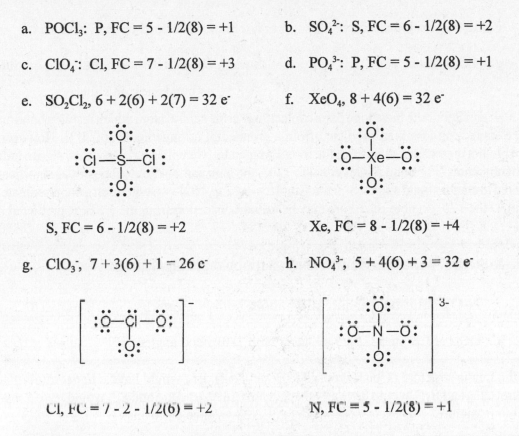

S, FC = 6 - 1/2(8) = +2   Xe, FC = 8 - 1/2(8) = +4

g. ClO_3^-, 7 + 3(6) + 1 = 26 e⁻   h. NO_4^{3-}, 5 + 4(6) + 3 = 32 e⁻

Cl, FC = 7 - 2 - 1/2(6) = +2   N, FC = 5 - 1/2(8) = +1

75. O_2F_2 has 2(6) + 2(7) = 26 valence e⁻. The formal charge and oxidation number of each atom is
      below the Lewis structure of O_2F_2.

Formal Charge	0	0	0	0
Oxid. Number	-1	+1	+1	-1

Oxidation numbers are more useful when accounting for the reactivity of O_2F_2. We are forced to
assign +1 as the oxidation number for oxygen. Oxygen is very electronegative, and +1 is not a stable
oxidation state for this element.

Molecular Geometry and Polarity

77.   The first step always is to draw a valid Lewis structure when predicting molecular structure. When
      resonance is possible, only one of the possible resonance structures is necessary to predict the
      correct structure since all resonance structures give the same structure. The Lewis structures are
      in Exercises 8.61 and 8.65. The structures and bond angles for each follow.

      8.61   a. HCN: linear, 180°   b. PH_3: trigonal pyramid, < 109.5°

             c. $CHCl_3$: tetrahedral, 109.5°   d. NH_4^+: tetrahedral, 109.5°

e. H_2CO: trigonal planar, 120°   f. SeF_2: V-shaped or bent, < 109.5°

g. CO_2: linear, 180°   h and i. O_2 and HBr are both linear, but there is

no bond angle in either.

Note: PH_3 and SeF_2 both have lone pairs of electrons on the central atom which result in bond angles that are something less than predicted from a tetrahedral arrangement (109.5°). However, we cannot predict the exact number. For the solutions manual, we will insert a less than sign to indicate this phenomenon. For bond angles equal to 120°, the lone pair phenomenon isn't as significant as compared to smaller bond angles. For these molecules, e.g., NO_2^-, we will insert an approximate sign in front of the 120° to note that there may be a slight distortion from the VSEPR predicted bond angle.

8.65   a. NO_2^-: V-shaped, ≈ 120°; NO_3^-: trigonal planar, 120°;

N_2O_4: trigonal planar, 120° about both N atoms

b. OCN^-, SCN^- and N_3^- are all linear with 180° bond angles.

79.   From the Lewis structures (see Exercises 8.63 and 8.64), Br_3^- would have a linear molecular structure, ClF_3 and BrF_3 would have a T-shaped molecular structure and SF_4 would have a see-saw molecular structure. For example, consider ClF_3 (28 valence electrons):

The central Cl atom is surrounded by 5 electron pairs, which requires a trigonal bipyramid geometry. Since there are 3 bonded atoms and 2 lone pairs of electrons about Cl, we describe the molecular structure of ClF_3 as T-shaped with predicted bond angles of about 90°. The actual bond angles would be slightly less than 90° due to the stronger repulsive effect of the lone pair electrons as compared to the bonding electrons.

81.   a. SeO_3, 6 + 3(6) = 24 e^-

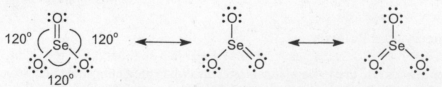

SeO_3 has a trigonal planar molecular structure with all bond angles equal to 120°. Note that any one of the resonance structures could be used to predict molecular structure and bond angles.

b. SeO_2, 6 + 2(6) = 18 e^-

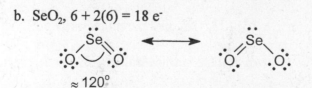

≈ 120°

SeO$_2$ has a V-shaped molecular structure. We would expect the bond angle to be approximately 120° as expected for trigonal planar geometry.

Note: Both of these structures have three effective pairs of electrons about the central atom. All of the structures are based on a trigonal planar geometry, but only SeO$_3$ is described as having a trigonal planar structure. Molecular structure always describes the relative positions of the atoms.

83.   a.   XeCl$_2$ has 8 + 2(7) = 22 valence electrons.

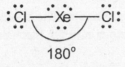

180°

There are 5 pairs of electrons about the central Xe atom. The structure will be based on a trigonal bipyramid geometry. The most stable arrangement of the atoms in XeCl$_2$ is a linear molecular structure with a 180° bond angle.

b.   ICl$_3$ has 7 + 3(7) = 28 valence electrons.

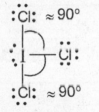

T-shaped; The ClICl angles are ≈ 90°. Since the lone pairs will take up more space, the ClICl bond angles will probably be slightly less than 90°.

c.   TeF$_4$ has 6 + 4(7) = 34 valence electrons.

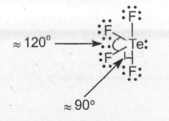

See-saw or teeter-totter or distorted tetrahedron

d.   PCl$_5$ has 5 + 5(7) = 40 valence electrons.

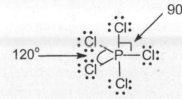

Trigonal bipyramid

All of the species in this exercise have 5 pairs of electrons around the central atom. All of the structures are based on a trigonal bipyramid geometry, but only in PCl$_5$ are all of the pairs bonding pairs. Thus, PCl$_5$ is the only one we describe as a trigonal bipyramid molecular structure. Still, we had to begin with the trigonal bipyramid geometry to get to the structures of the others.

85.   SeO$_3$ and SeO$_2$ both have polar bonds but only SeO$_2$ has a dipole moment. The three bond dipoles from the three polar Se – O bonds in SeO$_3$ will all cancel when summed together. Hence, SeO$_3$ is nonpolar since the overall molecule has no resulting dipole moment. In SeO$_2$, the two Se – O bond dipoles do not cancel when summed together, hence SeO$_2$ has a dipole moment (is polar). Since O is more electronegative than Se, the negative end of the dipole moment is between the two O atoms,

and the positive end is around the Se atom. The arrow in the following illustration represents the overall dipole moment in SeO_2. Note that to predict polarity for SeO_2, either of the two resonance structures can be used.

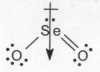

87.   All have polar bonds, but only TeF_4 and ICl_3 have dipole moments. The bond dipoles from the five P–Cl bonds in PCl_5 cancel each other when summed together, so PCl_5 has no dipole moment. The bond dipoles in $XeCl_2$ also cancel:

$$:\ddot{C}l \xrightleftharpoons{\quad} :\ddot{X}e: \xrightleftharpoons{\quad} \ddot{C}l:$$

Since the bond dipoles from the two Xe – Cl bonds are equal in magnitude but point in opposite directions, they cancel each other and $XeCl_2$ has no dipole moment (is nonpolar). For TeF_4 and ICl_3, the arrangement of these molecules is such that the individual bond dipoles do not all cancel, so each has an overall dipole moment.

89.   Molecules which have an overall dipole moment are called polar molecules, and molecules which do not have an overall dipole moment are called nonpolar molecules.

    a. OCl_2, 6 + 2(7) = 20 e⁻ KrF_2, 8 + 2(7) = 22 e⁻

V-shaped, polar; OCl_2 is polar because the two O – Cl bond dipoles don't cancel each other. The resultant dipole moment is shown in the drawing.

Linear, nonpolar; The molecule is nonpolar because the two Kr – F bond dipoles cancel each other.

BeH_2, 2 + 2(1) = 4 e⁻ SO_2, 6 + 2(6) = 18 e⁻

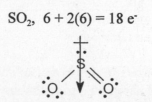

Linear, nonpolar; Be – H bond dipoles are equal and point in opposite directions. They cancel each other. BeH_2 is nonpolar.

V-shaped, polar; The S – O bond dipoles do not cancel, so SO_2 is polar (has a dipole moment). Only one resonance structure is shown.

Note: All four species contain three atoms. They have different structures because the number of lone pairs of electrons around the central atom is different in each case.

b. SO_3, $6 + 3(6) = 24$ e⁻ NF_3, $5 + 3(7) = 26$ e⁻

                                    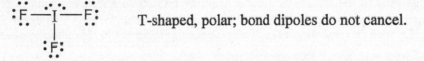

Trigonal planar, nonpolar;   Trigonal pyramid, polar;
Bond dipoles cancel. Only one   Bond dipoles do not cancel.
resonance structure is shown.

IF_3 has $7 + 3(7) = 28$ valence electrons.

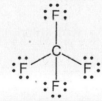   T-shaped, polar; bond dipoles do not cancel.

Note: Each molecule has the same number of atoms, but the structures are different because of differing numbers of lone pairs around each central atom.

c. CF_4, $4 + 4(7) = 32$ e⁻ SeF_4, $6 + 4(7) = 34$ e⁻

Tetrahedral, nonpolar;   See-saw, polar;
Bond dipoles cancel.   Bond dipoles do not cancel.

KrF_4, $8 + 4(7) = 36$ valence electrons

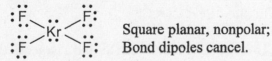   Square planar, nonpolar;
                             Bond dipoles cancel.

Again, each molecule has the same number of atoms, but a different structure because of differing numbers of lone pairs around the central atom.

d. IF_5, $7 + 5(7) = 42$ e$^-$ AsF_5, $5 + 5(7) = 40$ e$^-$

Square pyramid, polar;   Trigonal bipyramid, nonpolar;
Bond dipoles do not cancel.   Bond dipoles cancel.

Yet again, the molecules have the same number of atoms, but different structures because of the presence of differing numbers of lone pairs.

91. EO_3^- is the formula of the ion. The Lewis structure has 26 valence electrons. Let x = number of valence electrons of element E.

       $26 = x + 3(6) + 1$, $x = 7$ valence electrons

Element E is a halogen because halogens have 7 valence electrons. Some possible identities are F, Cl, Br and I. The EO_3^- ion has a trigonal pyramid molecular structure with bond angles < 109.5°.

93.   All these molecules have polar bonds that are symmetrically arranged about the central atoms. In each molecule, the individual bond dipoles cancel to give no net overall dipole moment.

Additional Exercises

95.   a. Radius: $N^+ < N < N^-$; IE: $N^- < N < N^+$

       N^+ has the fewest electrons held by the 7 protons in the nucleus while N^- has the most electrons held by the 7 protons. The 7 protons in the nucleus will hold the electrons most tightly in N^+ and least tightly in N^-. Therefore, N^+ has the smallest radius with the largest ionization energy (IE) and N^- is the largest species with the smallest IE.

       b. Radius: $Cl^+ < Cl < Se < Se^-$; IE: $Se^- < Se < Cl < Cl^+$

       The general trends tell us that Cl has a smaller radius than Se and a larger IE than Se. Cl^+, with fewer electron-electron repulsions than Cl, will be smaller than Cl and have a larger IE. Se^-, with more electron-electron repulsions than Se, will be larger than Se and have a smaller IE.

       c. Radius: $Sr^{2+} < Rb^+ < Br^-$; IE: $Br^- < Rb^+ < Sr^{2+}$

       These ions are isoelectronic. The species with the most protons (Sr^{2+}) will hold the electrons most tightly and will have the smallest radius and largest IE. The ion with the fewest protons (Br^-) will hold the electrons least tightly and will have the largest radius and smallest IE.

97. a. $HF(g) \rightarrow H(g) + F(g)$ $\Delta H = 565$ kJ
           $H(g) \rightarrow H^+(g) + e^-$ $\Delta H = 1312$ kJ
           $F(g) + e^- \rightarrow F^-(g)$ $\Delta H = -327.8$ kJ
           _____
           $HF(g) \rightarrow H^+(g) + F^-(g)$ $\Delta H = 1549$ kJ

     b. $HCl(g) \rightarrow H(g) + Cl(g)$ $\Delta H = 427$ kJ
           $H(g) \rightarrow H^+(g) + e^-$ $\Delta H = 1312$ kJ
           $Cl(g) + e^- \rightarrow Cl^-(g)$ $\Delta H = -348.7$ kJ
           _____
           $HCl(g) \rightarrow H^+(g) + Cl^-(g)$ $\Delta H = 1390.$ kJ

     c. $HI(g) \rightarrow H(g) + I(g)$ $\Delta H = 295$ kJ
           $H(g) \rightarrow H^+(g) + e^-$ $\Delta H = 1312$ kJ
           $I(g) + e^- \rightarrow I^-(g)$ $\Delta H = -295.2$ kJ
           _____
           $HI(g) \rightarrow H^+(g) + I^-(g)$ $\Delta H = 1312$ kJ

     d. $H_2O(g) \rightarrow OH(g) + H(g)$ $\Delta H = 467$ kJ
           $H(g) \rightarrow H^+(g) + e^-$ $\Delta H = 1312$ kJ
           $OH(g) + e^- \rightarrow OH^-(g)$ $\Delta H = -180.$ kJ
           _____
           $H_2O(g) \rightarrow H^+(g) + OH^-(g)$ $\Delta H = 1599$ kJ

99. The stable species are:

     a. NaBr: In $NaBr_2$, the sodium ion would have a + 2 charge assuming each bromine has a -1 charge. Sodium doesn't form stable Na^{2+} compounds.

     b. ClO_4^-: ClO_4 has 31 valence electrons so it is impossible to satisfy the octet rule for all atoms in ClO_4. The extra electron from the -1 charge in ClO_4^- allows for complete octets for all atoms.

     c. XeO_4: We can't draw a Lewis structure that obeys the octet rule for SO_4 (30 electrons), unlike XeO_4 (32 electrons).

     d. SeF_4: Both compounds require the central atom to expand its octet. O is too small and doesn't have low energy d orbitals to expand its octet (which is true for all row 2 elements).

101. The general structure of the trihalide ions is: $\left[:\ddot{X} - \ddot{X} - \ddot{X}: \right]^-$

      Bromine and iodine are large enough and have low energy, empty d-orbitals to accommodate the expanded octet. Fluorine is small, its valence shell contains only 2s and 2p orbitals (4 orbitals) and it does not expand its octet. The lowest energy d orbitals in F are 3d orbitals; they are too high in energy as compared to 2s and 2p to be used in bonding.

103. Yes, each structure has the same number of effective pairs around the central atom. (A multiple bond is counted as a single group of electrons.)

105.   TeF$_5^-$ has $6 + 5(7) + 1 = 42$ valence electrons.

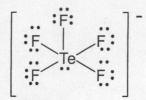

The lone pair of electrons around Te exerts a stronger repulsion than the bonding pairs, pushing the four square planar F's away from the lone pair and thus reducing the bond angles between the axial F atom and the square planar F atoms.

Challenge Problems

107.   As the halogen atoms get larger, it becomes more difficult to fit three halogen atoms around the small nitrogen atom, and the NX$_3$ molecule becomes less stable.

109.   a.   i. $C_6H_6N_{12}O_{12} \rightarrow 6\ CO + 6\ N_2 + 3\ H_2O + 3/2\ O_2$

The NO$_2$ groups are assumed to have one $N - O$ single bond and one $N = O$ double bond and each carbon atom has one $C - H$ single bond. We must break and form all bonds.

Bonds broken: Bonds formed:

   3 C $-$ C   (347 kJ/mol)   6 C $\equiv$ O   (1072 kJ/mol)
   6 C $-$ H   (413 kJ/mol)   6 N $\equiv$ N   (941 kJ/mol)
 12 C $-$ N   (305 kJ/mol)   6 H $-$ O   (467 kJ/mol)
   6 N $-$ N   (160. kJ/mol)   3/2 O $=$ O   (495 kJ/mol)
   6 N $-$ O   (201 kJ/mol)   _____
   6 N $=$ O   (607 kJ/mol) $\Sigma D_{formed} = 15{,}623$ kJ
_____
$\Sigma D_{broken} = 12{,}987$ kJ

$\Delta H = \Sigma D_{broken} - \Sigma D_{formed} = 12{,}987$ kJ - 15,623 kJ = -2636 kJ

ii. $C_6H_6N_{12}O_{12} \rightarrow 3\ CO + 3\ CO_2 + 6\ N_2 + 3\ H_2O$

Note: The bonds broken will be the same for all three reactions.

Bonds formed:

   3 C $\equiv$ O   (1072 kJ/mol)
   6 C $=$ O   (799 kJ/mol)
   6 N $\equiv$ N   (941 kJ/mol)
   6 H $-$ O   (467 kJ/mol)
_____
$\Sigma D_{formed} = 16{,}458$ kJ

$\Delta H = 12{,}987$ kJ - 16,458 kJ = -3471 kJ

iii. $C_6H_6N_{12}O_{12} \rightarrow 6\ CO_2 + 6\ N_2 + 3\ H_2$

Bonds formed:

$12\ C = O$ (799 kJ/mol)
$6\ N \equiv N$ (941 kJ/mol)
$3\ H - H$ (432 kJ/mol)
_____

$\Sigma D_{formed} = 16,530.\ kJ$

$\Delta H = 12,987\ kJ - 16,530.\ kJ = -3543\ kJ$

b. Reaction iii yields the most energy per mole of CL-20 so it will yield the most energy per kg.

$$\frac{-3543\ kJ}{mol} \times \frac{1\ mol}{438.23\ g} \times \frac{1000\ g}{kg} = -8085\ kJ/kg$$

111.   PAN ($H_3C_2NO_5$) has $3(1) + 2(4) + 5 + 5(6) = 46$ valence electrons.

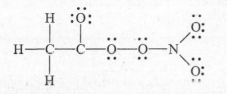

Skeletal structure with complete octets about oxygen atoms (46 electrons used).

This structure has used all 46 electrons, but there are only six electrons around one of the carbon atoms and the nitrogen atom. Two unshared pairs must become shared, that is we must form two double bonds.

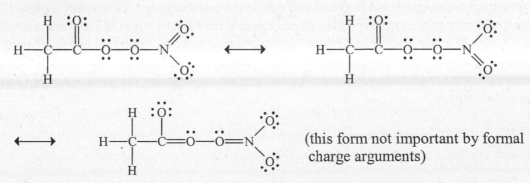

  (this form not important by formal charge arguments)

113.   a. $BrFI_2$, $7 + 7 + 2(7) = 28\ e^-$; Two possible structures exist; each has a T-shaped molecular structure.

90° bond angles between I atoms   180° bond angles between I atoms

b. XeO_2F_2, $8 + 2(6) + 2(7) = 34$ e⁻; Three possible structures exist; each has a see-saw molecular structure.

90° bond angle
between O atoms

180° bond angle
between O atoms

120° bond angle
between O atoms

c. $TeF_2Cl_3^-$; $6 + 2(7) + 3(7) + 1 = 42$ e⁻; Three possible structures exist; each has a square pyramid molecular structure.

One F is 180° from
lone pair.

Both F atoms are 90°
from lone pair and 90°
from each other.

Both F atoms are 90°
from lone pair and 180°
from each other.

115. The nitrogen-nitrogen bond length of 112 pm is between a double (120 pm) and a triple (110 pm) bond. The nitrogen-oxygen bond length of 119 pm is between a single (147 pm) and a double bond (115 pm). The last resonance structure doesn't appear to be as important as the other two since there is no evidence from bond lengths for a nitrogen-oxygen triple bond or a nitrogen-nitrogen single bond as in the third resonance form. We can adequately describe the structure of N_2O using the resonance forms:

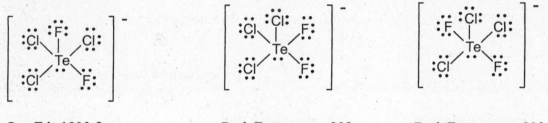

Assigning formal charges for all 3 resonance forms:

For:

$\left(\ddot{\overset{\cdot\cdot}{N}} = \right)$, FC = 5 - 4 - 1/2(4) = -1

$\left(= N = \right)$, FC = 5 - 1/2(8) = +1 , Same for $\left(\equiv N - \right)$ and $\left(- N \equiv \right)$

$\left(:\ddot{N} - \right)$, FC = 5 - 6 - 1/2(2) = -2 ; $\left(:N \equiv \right)$, FC = 5 - 2 - 1/2(6) = 0

$\left(= \ddot{\overset{\cdot\cdot}{O}} \right)$, FC = 6 - 4 - 1/2(4) = 0 ; $\left(- \ddot{\overset{\cdot\cdot}{O}}: \right)$, FC = 6 - 6 - 1/2(2) = -1

$\left(\equiv O: \right)$, FC = 6 - 2 - 1/2(6) = +1

We should eliminate $N - N \equiv O$ since it has a formal charge of +1 on the most electronegative element (O). This is consistent with the observation that the $N - N$ bond is between a double and triple bond and that the $N - O$ bond is between a single and double bond.

CHAPTER NINE

COVALENT BONDING: ORBITALS

Questions

7.   Bond energy is directly proportional to bond order. Bond length is inversely proportional to bond order. Bond energy and bond length can be measured.

9.   Paramagnetic: Unpaired electrons are present. Measure the mass of a substance in the presence and absence of a magnetic field. A substance with unpaired electrons will be attracted by the magnetic field, giving an apparent increase in mass in the presence of the field. A greater number of unpaired electrons will give a greater attraction and a greater observed mass increase.

Exercises

The Localized Electron Model and Hybrid Orbitals

11. H_2O has $2(1) + 6 = 8$ valence electrons.

H_2O has a tetrahedral arrangement of the electron pairs about the O atom that requires sp^3 hybridization. Two of the four sp^3 hybrid orbitals are used to form bonds to the two hydrogen atoms and the other two sp^3 hybrid orbitals hold the two lone pairs of oxygen. The two O – H bonds are formed from overlap of the sp^3 hybrid orbitals on oxygen with the 1s atomic orbitals on the hydrogen atoms.

13. H_2CO has $2(1) + 4 + 6 = 12$ valence electrons.

The central carbon atom has a trigonal planar arrangement of the electron pairs which requires sp^2 hybridization. The two C – H sigma bonds are formed from overlap of the sp^2 hybrid orbitals on carbon with the hydrogen 1s atomic orbitals. The double bond between carbon and oxygen consists of one σ and one π bond. The oxygen atom, like the carbon atom, also has a trigonal planar arrangement of the electrons which requires sp^2 hybridization. The σ bond in the double bond is formed from overlap of a carbon sp^2 hybrid orbital with an oxygen sp^2 hybrid orbital. The π bond in

116

the double bond is formed from overlap of the unhybridized p atomic orbitals. Carbon and oxygen each have one unhybridized p atomic orbital which are parallel to each other. When two parallel p atomic orbitals overlap, a π bond results.

15.   See Exercises 8.61 and 8.65 for the Lewis structures. To predict the hybridization, first determine the arrangement of electron pairs about each central atom using the VSEPR model; then utilize the information in Figure 9.24 of the text to deduce the hybridization required for that arrangement of electron pairs.

8.61   a.   HCN; C is sp hybridized.   b. PH_3; P is sp^3 hybridized.

c. $CHCl_3$; C is sp^3 hybridized.   d. NH_4^+; N is sp^3 hybridized.

e. H_2CO; C is sp^2 hybridized.   f. SeF_2; Se is sp^3 hybridized.

g. CO_2; C is sp hybridized.   h. O_2; Each O atom is sp^2 hybridized.

i.   HBr; Br is sp^3 hybridized.

8.65   a.   The central N atom is sp^2 hybridized in NO_2^- and NO_3^-. In N_2O_4, both central N atoms are sp^2 hybridized.

b.   In OCN^- and SCN^-, the central carbon atoms in each ion are sp hybridized and in N_3^-, the central N atom is also sp hybridized.

17.   See Exercise 8.63 for the Lewis structures.

PF_5:   P is dsp^3 hybridized. BeH_2:   Be is sp hybridized.

BH_3:   B is sp^2 hybridized. Br_3^-:   Br is dsp^3 hybridized.

SF_4:   S is dsp^3 hybridized. XeF_4:   Xe is d^2sp^3 hybridized.

ClF_6^-:   Cl is d^2sp^3 hybridized. SF_6:   S is d^2sp^3 hybridized.

19.   The molecules in Exercise 8.81 all have a trigonal planar arrangement of electron pairs about the central atom so all have central atoms with sp^2 hybridization. The molecules in Exercise 8.82 all have a tetrahedral arrangement of electron pairs about the central atom so all have central atoms with sp^3 hybridization. See Exercises 8.81 and 8.82 for the Lewis structures.

21.   a.   b.

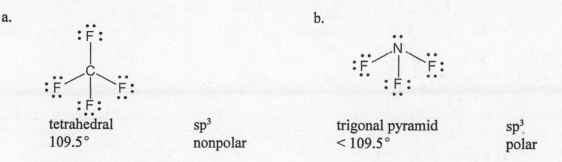

tetrahedral sp^3 trigonal pyramid sp^3
109.5°   nonpolar   < 109.5°   polar

The angles in NF_3 should be slightly less than 109.5° because the lone pair requires more space than the bonding pairs.

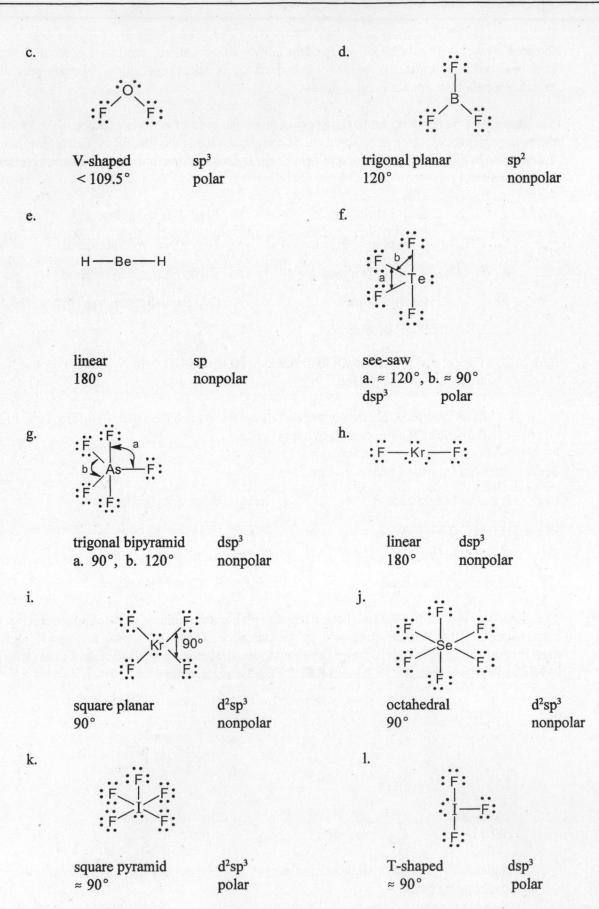

c.

V-shaped   sp³
< 109.5°   polar

d.

trigonal planar   sp²
120°   nonpolar

e.

H—Be—H

linear   sp
180°   nonpolar

f.

see-saw
a. ≈ 120°, b. ≈ 90°
dsp³   polar

g.

trigonal bipyramid dsp³
a. 90°, b. 120°   nonpolar

h.

linear   dsp³
180°   nonpolar

i.

square planar   d²sp³
90°   nonpolar

j.

octahedral   d²sp³
90°   nonpolar

k.

square pyramid   d²sp³
≈ 90°   polar

l.

T-shaped   dsp³
≈ 90°   polar

23.

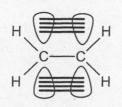

For the p-orbitals to properly line up to form the π bond, all six atoms are forced into the same plane. If the atoms were not in the same plane, the π bond could not form since the p-orbitals would no longer be parallel to each other.

25.   To complete the Lewis structures, just add lone pairs of electrons to satisfy the octet rule for the atoms with fewer than eight electrons.

Biacetyl ($C_4H_6O_2$) has $4(4) + 6(1) + 2(6) = 34$ valence electrons.

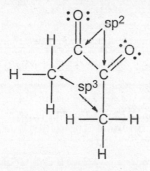

All CCO angles are 120°. The six atoms are not in the same plane because of free rotation about the carbon-carbon single (sigma) bonds. There are 11 σ and 2 π bonds in biacetyl.

Acetoin ($C_4H_8O_2$) has $4(4) + 8(1) + 2(6) = 36$ valence electrons.

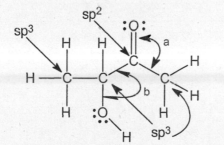

The carbon with the doubly-bonded O is sp^2 hybridized. The other 3 C atoms are sp^3 hybridized. Angle a = 120° and angle b = 109.5°. There are 13 σ and 1 π bonds in acetoin.

Note: All single bonds are σ bonds, all double bonds are one σ and one π bond, and all triple bonds are one σ and two π bonds.

27.   To complete the Lewis structure, just add lone pairs of electrons to satisfy the octet rule for the atoms that have fewer than eight electrons.

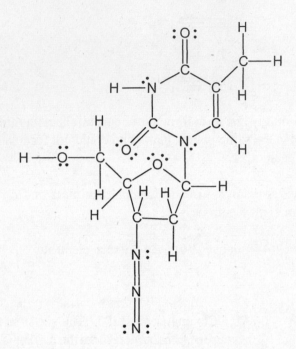

a. 6   b. 4   c.   The center N in – N = N = N group

d. 33 σ   e. 5 π bonds   f.   180°

g. < 109.5°   h. sp³

29.

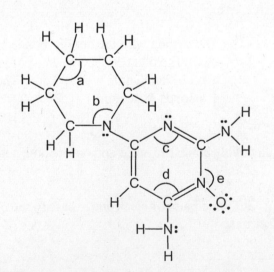

a. The two nitrogens in the ring with double bonds are sp² hybridized. The other three nitrogens are sp³ hybridized.

b. The five carbon atoms in the ring with one nitrogen are all sp³ hybridized. The four carbon atoms in the other ring with double bonds are all sp² hybridized.

c. Angles a and b: ≈109.5°; angles c, d, and e: ≈120°

d.   31 sigma bonds

e.   3 pi bonds (Each double bond consists of one sigma and one pi bond.)

The Molecular Orbital Model

31.   If we calculate a non-zero bond order for a molecule, then we predict that it can exist (is stable).

a. H_2^+: $(\sigma_{1s})^1$ B.O. = (1-0)/2 = 1/2, stable

   H_2: $(\sigma_{1s})^2$ B.O. = (2-0)/2 = 1, stable

   H_2^-: $(\sigma_{1s})^2(\sigma_{1s}*)^1$ B.O. = (2-1)/2 = 1/2, stable

   H_2^{2-}: $(\sigma_{1s})^2(\sigma_{1s}*)^2$ B.O. = (2-2)/2 = 0, not stable

b. He_2^{2+}: $(\sigma_{1s})^2$ B.O. = (2-0)/2 = 1, stable

   He_2^+: $(\sigma_{1s})^2(\sigma_{1s}*)^1$ B.O. = (2-1)/2 = 1/2, stable

   He_2: $(\sigma_{1s})^2(\sigma_{1s}*)^2$ B.O. = (2-2)/2 = 0, not stable

33.   The electron configurations are:

a. Li_2: $(\sigma_{2s})^2$ B.O. = (2-0)/2 = 1, diamagnetic (0 unpaired e⁻)

b. C_2: $(\sigma_{2s})^2(\sigma_{2s}*)^2(\pi_{2p})^4$ B.O. = (6-2)/2 = 2, diamagnetic (0 unpaired e⁻)

c. S_2: $(\sigma_{3s})^2(\sigma_{3s}*)^2(\sigma_{3p})^2(\pi_{3p})^4(\pi_{3p}*)^2$ B.O. = (8-4)/2 = 2, paramagnetic (2 unpaired e⁻)

35.   The electron configurations are:

O_2^+: $(\sigma_{2s})^2(\sigma_{2s}*)^2(\sigma_{2p})^2(\pi_{2p})^4(\pi_{2p}*)^1$

O_2: $(\sigma_{2s})^2(\sigma_{2s}*)^2(\sigma_{2p})^2(\pi_{2p})^4(\pi_{2p}*)^2$

O_2^-: $(\sigma_{2s})^2(\sigma_{2s}*)^2(\sigma_{2p})^2(\pi_{2p})^4(\pi_{2p}*)^3$

O_2^{2-}: $(\sigma_{2s})^2(\sigma_{2s}*)^2(\sigma_{2p})^2(\pi_{2p})^4(\pi_{2p}*)^4$

	O_2^+	O_2	O_2^-	O_2^{2-}
Bond order	2.5	2	1.5	1
# of unpaired electrons	1	2	1	0

Bond energy: $O_2^{2-} < O_2^- < O_2 < O_2^+$; Bond length: $O_2^+ < O_2 < O_2^- < O_2^{2-}$

Bond energy is directly proportional to bond order, and bond length is inversely proportional to bond order.

37.   The electron configurations are (assuming the same orbital order as that for N_2):

a.   CO: $(\sigma_{2s})^2(\sigma_{2s}*)^2(\pi_{2p})^4(\sigma_{2p})^2$ B.O. = (8-2)/2 = 3, diamagnetic

b. CO^+: $(\sigma_{2s})^2(\sigma_{2s}*)^2(\pi_{2p})^4(\sigma_{2p})^1$ B.O. = (7-2)/2 = 2.5, paramagnetic

c. CO^{2+}: $(\sigma_{2s})^2(\sigma_{2s}*)^2(\pi_{2p})^4$ B.O. = (6-2)/2 = 2, diamagnetic

Since bond order is directly proportional to bond energy and inversely proportional to bond length, then:

   shortest → longest bond length: $CO < CO^+ < CO^{2+}$

   smallest → largest bond energy: $CO^{2+} < CO^+ < CO$

39. H_2: $(\sigma_{1s})^2$
       B_2: $(\sigma_{2s})^2(\sigma_{2s}*)^2(\pi_{2p})^2$
       N_2: $(\sigma_{2s})^2(\sigma_{2s}*)^2(\pi_{2p})^4(\sigma_{2p})^2$
       OF: $(\sigma_{2s})^2(\sigma_{2s}*)^2(\sigma_{2p})^2(\pi_{2p})^4(\pi_{2p}*)^3$

The bond strength will weaken if the electron removed comes from a bonding orbital. Of the molecules listed, H_2, B_2, and N_2 would be expected to have their bond strength weaken as an electron is removed. OF has the electron removed from an antibonding orbital, so its bond strength increases.

41.   The two types of overlap that result in bond formation for p orbitals are side to side overlap (π bond) and head to head overlap (σ bond).

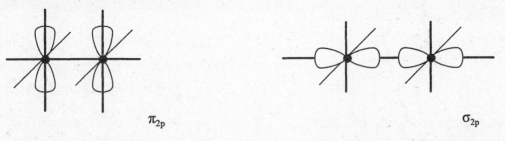

              π_{2p} σ_{2p}

43.   a.   The electron density would be closer to F on the average. The F atom is more electronegative than the H atom, and the 2p orbital of F is lower in energy than the 1s orbital of H.

       b.   The bonding MO would have more fluorine 2p character since it is closer in energy to the fluorine 2p atomic orbital.

       c.   The antibonding MO would place more electron density closer to H and would have a greater contribution from the higher energy hydrogen 1s atomic orbital.

45. O_3 and NO_2^- are isoelectronic, so we only need consider one of them since the same bonding ideas apply to both. The Lewis structures for O_3 are:

For each of the two resonance forms, the central O atom is sp^2 hybridized with one unhybridized p atomic orbital. The sp^2 hybrid orbitals are used to form the two sigma bonds to the central atom. The localized electron view of the π bond utilizes unhybridized p atomic orbitals. The π bond resonates between the two positions in the Lewis structures:

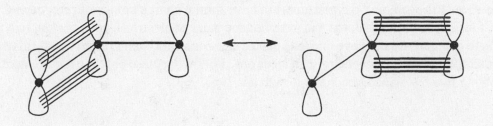

In the MO picture of the π bond, all three unhybridized p-orbitals overlap at the same time, resulting in π electrons that are delocalized over the entire surface of the molecule. This is represented as:

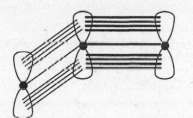

or

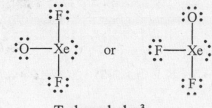

Additional Exercises

47.   a. XeO_3, $8 + 3(6) = 26$ e⁻   b. XeO_4, $8 + 4(6) = 32$ e⁻

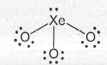

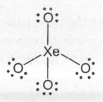

trigonal pyramid; sp^3 tetrahedral; sp^3

c. $XeOF_4$, $8 + 6 + 4(7) = 42$ e⁻   d. $XeOF_2$, $8 + 6 + 2(7) = 28$ e⁻

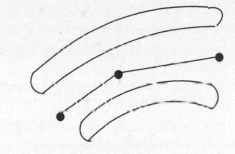

square pyramid; d^2sp^3 T-shaped; dsp^3

e. XeO_3F_2 has $8 + 3(6) + 2(7) = 40$ valence electrons.

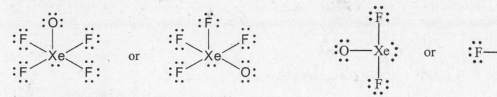

trigonal
bipyramid;
dsp^3

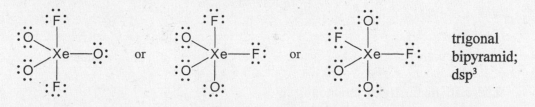

49.   For carbon, nitrogen, and oxygen atoms to have formal charge values of zero, each C atom will
      form four bonds to other atoms and have no lone pairs of electrons, each N atom will form three
      bonds to other atoms and have one lone pair of electrons, and each O atom will form two bonds to
      other atoms and have two lone pairs of electrons. Following these bonding requirements gives the
      following two resonance structures for vitamin B_6:

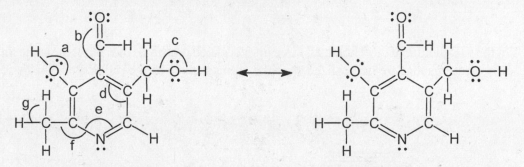

      a.   21 σ bonds; 4 π bonds (The electrons in the 3 π bonds in the ring are delocalized.)

      b.   angles a, c, and g: ≈ 109.5°; angles b, d, e and f: ≈ 120°

      c.   6 sp^2 carbons; the 5 carbon atoms in the ring are sp^2 hybridized, as is the carbon with the double
           bond to oxygen.

      d.   4 sp^3 atoms; the 2 carbons which are not sp^2 hybridized are sp^3 hybridized, and the oxygens
           marked with angles a and c are sp^3 hybridized.

      e.   Yes, the π electrons in the ring are delocalized. The atoms in the ring are all sp^2 hybridized. This
           leaves a p orbital perpendicular to the plane of the ring from each atom. Overlap of all six of
           these p orbitals results in a π molecular orbital system where the electrons are delocalized above
           and below the plane of the ring (similar to benzene in Figure 9.48 of the text).

51.   a. $COCl_2$ has 4 + 6 + 2(7) = 24 valence electrons.

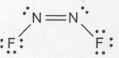

           trigonal planar
           polar
           120°
           sp^2

      b. N_2F_2 has 2(5) + 2(7) = 24 valence electrons.

                                                                            Can also be:

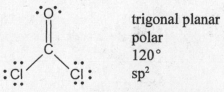   V-shaped about both Ns;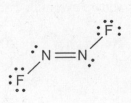
                              ≈ 120° about both Ns;
                              Both Ns: sp^2

           polar   nonpolar

      These are distinctly different molecules.

c.   COS has $4 + 6 + 6 = 16$ valence electrons.

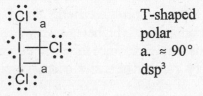   linear, polar, 180°, sp

d. ICl_3 has $7 + 3(7) = 28$ valence electrons.

:Cl: T-shaped
  a   polar
:I—Cl: a. ≈ 90°
  a dsp^3
:Cl:

53.   a.   The Lewis structures for NNO and NON are:

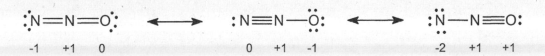

The NNO structure is correct. From the Lewis structures, we would predict both NNO and NON to be linear. However, we would predict NNO to be polar and NON to be nonpolar. Since experiments show N_2O to be polar, NNO is the correct structure.

b.   Formal charge = number of valence electrons of atoms - [(number of lone pair electrons) + 1/2 (number of shared electrons)].

:N═N═O: ←→ :N≡N—O: ←→ :N—N≡O:
 -1 +1 0   0 +1 -1   -2 +1 +1

The formal charges for the atoms in the various resonance structures are below each atom. The central N is sp hybridized in all of the resonance structures. We can probably ignore the third resonance structure on the basis of the relatively large formal charges as compared to the first two resonance structures.

c.   The sp hybrid orbitals on the center N overlap with atomic orbitals (or hybrid orbitals) on the other two atoms to form the two sigma bonds. The remaining two unhybridized p orbitals on the center N overlap with two p orbitals on the peripheral N to form the two π bonds.

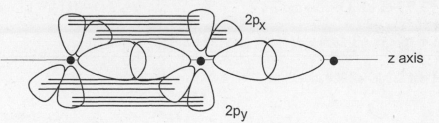

55. N_2 (ground state): $(\sigma_{2s})^2(\sigma_{2s}*)^2(\pi_{2p})^4(\sigma_{2p})^2$, B.O. = 3, diamagnetic (0 unpaired e⁻)

       N_2 (1st excited state): $(\sigma_{2s})^2(\sigma_{2s}*)^2(\pi_{2p})^4(\sigma_{2p})^1(\pi_{2p}*)^1$

       B.O. = (7-3)/2 = 2, paramagnetic (2 unpaired e⁻)

       The first excited state of N_2 should have a weaker bond and should be paramagnetic.

57. F_2: $(\sigma_{2s})^2(\sigma_{2s}*)^2(\sigma_{2p})^2(\pi_{2p})^4(\pi_{2p}*)^4$; F_2 should have a lower ionization energy than F. The electron removed from F_2 is in a $\pi_{2p}*$ antibonding molecular orbital that is higher in energy than the 2p atomic orbitals from which the electron in atomic fluorine is removed. Since the electron removed from F_2 is higher in energy than the electron removed from F, it should be easier to remove an electron from F_2 than from F.

Challenge Problems

59.   a. No, some atoms are in different places. Thus, these are not resonance structures; they are different compounds.

       b. For the first Lewis structure, all nitrogens are sp^3 hybridized and all carbons are sp^2 hybridized. In the second Lewis structure, all nitrogens and carbons are sp^2 hybridized.

       c. For the reaction:

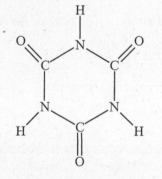   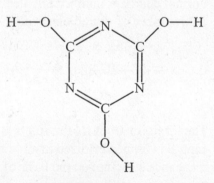

       Bonds broken: Bonds formed:

          3 C=O (745 kJ/mol)   3 C=N (615 kJ/mol)

          3 C−N (305 kJ/mol)   3 C−O (358 kJ/mol)

          3 N−H (391 kJ/mol)   3 O−H (467 kJ/mol)

       ΔH = 3(745) + 3(305) + 3(391) - [3(615) + 3(358) + 3(467)]

       ΔH = 4323 kJ - 4320 kJ = 3 kJ

The bonds are slightly stronger in the first structure with the carbon-oxygen double bonds since ΔH for the reaction is positive. However, the value of ΔH is so small that the best conclusion is that the bond strengths are comparable in the two structures.

61.   a. NCN^{2-} has $5 + 4 + 5 + 2 = 16$ valence electrons.

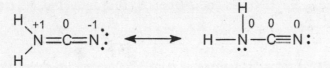

H_2NCN has $2(1) + 5 + 4 + 5 = 16$ valence electrons.

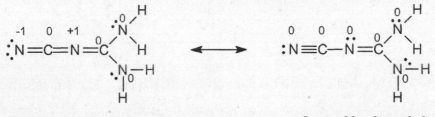

favored by formal charge

$NCNC(NH_2)_2$ has $5 + 4 + 5 + 4 + 2(5) + 4(1) = 32$ valence electrons.

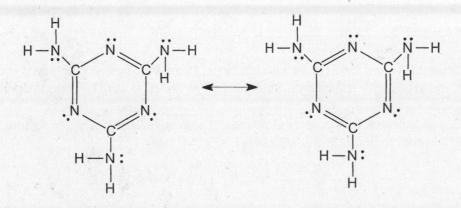

favored by formal charge

Melamine $(C_3N_6H_6)$ has $3(4) + 6(5) + 6(1) = 48$ valence electrons.

b. NCN^{2-}: C is sp hybridized. Depending on the resonance form, N can be sp, sp^2, or sp^3 hybridized. For the remaining compounds, we will give hybrids for the favored resonance structures as predicted from formal charge considerations.

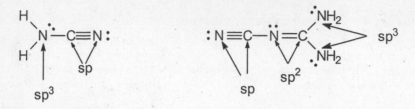

Melamine: N in NH_2 groups are all sp^3 hybridized. Atoms in ring are all sp^2 hybridized.

c. NCN^{2-}: 2 σ and 2 π bonds; H_2NCN: 4 σ and 2 π bonds; dicyandiamide: 9 σ and 3 π bonds; melamine: 15 σ and 3 π bonds

d. The π-system forces the ring to be planar just as the benzene ring is planar.

e. The structure:

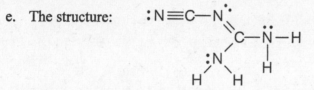

is the most important since it has three different CN bonds. This structure is also favored on the basis of formal charge.

63.   a. $E = \dfrac{hc}{\lambda} = \dfrac{(6.626 \times 10^{-34} \text{ J s}) \, (2.998 \times 10^8 \text{ m/s})}{25 \times 10^{-9} \text{ m}} = 7.9 \times 10^{-18} \text{ J}$

$7.9 \times 10^{-18} \text{ J} \times \dfrac{6.022 \times 10^{23}}{\text{mol}} \times \dfrac{1 \text{ kJ}}{1000 \text{ J}} = 4800 \text{ kJ/mol}$

Using ΔH values from the various reactions, 25 nm light has sufficient energy to ionize N_2 and N and to break the triple bond. Thus, N_2, N_2^+, N, and N^+ will all be present, assuming excess N_2.

b. To produce atomic nitrogen but no ions, the range of energies of the light must be from 941 kJ/mol to just below 1402 kJ/mol.

$\dfrac{941 \text{ kJ}}{\text{mol}} \times \dfrac{1 \text{ mol}}{6.022 \times 10^{23}} \times \dfrac{1000 \text{ J}}{\text{kJ}} = 1.56 \times 10^{-18} \text{ J/photon}$

$\lambda = \dfrac{hc}{E} = \dfrac{(6.626 \times 10^{-34} \text{ J s}) \, (2.998 \times 10^8 \text{ m/s})}{1.56 \times 10^{-18} \text{ J}} = 1.27 \times 10^{-7} \text{ m} = 127 \text{ nm}$

$$\frac{1402 \text{ kJ}}{\text{mol}} \times \frac{1 \text{ mol}}{6.0221 \times 10^{23}} \times \frac{1000 \text{ J}}{\text{kJ}} = 2.328 \times 10^{-18} \text{ J/photon}$$

$$\lambda = \frac{hc}{E} = \frac{(6.6261 \times 10^{-34} \text{ J s}) \, (2.9979 \times 10^8 \text{ m/s})}{2.328 \times 10^{-18} \text{ J}} = 8.533 \times 10^{-8} \text{ m} = 85.33 \text{ nm}$$

Light with wavelengths in the range of 85.33 nm $< \lambda \le$ 127 nm will produce N but no ions.

c. N_2: $(\sigma_{2s})^2(\sigma_{2s}*)^2(\pi_{2p})^4(\sigma_{2p})^2$; The electron removed from N_2 is in the σ_{2p} molecular orbital which is lower in energy than the 2p atomic orbital from which the electron in atomic nitrogen is removed. Since the electron removed from N_2 is lower in energy than the electron in N, the ionization energy of N_2 is greater than for N.

65.   O=N–Cl: The bond order of the NO bond in NOCl is 2 (a double bond).

NO: From molecular orbital theory, the bond order of this NO bond is 2.5.

Both reactions apparently involve only the breaking of the N–Cl bond. However, in the reaction ONCl → NO + Cl, some energy is released in forming the stronger NO bond, lowering the value of ΔH. Therefore, the apparent N–Cl bond energy is artificially low for this reaction. The first reaction involves only the breaking of the N–Cl bond.

67.   a.   The CO bond is polar with the negative end around the more electronegative oxygen atom. We would expect metal cations to be attracted to and bond to the oxygen end of CO on the basis of electronegativity.

b.   :C≡O: FC (carbon) = 4 - 2 - 1/2(6) = -1

FC (oxygen) = 6 - 2 - 1/2(6) = +1

From formal charge, we would expect metal cations to bond to the carbon (with the negative formal charge).

c.   In molecular orbital theory, only orbitals with proper symmetry overlap to form bonding orbitals. The metals that form bonds to CO are usually transition metals, all of which have outer electrons in the d orbitals. The only molecular orbitals of CO that have proper symmetry to overlap with d orbitals are the $\pi_{2p}*$ orbitals, whose shape is similar to the d orbitals (see Figure 9.34). Since the antibonding molecular orbitals have more carbon character (carbon is less electronegative than oxygen), one would expect the bond to form through carbon.

CHAPTER TEN

LIQUIDS AND SOLIDS

Questions

13. London dispersion (LD) < dipole-dipole < H bonding < metallic bonding, covalent network, ionic.

    Yes, there is considerable overlap. Consider some of the examples in Exercise 10.92. Benzene (only LD forces) has a higher boiling point than acetone (dipole-dipole forces). Also, there is even more overlap among the stronger forces (metallic, covalent, and ionic).

15. a. Polarizability of an atom refers to the ease of distorting the electron cloud. It can also refer to distorting the electron clouds in molecules or ions. Polarity refers to the presence of a permanent dipole moment in a molecule.

    b. London dispersion (LD) forces are present in all substances. LD forces can be referred to as accidental dipole-induced dipole forces. Dipole-dipole forces involve the attraction of molecules with permanent dipoles for each other.

    c. inter: between; intra: within; For example, in Br_2 the covalent bond is an intramolecular force holding the two Br atoms together in the molecule. The much weaker London dispersion forces are the intermolecular forces of attraction which hold different molecules of Br_2 together in the liquid phase.

17. Atoms have an approximately spherical shape (on the average). It is impossible to pack spheres together without some empty space among the spheres.

19. As the intermolecular forces increase, the critical temperature increases.

21. a. Crystalline solid: Regular, repeating structure

       Amorphous solid: Irregular arrangement of atoms or molecules

    b. Ionic solid: Made up of ions held together by ionic bonding.

       Molecular solid: Made up of discrete covalently bonded molecules held together in the solid phase by weaker forces (LD, dipole or hydrogen bonds).

    c. Molecular solid: Discrete, individual molecules

        Network solid: No discrete molecules; A network solid is one large molecule. The intermolecular forces are the covalent bonds between atoms.

    d. Metallic solid: Completely delocalized electrons, conductor of electricity (ions in a sea of electrons)

        Network solid: Localized electrons; Insulator or semiconductor

23.   Conductor: The energy difference between the filled and unfilled molecular orbitals is minimal. We call this energy difference the band gap. Since the band gap is minimal, electrons can easily move into the conduction bands (the unfilled molecular orbitals).

    Insulator: Large band gap; Electrons do not move from the filled molecular orbitals to the conduction bands since the energy difference is large.

    Semiconductor: Small band gap; Since the energy difference between the filled and unfilled molecular orbitals is smaller than in insulators, some electrons can jump into the conduction bands. The band gap, however, is not as small as with conductors, so semiconductors have intermediate conductivity.

    a. As the temperature is increased, more electrons in the filled molecular orbitals have sufficient kinetic energy to jump into the conduction bands (the unfilled molecular orbitals).

    b. A photon of light is absorbed by an electron which then has sufficient energy to jump into the conduction bands.

    c. An impurity either adds electrons at an energy near that of the conduction bands (n-type) or creates holes (unfilled energy levels) at energies in the previously filled molecular orbitals (p-type)

25.   To produce an n-type semiconductor, dope Ge with a substance that has more than 4 valence electrons, e.g., a group 5A element. Phosphorus or arsenic are two substances which will produce n-type semiconductors when they are doped into germanium. To produce a p-type semiconductor, dope Ge with a substance that has fewer than 4 valence electrons, e.g., a group 3A element. Gallium or indium are two substances which will produce p-type semiconductors when they are doped into germanium.

27.   a. Condensation: vapor → liquid   b. Evaporation: liquid → vapor

    c. Sublimation: solid → vapor

    d. A supercooled liquid is a liquid which is at a temperature below its freezing point.

29.   a. As the intermolecular forces increase, the rate of evaporation decreases.

    b. As temperature increases, the rate of evaporation increases.

c.   As surface area increases, the rate of evaporation increases.

31. $C_2H_5OH(l) \rightarrow C_2H_5OH(g)$ is an endothermic process. Heat is absorbed when liquid ethanol vaporizes; the internal heat from the body provides this heat which results in the cooling of the body.

33.   The phase change, $H_2O(g) \rightarrow H_2O(l)$, releases heat that can cause additional damage. Also steam can be at a temperature greater than $100°C$.

Exercises

Intermolecular Forces and Physical Properties

35.   Ionic compounds have ionic forces. Covalent compounds all have London Dispersion (LD) forces, while polar covalent compounds have dipole forces and/or hydrogen bonding forces. For H bonding forces, the covalent compound must have either a N–H, O–H or F–H bond in the molecule.

    a.   LD only   b.   dipole, LD   c.   H bonding, LD

    d.   ionic   e.   LD only (CH_4 in a nonpolar covalent compound.)

    f.   dipole, LD   g.   ionic

37.   a.   OCS; OCS is polar and has dipole-dipole forces in addition to London dispersion (LD) forces. All polar molecules have dipole forces. CO_2 is nonpolar and only has LD forces. To predict polarity, draw the Lewis structure and deduce whether the individual bond dipoles cancel.

    b. SeO_2; Both SeO_2 and SO_2 are polar compounds, so they both have dipole forces as well as LD forces. However, SeO_2 is a larger molecule, so it would have stronger LD forces.

    c. $H_2NCH_2CH_2NH_2$; More extensive hydrogen bonding is possible.

    d. H_2CO; H_2CO is polar while CH_3CH_3 is nonpolar. H_2CO has dipole forces in addition to LD forces.

    e. CH_3OH; CH_3OH can form relatively strong H bonding interactions, unlike H_2CO.

39.   See Question 10.14 to review the dependence of some physical properties on the strength of the intermolecular forces.

    a.   HCl; HCl is polar while Ar and F_2 are nonpolar. HCl has dipole forces unlike Ar and F_2.

    b.   NaCl; Ionic forces are much stronger than molecular forces.

    c. I_2; All are nonpolar, so the largest molecule (I_2) will have the strongest LD forces and the lowest vapor pressure.

d. N_2; Nonpolar and smallest, so has the weakest intermolecular forces.

e. CH_4; Smallest, nonpolar molecule so has the weakest LD forces.

f. HF; HF can form relatively strong H bonding interactions unlike the others.

g. $CH_3CH_2CH_2OH$; H bonding, unlike the others, so has strongest intermolecular forces.

Properties of Liquids

41. The attraction of H_2O for glass is stronger than the $H_2O - H_2O$ attraction. The miniscus is concave to increase the area of contact between glass and H_2O. The $Hg - Hg$ attraction is greater than the $Hg - glass$ attraction. The miniscus is convex to minimize the $Hg - glass$ contact.

43. The structure of H_2O_2 is $H - O - O - H$, which produces greater hydrogen bonding than water. Long chains of hydrogen bonded H_2O_2 molecules then get tangled together.

Structures and Properties of Solids

45. $n\lambda = 2d \sin \theta$, $d = \dfrac{n\lambda}{2 \sin \theta} = \dfrac{1 \times 1.54 \text{ Å}}{2 \times \sin 14.22°} = 3.13 \text{ Å} = 3.13 \times 10^{-10} \text{ m} = 313 \text{ pm}$

47. A cubic closest packed structure has a face-centered cubic unit cell. In a face-centered cubic unit, there are:

$$8 \text{ corners} \times \frac{1/8 \text{ atom}}{\text{corner}} + 6 \text{ faces} \times \frac{1/2 \text{ atom}}{\text{face}} = 4 \text{ atoms}$$

The atoms in a face-centered cubic unit cell touch along the face diagonal of the cubic unit cell. Using the Pythagorean formula where l = length of the face diagonal and r = radius of the atom:

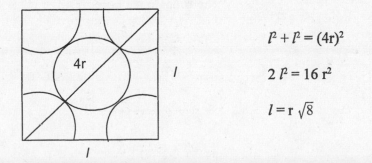

$$l^2 + l^2 = (4r)^2$$

$$2\, l^2 = 16\, r^2$$

$$l = r\, \sqrt{8}$$

$l = r \sqrt{8} = 197 \times 10^{-12} \text{ m} \times \sqrt{8} = 5.57 \times 10^{-10} \text{ m} = 5.57 \times 10^{-8} \text{ cm}$

Volume of a unit cell $= l^3 = (5.57 \times 10^{-8} \text{ cm})^3 = 1.73 \times 10^{-22} \text{ cm}^3$

Mass of a unit cell = 4 Ca atoms $\times \dfrac{1 \text{ mol Ca}}{6.022 \times 10^{23} \text{ atoms}} \times \dfrac{40.08 \text{ g Ca}}{\text{mol Ca}} = 2.662 \times 10^{-22}$ g Ca

density = $\dfrac{\text{mass}}{\text{volume}} = \dfrac{2.662 \times 10^{-22} \text{ g}}{1.73 \times 10^{-22} \text{ cm}^3} = 1.54$ g/cm^3

49.   The volume of a unit cell is:

$$V = l^3 = (383.3 \times 10^{-10} \text{ cm})^3 = 5.631 \times 10^{-23} \text{ cm}^3$$

There are 4 Ir atoms in the unit cell, as is the case for all face-centered cubic unit cells. The mass of atoms in a unit cell is:

mass = 4 Ir atoms $\times \dfrac{1 \text{ mol Ir}}{6.022 \times 10^{23} \text{ atoms}} \times \dfrac{192.2 \text{ g Ir}}{\text{mol Ir}} = 1.277 \times 10^{-21}$ g

density = $\dfrac{\text{mass}}{\text{volume}} = \dfrac{1.277 \times 10^{-21} \text{ g}}{5.631 \times 10^{-23} \text{ cm}^3} = 22.68$ g/cm^3

51.   For a body-centered unit cell: 8 corners $\times \dfrac{1/8 \text{ Ti}}{\text{corner}}$ + Ti at body center = 2 Ti atoms

All body-centered unit cells have 2 atoms per unit cell. For a unit cell:

density = 4.50 g/cm^3 = $\dfrac{2 \text{ atoms Ti} \times \dfrac{1 \text{ mol Ti}}{6.022 \times 10^{23} \text{ atoms}} \times \dfrac{47.88 \text{ g Ti}}{\text{mol Ti}}}{l^3}$, l = cube edge length

Solving: l = edge length of unit cell = 3.28×10^{-8} cm = 328 pm

Assume Ti atoms just touch along the body diagonal of the cube, so body diagonal = 4 $\times$ radius of atoms = 4r.

The triangle we need to solve is:

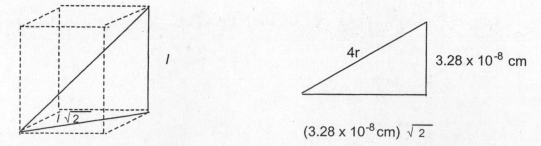

$(4r)^2 = (3.28 \times 10^{-8} \text{ cm})^2 + [(3.28 \times 10^{-8} \text{ cm}) \sqrt{2} \ ]^2$, r = 1.42×10^{-8} cm = 142 pm

For a body-centered unit cell (bcc), the radius of the atom is related to the cube edge length by $4r = l\sqrt{3}$ or $l = 4r/\sqrt{3}$.

53.   In a face-centered unit cell (ccp structure), the atoms touch along the face diagonal:

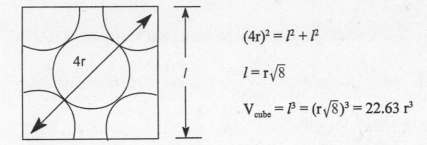

$$(4r)^2 = l^2 + l^2$$

$$l = r\sqrt{8}$$

$$V_{cube} = l^3 = (r\sqrt{8})^3 = 22.63\ r^3$$

There are four atoms in a face-centered cubic cell (see Exercise 10.47). Each atom has a volume of 4/3 πr^3.

$$V_{atoms} = 4 \times \frac{4}{3}\ \pi r^3 = 16.76\ r^3$$

So, $\dfrac{V_{atom}}{V_{cube}} - \dfrac{16.76\ r^3}{22.63\ r^3} = 0.7406$ or 74.06% of the volume of each unit cell is occupied by atoms.

In a simple cubic unit cell, the atoms touch along the cube edge (l):

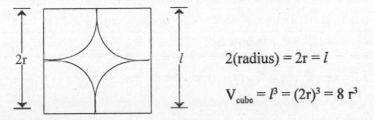

$$2(radius) = 2r = l$$

$$V_{cube} = l^3 = (2r)^3 = 8\ r^3$$

There is one atom per simple cubic cell (8 corner atoms × 1/8 atom per corner = 1 atom/unit cell). Each atom has an assumed volume of 4/3 πr^3 = volume of a sphere.

$$V_{atom} = \frac{4}{3}\ \pi r^3 = 4.189\ r^3$$

So, $\dfrac{V_{atom}}{V_{cube}} = \dfrac{4.189\ r^3}{8\ r^3} = 0.5236$ or 52.36% of the volume of each unit cell is occupied by atoms.

A cubic closest packed structure packs the atoms much more efficiently than a simple cubic structure.

55.   In has fewer valence electrons than Se, thus, Se doped with In would be a p-type semiconductor.

57. $E_{gap} = 2.5\ eV \times 1.6 \times 10^{-19}\ J/eV = 4.0 \times 10^{-19}\ J$; We want $E_{gap} = E_{light}$, so:

$$E_{light} = \frac{hc}{\lambda},\ \lambda = \frac{hc}{E} = \frac{(6.63 \times 10^{-34}\ J\ s)\ (3.00 \times 10^8\ m/s)}{4.0 \times 10^{-19}\ J} = 5.0 \times 10^{-7}\ m = 5.0 \times 10^2\ nm$$

59.   a. $8 \text{ corners} \times \dfrac{1/8 \text{ Cl}}{\text{corner}} + 6 \text{ faces} \times \dfrac{1/2 \text{ Cl}}{\text{face}} = 4 \text{ Cl ions}$

$12 \text{ edges} \times \dfrac{1/4 \text{ Na}}{\text{edge}} + 1 \text{ Na at body center} = 4 \text{ Na ions}; \quad \text{NaCl is the formula.}$

   b.   1 Cs ion at body center; $8 \text{ corners} \times \dfrac{1/8 \text{ Cl}}{\text{corner}} = 1 \text{ Cl ion}; \quad \text{CsCl is the formula.}$

   c.   There are 4 Zn ions inside the cube.

$8 \text{ corners} \times \dfrac{1/8 \text{ S}}{\text{corner}} + 6 \text{ faces} \times \dfrac{1/2 \text{ S}}{\text{face}} = 4 \text{ S ions}; \quad \text{ZnS is the formula.}$

   d. $8 \text{ corners} \times \dfrac{1/8 \text{ Ti}}{\text{corner}} + 1 \text{ Ti at body center} = 2 \text{ Ti ions}$

$4 \text{ faces} \times \dfrac{1/2 \text{ O}}{\text{face}} + 2 \text{ O inside cube} = 4 \text{ O ions}; \quad \text{TiO}_2 \text{ is the formula.}$

61.   There is one octahedral hole per closest packed anion in a closest packed structure. If half of the octahedral holes are filled, there is a 2:1 ratio of fluoride ions to cobalt ions in the crystal. The formula is CoF_2.

63. $8 \text{ F}^- \text{ ions at corners} \times \dfrac{1/8 \text{ F}^-}{\text{corner}} = 1 \text{ F}^- \text{ ion per unit cell};$ Since there is one cubic hole per cubic unit cell, there is a 2:1 ratio of F$^-$ ions to metal ions in the crystal. The formula is MF_2 where M^{2+} is the metal ion.

65.   Since magnesium oxide has the same structure as NaCl, each unit cell contains 4 Mg^{2+} ions and 4 O^{2-} ions. The mass of a unit cell is:

$$4 \text{ MgO formula units} \left(\frac{1 \text{ mol MgO}}{6.022 \times 10^{23} \text{ formula units}} \right) \left(\frac{40.31 \text{ g MgO}}{1 \text{ mol MgO}} \right) = 2.678 \times 10^{-22} \text{ g MgO}$$

$$\text{Volume of unit cell} = 2.678 \times 10^{-22} \text{ g MgO} \left(\frac{1 \text{ cm}^3}{3.58 \text{ g}} \right) = 7.48 \times 10^{-23} \text{ cm}^3$$

Volume of unit cell $= l^3$, $l =$ cube edge length; $l = (7.48 \times 10^{-23} \text{ cm}^3)^{1/3} = 4.21 \times 10^{-8} \text{ cm} = 421 \text{ pm}$

From the NaCl structure in Figure 10.35 of the text, Mg^{2+} and O^{2-} ions should touch along the cube edge, l:

$l = 2 \, r_{Mg^{2+}} + 2 \, r_{O^{2-}} = 2 \, (65 \text{ pm}) + 2 \, (140. \text{ pm}) = 410. \text{ pm}$

The two values agree within 3%. In the actual crystals, the Mg^{2+} and O^{2-} ions may not touch, which is assumed in calculating the 410. pm value.

67. a. CO_2: molecular   b. SiO_2: network   c. Si: atomic, network

     d. CH_4: molecular   e. Ru: atomic, metallic   f. I_2: molecular

     g. KBr: ionic   h. H_2O: molecular   i. NaOH: ionic

     j. U: atomic, metallic   k. $CaCO_3$: ionic   l. PH_3: molecular

69. a. The unit cell consists of Ni at the cube corners and Ti at the body center, or Ti at the cube corners and Ni at the body center.

     b. $8 \times 1/8 = 1$ atom from corners + 1 atom at body center; Empirical formula = NiTi

     c. Both have a coordination number of 8 (both are surrounded by 8 atoms).

71.   Structure 1   Structure 2

     $8 \text{ corners} \times \dfrac{1/8 \text{ Ca}}{\text{corner}} = 1 \text{ Ca atom}$ $8 \text{ corners} \times \dfrac{1/8 \text{ Ti}}{\text{corner}} = 1 \text{ Ti atom}$

     $6 \text{ faces} \times \dfrac{1/2 \text{ O}}{\text{face}} = 3 \text{ O atoms}$ $12 \text{ edges} \times \dfrac{1/4 \text{ O}}{\text{edge}} = 3 \text{ O atoms}$

     1 Ti at body center. Formula = $CaTiO_3$ 1 Ca at body center. Formula = $CaTiO_3$

     In the extended lattice of both structures, each Ti atom is surrounded by six O atoms.

73. a. Y: 1 Y in center; Ba: 2 Ba in center

     Cu: $8 \text{ corners} \times \dfrac{1/8 \text{ Cu}}{\text{corner}} = 1 \text{ Cu}$, $8 \text{ edges} \times \dfrac{1/4 \text{ Cu}}{\text{edge}} = 2 \text{ Cu}$, total = 3 Cu atoms

     O: $20 \text{ edges} \times \dfrac{1/4 \text{ O}}{\text{edge}} = 5 \text{ oxygen}$, $8 \text{ faces} \times \dfrac{1/2 \text{ O}}{\text{face}} = 4 \text{ oxygen}$, total = 9 O atoms

     Formula: $YBa_2Cu_3O_9$

     b. The structure of this superconductor material follows the second perovskite structure described in Exercise 10.71. The $YBa_2Cu_3O_9$ structure is three of these cubic perovskite unit cells stacked on top of each other. The oxygen atoms are in the same places, Cu takes the place of Ti, two of the calcium atoms are replaced by two barium atoms, and one Ca is replaced by Y.

     c. Y, Ba, and Cu are the same. Some oxygen atoms are missing.

     $12 \text{ edges} \times \dfrac{1/4 \text{ O}}{\text{edge}} = 3 \text{ O}$, $8 \text{ faces} \times \dfrac{1/2 \text{ O}}{\text{face}} = 4 \text{ O}$, total = 7 O atoms

     Superconductor formula is $YBa_2Cu_3O_7$.

Phase Changes and Phase Diagrams

75.   If we graph ln P_{vap} vs 1/T, the slope of the resulting straight line will be $-\Delta H_{vap}/R$.

P_{vap}	ln P_{vap}	T (Li)	1/T	T (Mg)	1/T
1 torr	0	1023 K	9.775×10^{-4} K^{-1}	893 K	11.2×10^{-4} K^{-1}
10.	2.3	1163	8.598×10^{-4}	1013	9.872×10^{-4}
100.	4.61	1353	7.391×10^{-4}	1173	8.525×10^{-4}
400.	5.99	1513	6.609×10^{-4}	1313	7.616×10^{-4}
760.	6.63	1583	6.317×10^{-4}	1383	7.231×10^{-4}

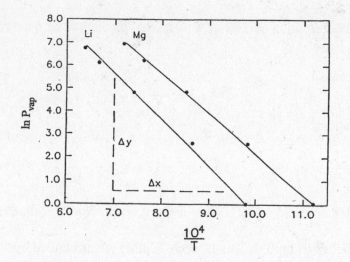

For Li:

We get the slope by taking two points (x, y) that are on the line we draw. For a line:

$$\text{slope} = \frac{\Delta y}{\Delta x} = \frac{y_2 - y_1}{x_2 - x_1}$$

or we can determine the straight line equation using a computer or calculator. The general straight line equation is $y = mx + b$ where m = slope and b = y-intercept.

The equation of the Li line is: ln $P_{vap} = -1.90 \times 10^4(1/T) + 18.6$, slope $= -1.90 \times 10^4$ K

Slope $= -\Delta H_{vap}/R$, $\Delta H_{vap} = -\text{slope} \times R = 1.90 \times 10^4$ K $\times 8.3145$ J/K•mol

$\Delta H_{vap} = 1.58 \times 10^5$ J/mol $= 158$ kJ/mol

For Mg:

The equation of the line is: ln $P_{vap} = -1.67 \times 10^4(1/T) + 18.7$, slope $= -1.67 \times 10^4$ K

$\Delta H_{vap} = -\text{slope} \times R = 1.67 \times 10^4$ K $\times 8.3145$ J/K•mol, $\Delta H_{vap} = 1.39 \times 10^5$ J/mol $= 139$ kJ/mol

The bonding is stronger in Li since ΔH_{vap} is larger for Li.

77.   At 100.°C (373 K), the vapor pressure of H_2O is 1.00 atm = 760. torr.
        For water, $\Delta H_{vap} = 40.7$ kJ/mol.

$$\ln\left(\frac{P_1}{P_2}\right) = \frac{\Delta H_{vap}}{R}\left(\frac{1}{T_2} - \frac{1}{T_1}\right) \text{ or } \ln\left(\frac{P_2}{P_1}\right) = \frac{\Delta H_{vap}}{R}\left(\frac{1}{T_1} - \frac{1}{T_2}\right)$$

$$\ln\left(\frac{520.\text{ torr}}{760.\text{ torr}}\right) = \frac{40.7 \times 10^3 \text{ J/mol}}{8.3145 \text{ J/K}\bullet\text{mol}}\left(\frac{1}{373 \text{ K}} - \frac{1}{T_2}\right), \; -7.75 \times 10^{-5} = \left(\frac{1}{373} - \frac{1}{T_2}\right)$$

$$-7.75 \times 10^{-5} = 2.68 \times 10^{-3} - \frac{1}{T_2}, \; \frac{1}{T_2} = 2.76 \times 10^{-3}, \; T_2 = \frac{1}{2.76 \times 10^{-3}} = 362 \text{ K or } 89°C$$

79. $$\ln\left(\frac{P_1}{P_2}\right) = \frac{\Delta H_{vap}}{R}\left(\frac{1}{T_2} - \frac{1}{T_1}\right); \; P_1 = 760.\text{ torr}, T_1 = 630.\text{ K}; \; P_2 = ?, \; T_2 = 298 \text{ K}$$

$$\ln\left(\frac{760.}{P_2}\right) = \frac{59.1 \times 10^3 \text{ J/mol}}{8.3145 \text{ J/K}\bullet\text{mol}}\left(\frac{1}{298 \text{ K}} - \frac{1}{630.\text{ K}}\right) = 12.6$$

$$760./P_2 = e^{12.6}, \; P_2 = 760./(2.97 \times 10^5) = 2.56 \times 10^{-3} \text{ torr}$$

81.

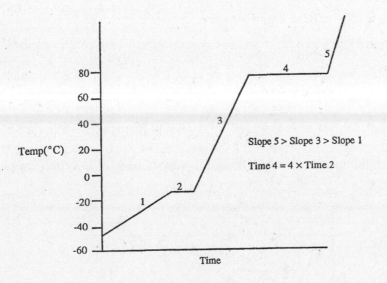

        Temp(°C)

        Slope 5 > Slope 3 > Slope 1

        Time 4 = 4 × Time 2

        Time

83. $H_2O(s, -20.°C) \rightarrow H_2O(s, 0°C), \; \Delta T = 20.°C$

$$q_1 = s_{ice} \times m \times \Delta T = \frac{2.1 \text{ J}}{g \, °C} \times 5.00 \times 10^2 \text{ g} \times 20.°C = 2.1 \times 10^4 \text{ J} = 21 \text{ kJ}$$

$$H_2O(s, 0°C) \rightarrow H_2O(l, 0°C), \; q_2 = 5.00 \times 10^2 \text{ g } H_2O \times \frac{1 \text{ mol}}{18.02 \text{ g}} \times \frac{6.02 \text{ kJ}}{\text{mol}} = 167 \text{ kJ}$$

$H_2O(l, 0°C) \rightarrow H_2O(l, 100°C)$, $q_3 = \dfrac{4.2\ J}{g\ °C} \times 5.00 \times 10^2\ g \times 100.°C = 2.1 \times 10^5\ J = 210\ kJ$

$H_2O(l, 100°C) \rightarrow H_2O(g, 100°C)$, $q_4 = 5.00 \times 10^2\ g \times \dfrac{1\ mol}{18.02\ g} \times \dfrac{40.7\ kJ}{mol} = 1130\ kJ$

$H_2O(g, 100°C) \rightarrow H_2O(g, 250°C)$, $q_5 = \dfrac{2.0\ J}{g\ °C} \times 5.00 \times 10^2\ g \times 150.°C = 1.5 \times 10^5\ J = 150\ kJ$

$q_{total} = q_1 + q_2 + q_3 + q_4 + q_5 = 21 + 167 + 210 + 1130 + 150 = 1680\ kJ$

85.   Total mass $H_2O = 18$ cubes $\times \dfrac{30.0\ g}{cube} = 540.\ g$; $540.\ g\ H_2O \times \dfrac{1\ mol\ H_2O}{18.02\ g} = 30.0\ mol\ H_2O$

Heat removed to produce ice at -5.0°C:

$\dfrac{4.18\ J}{g\ °C} \times 540.\ g \times 22.0\ °C + \dfrac{6.02 \times 10^3\ J}{mol} \times 30.0\ mol + \dfrac{2.08\ J}{g\ °C} \times 540.\ g \times 5.0\ °C$

$= 4.97 \times 10^4\ J + 1.81 \times 10^5\ J + 5.6 \times 10^3\ J = 2.36 \times 10^5\ J$

$2.36 \times 10^5\ J \times \dfrac{1\ g\ CF_2Cl_2}{158\ J} = 1.49 \times 10^3\ g\ CF_2Cl_2$ must be vaporized.

87.   A: solid   B: liquid   C: vapor

D: solid + vapor   E: solid + liquid + vapor

F: liquid + vapor   G: liquid + vapor   H: vapor

triple point: E   critical point: G

normal freezing point: temperature at which solid - liquid line is at 1.0 atm (see plot below).

normal boiling point: temperature at which liquid - vapor line is at 1.0 atm (see plot below).

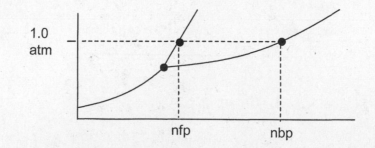

Since the solid-liquid line has a positive slope, the solid phase is denser than the liquid phase.

89.   a.   two

      b.   Higher pressure triple point: graphite, diamond and liquid; Lower pressure triple point: graphite, liquid and vapor

      c.   It is converted to diamond (the more dense solid form).

      d.   Diamond is more dense, which is why graphite can be converted to diamond by applying pressure.

Additional Exercises

91. $C_{25}H_{52}$ has the stronger intermolecular forces because it has the higher boiling point. Even though $C_{25}H_{52}$ is nonpolar, it is so large that its London dispersion forces are much stronger than the sum of the London dispersion and hydrogen bonding interactions found in H_2O.

93.   At any temperature, the plot tells us that substance A has a higher vapor pressure than substance B, with substance C having the lowest vapor pressure.   Therefore, the substance with the weakest intermolecular forces is A, and the substance with the strongest intermolecular forces is C.

      NH_3 can form hydrogen bonding interactions while the others cannot. Substance C is NH_3. The other two are nonpolar compounds with only London dispersion forces. Since CH_4 is smaller than SiH_4, CH_4 will have weaker LD forces and is substance A.   Therefore, substance B is SiH_4.

95. $n\lambda = 2d \sin \theta$, $\lambda = \dfrac{2d \sin \theta}{n} = \dfrac{2 \times 201 \text{ pm} \times \sin 34.68°}{1}$, $\lambda = 229 \text{ pm} = 2.29 \times 10^{-10} \text{ m} = 0.229 \text{ nm}$

97.   If TiO_2 conducts electricity as a liquid, then it is an ionic solid; if not, then TiO_2 is a network solid.

99. B_2H_6:   This compound contains only nonmetals so it is probably a molecular solid with covalent bonding. The low boiling point confirms this.

      SiO_2:   This is the empirical formula for quartz, which is a network solid.

      CsI: This is a metal bonded to a nonmetal, which generally form ionic solids. The electrical conductivity in aqueous solution confirms this.

      W: Tungsten is a metallic solid as the conductivity data confirms.

101. $1.00 \text{ lb} \times \dfrac{454 \text{ g}}{\text{lb}} = 454 \text{ g } H_2O$; A change of $1.00°F$ is equal to a change of $5/9°C$.

      The amount of heat in J in 1 Btu is: $\dfrac{4.18 \text{ J}}{\text{g }°C} \times 454 \text{ g} \times \dfrac{5}{9}°C = 1.05 \times 10^3 \text{ J}$ or 1.05 kJ

It takes 40.7 kJ to vaporize 1 mol H_2O (ΔH_{vap}). Combining these:

$$\frac{1.00 \times 10^4 \text{ Btu}}{\text{hr}} \times \frac{1.05 \text{ kJ}}{\text{Btu}} \times \frac{1 \text{ mol } H_2O}{40.7 \text{ kJ}} = 258 \text{ mol/hr}$$

or: $\dfrac{258 \text{ mol}}{\text{hr}} \times \dfrac{18.02 \text{ g } H_2O}{\text{mol}} = 4650 \text{ g/hr} = 4.65 \text{ kg/hr}$

Challenge Problems

103.   A single hydrogen bond in H_2O has a strength of 21 kJ/mol. Each H_2O molecule forms two H bonds. Thus, it should take 42 kJ/mol of energy to break all of the H bonds in water. Consider the phase transitions:

$$\text{solid} \quad \overset{6.0 \text{ kJ}}{\rightarrow} \quad \text{liquid} \quad \overset{40.7 \text{ kJ}}{\rightarrow} \quad \text{vapor} \qquad \Delta H_{sub} = \Delta H_{fus} + \Delta H_{vap}$$

It takes a total of 46.7 kJ/mol to convert solid H_2O to vapor (ΔH_{sub}). This would be the amount of energy necessary to disrupt all of the intermolecular forces in ice. Thus, $(42 \div 46.7) \times 100 = 90\%$ of the attraction in ice can be attributed to H bonding.

105.   NaCl, $MgCl_2$, NaF, MgF_2 AlF_3 all have very high melting points indicative of strong intermolecular forces. They are all ionic solids. $SiCl_4$, SiF_4, F_2, Cl_2, PF_5 and SF_6 are nonpolar covalent molecules. Only LD forces are present. PCl_3 and SCl_2 are polar molecules. LD forces and dipole forces are present. In these 8 molecular substances, the intermolecular forces are weak and the melting points low. $AlCl_3$ doesn't seem to fit in as well. From the melting point, there are much stronger forces present than in the nonmetal halides, but they aren't as strong as we would expect for an ionic solid. $AlCl_3$ illustrates a gradual transition from ionic to covalent bonding, from an ionic solid to discrete molecules.

107.   Out of 100.00 g: $28.31 \text{ g O} \times \dfrac{1 \text{ mol}}{16.00 \text{ g}} = 1.769 \text{ mol O}$; $71.69 \text{ g Ti} \times \dfrac{1 \text{ mol}}{47.88 \text{ g}} = 1.497 \text{ mol Ti}$

$\dfrac{1.769}{1.497} = 1.182$; $\dfrac{1.497}{1.769} = 0.8462$; The formula is $TiO_{1.182}$ or $Ti_{0.8462}O$.

For $Ti_{0.8462}O$, let $x = Ti^{2+}$ per mol O^{2-} and $y = Ti^{3+}$ per mol O^{2-}. Setting up two equations and solving:

$x + y = 0.8462$ (mass balance) and $2x + 3y = 2$ (charge balance); $2x + 3(0.8462 - x) = 2$

$x = 0.539 \text{ mol } Ti^{2+}/\text{mol } O^{2-}$ and $y = 0.307 \text{ mol } Ti^{3+}/\text{mol } O^{2-}$

$\dfrac{0.539}{0.8462} \times 100 = 63.7\%$ of the titanium ions is Ti^{2+} and 36.3% is Ti^{3+} (a 1.75:1 ion ratio).

109. $$\frac{density_{Mn}}{density_{Cu}} = \frac{mass_{Mn} \times volume_{Cu}}{volume_{Mn} \times mass_{Cu}} = \frac{mass_{Mn}}{mass_{Cu}} \times \frac{volume_{Cu}}{volume_{Mn}}$$

The type of cubic cell formed is not important; only that Cu and Mn crystallize in the same type of cubic unit cell is important. Each cubic unit cell has a specific relationship between the cube edge length, l, and the radius, r. In all cases $l \propto r$. Therefore, $V \propto l^3 \propto r^3$. For the mass ratio, we can use the molar masses of Mn and Cu since each unit cell must contain the same number of Mn and Cu atoms. Solving:

$$\frac{density_{Mn}}{density_{Cu}} = \frac{mass_{Mn}}{mass_{Cu}} \times \frac{volume_{Cu}}{volume_{Mn}} = \frac{54.94 \text{ g/mol}}{63.55 \text{ g/mol}} \times \frac{(r_{Cu})^3}{(1.056 \, r_{Cu})^3}$$

$$\frac{density_{Mn}}{density_{Cu}} = 0.8645 \times \left(\frac{1}{1.056}\right)^3 = 0.7341$$

$$density_{Mn} = 0.7341 \times density_{Cu} = 0.7341 \times 8.96 \text{ g/cm}^3 = 6.58 \text{ g/cm}^3$$

111.

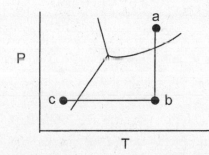

As P is lowered, we go from a to b on the phase diagram. The water boils. The boiling of water is endothermic and the water is cooled (b →c), forming some ice. If the pump is left on, the ice will sublime until none is left. This is the basis of freeze drying.

CHAPTER ELEVEN

PROPERTIES OF SOLUTIONS

Solution Review

9. $125 \text{ g } C_{12}H_{22}O_{11} \times \dfrac{1 \text{ mol}}{342.30 \text{ g}} = 0.365 \text{ mol}; \quad M = \dfrac{0.365 \text{ mol}}{1.00 \text{ L}} = \dfrac{0.365 \text{ mol sucrose}}{L}$

11. $25.00 \times 10^{-3} \text{ L} \times \dfrac{0.308 \text{ mol}}{L} = 7.70 \times 10^{-3} \text{ mol}; \quad \dfrac{7.70 \times 10^{-3} \text{ mol}}{0.500 \text{ L}} = \dfrac{1.54 \times 10^{-2} \text{ mol NiCl}_2}{L}$

   $NiCl_2(s) \rightarrow Ni^{2+}(aq) + 2 Cl^-(aq); \quad M_{Ni^{2+}} = \dfrac{1.54 \times 10^{-2} \text{ mol}}{L}; \quad M_{Cl^-} = \dfrac{3.08 \times 10^{-2} \text{ mol}}{L}$

Questions

13. $\text{Molarity} = \dfrac{\text{moles solute}}{\text{L solution}}; \quad \text{Molality} = \dfrac{\text{moles solute}}{\text{kg solvent}}$

   Volume is temperature dependent and mass is not. Therefore, molarity is temperature dependent and molality is temperature independent. In determining ΔT_f and ΔT_b, we are interested in how some temperature depends on composition. Thus, we don't want our expression of composition to also depend on temperature.

15.   hydrophobic: water hating;   hydrophilic: water loving

17.   Since the solute is volatile, then both the water and solute will transfer back and forth between the two beakers. The volume in each beaker will become constant when the concentrations of solute in the beakers are equal to each other. Since the solute is less volatile than water, one would expect there to be a larger net transfer of water molecules into the right beaker than the net transfer of solute molecules into the left beaker. This results in a larger solution volume in the right beaker when equilibrium is reached, i.e., when the solute concentration is identical in each beaker.

19.   No, the solution is not ideal. For an ideal solution, the strength of intermolecular forces in solution is the same as in pure solute and pure solvent. This results in $\Delta H_{soln} = 0$ for an ideal solution. ΔH_{soln} for methanol/water is not zero. Since $\Delta H_{soln} < 0$, this solution shows a negative deviation from Raoult's law.

21.   With addition of salt or sugar, the osmotic pressure inside the fruit cells (and bacteria) is less than outside the cell. Water will leave the cells which will dehydrate bacteria present, causing them to die.

23.   Both solutions and colloids have suspended particles in some medium. The major difference between the two is the size of the particles. A colloid is a suspension of relatively large particles as compared to a solution. Because of this, colloids will scatter light while solutions will not. The scattering of light by a colloidal suspension is called the Tyndall effect.

Exercises

Concentration of Solutions

25. $1.00 \text{ L} \times \dfrac{1000 \text{ mL}}{\text{L}} \times \dfrac{1.00 \text{ g}}{\text{mL}} = 1.00 \times 10^3 \text{ g solution}$

mass % NaCl $= \dfrac{25 \text{ g NaCl}}{1.00 \times 10^3 \text{ g solution}} \times 100 = 2.5\%$

molarity $= \dfrac{25 \text{ g NaCl}}{1.00 \text{ L solution}} \times \dfrac{1 \text{ mol NaCl}}{58.44 \text{ g NaCl}} = 0.43 \; M$

1.00×10^3 g solution contains 25 g NaCl and 975 g $H_2O \approx 980$ g H_2O.

molality $= \dfrac{0.43 \text{ mol NaCl}}{0.98 \text{ kg solvent}} = 0.44 \text{ molal}$

0.43 mol NaCl; $980 \text{ g} \times \dfrac{1 \text{ mol}}{18.0 \text{ g}} = 54 \text{ mol } H_2O$; $\chi_{NaCl} = \dfrac{0.43}{0.43 + 54} = 7.9 \times 10^{-3}$

27.   Hydrochloric acid:

molarity $= \dfrac{38 \text{ g HCl}}{100. \text{ g soln}} \times \dfrac{1.19 \text{ g soln}}{\text{cm}^3 \text{ soln}} \times \dfrac{1000 \text{ cm}^3}{\text{L}} \times \dfrac{1 \text{ mol HCl}}{36.46 \text{ g}} = 12 \text{ mol/L}$

molality $= \dfrac{38 \text{ g HCl}}{62 \text{ g solvent}} \times \dfrac{1000 \text{ g}}{\text{kg}} \times \dfrac{1 \text{ mol HCl}}{36.46 \text{ g}} = 17 \text{ mol/kg}$

$38 \text{ g HCl} \times \dfrac{1 \text{ mol}}{36.46 \text{ g}} = 1.0 \text{ mol HCl}$; $62 \text{ g } H_2O \times \dfrac{1 \text{ mol}}{18.02 \text{ g}} = 3.4 \text{ mol } H_2O$

mole fraction of HCl $= \chi_{HCl} = \dfrac{1.0}{3.4 + 1.0} = 0.23$

Nitric acid:

$\dfrac{70. \text{ g HNO}_3}{100. \text{ g soln}} \times \dfrac{1.42 \text{ g soln}}{\text{cm}^3 \text{ soln}} \times \dfrac{1000 \text{ cm}^3}{\text{L}} \times \dfrac{1 \text{ mol HNO}_3}{63.02 \text{ g}} = 16 \text{ mol/L}$

$$\frac{70. \text{ g HNO}_3}{30. \text{ g solvent}} \times \frac{1000 \text{ g}}{\text{kg}} \times \frac{1 \text{ mol HNO}_3}{63.02 \text{ g}} = 37 \text{ mol/kg}$$

$$\textbf{70. g HNO}_3 \times \frac{1 \text{ mol}}{63.02 \text{ g}} = 1.1 \text{ mol HNO}_3; \quad 30. \text{ g H}_2\text{O} \times \frac{1 \text{ mol}}{18.02 \text{ g}} = 1.7 \text{ mol H}_2\text{O}$$

$$\chi_{\text{HNO}_3} = \frac{1.1}{1.7 + 1.1} = 0.39$$

Sulfuric acid:

$$\frac{95 \text{ g H}_2\text{SO}_4}{100. \text{ g soln}} \times \frac{1.84 \text{ g soln}}{\text{cm}^3 \text{ soln}} \times \frac{1000 \text{ cm}^3}{\text{L}} \times \frac{1 \text{ mol H}_2\text{SO}_4}{98.09 \text{ g H}_2\text{SO}_4} = 18 \text{ mol/L}$$

$$\frac{95 \text{ g H}_2\text{SO}_4}{5 \text{ g H}_2\text{O}} \times \frac{1000 \text{ g}}{\text{kg}} \times \frac{1 \text{ mol}}{98.09 \text{ g}} = 194 \text{ mol/kg} \approx 200 \text{ mol/kg}$$

$$95 \text{ g H}_2\text{SO}_4 \times \frac{1 \text{ mol}}{98.09 \text{ g}} = 0.97 \text{ mol H}_2\text{SO}_4; \quad 5 \text{ g H}_2\text{O} \times \frac{1 \text{ mol}}{18.02 \text{ g}} = 0.3 \text{ mol H}_2\text{O}$$

$$\chi_{\text{H}_2\text{SO}_4} = \frac{0.97}{0.97 + 0.3} = 0.76$$

Acetic Acid:

$$\frac{99 \text{ g HC}_2\text{H}_3\text{O}_2}{100. \text{ g soln}} \times \frac{1.05 \text{ g soln}}{\text{cm}^3 \text{ soln}} \times \frac{1000 \text{ cm}^3}{\text{L}} \times \frac{1 \text{ mol}}{60.05 \text{ g}} = 17 \text{ mol/L}$$

$$\frac{99 \text{ g HC}_2\text{H}_3\text{O}_2}{1 \text{ g H}_2\text{O}} \times \frac{1000 \text{ g}}{\text{kg}} \times \frac{1 \text{ mol}}{60.05 \text{ g}} = 1600 \text{ mol/kg} \approx 2000 \text{ mol/kg}$$

$$99 \text{ g HC}_2\text{H}_3\text{O}_2 \times \frac{1 \text{ mol}}{60.05 \text{ g}} = 1.6 \text{ mol HC}_2\text{H}_3\text{O}_2; \quad 1 \text{ g H}_2\text{O} \times \frac{1 \text{ mol}}{18.02 \text{ g}} = 0.06 \text{ mol H}_2\text{O}$$

$$\chi_{\text{HC}_2\text{H}_3\text{O}_2} = \frac{1.6}{1.6 + 0.06} = 0.96$$

Ammonia:

$$\frac{28 \text{ g NH}_3}{100. \text{ g soln}} \times \frac{0.90 \text{ g}}{\text{cm}^3} \times \frac{1000 \text{ cm}^3}{\text{L}} \times \frac{1 \text{ mol}}{17.03 \text{ g}} = 15 \text{ mol/L}$$

$$\frac{28 \text{ g NH}_3}{72 \text{ g H}_2\text{O}} \times \frac{1000 \text{ g}}{\text{kg}} \times \frac{1 \text{ mol}}{17.03 \text{ g}} = 23 \text{ mol/kg}$$

$$28 \text{ g NH}_3 \times \frac{1 \text{ mol}}{17.03 \text{ g}} = 1.6 \text{ mol NH}_3; \quad 72 \text{ g H}_2\text{O} \times \frac{1 \text{ mol}}{18.02 \text{ g}} = 4.0 \text{ mol H}_2\text{O}$$

$$\chi_{\text{NH}_3} = \frac{1.6}{4.0 + 1.6} = 0.29$$

29. $25 \text{ mL } C_5H_{12} \times \dfrac{0.63 \text{ g}}{\text{mL}} = 16 \text{ g } C_5H_{12}$; $25 \text{ mL} \times \dfrac{0.63}{\text{mL}} \times \dfrac{1 \text{ mol}}{72.15 \text{ g}} = 0.22 \text{ mol } C_5H_{12}$

$45 \text{ mL } C_6H_{14} \times \dfrac{0.66 \text{ g}}{\text{mL}} = 30. \text{ g } C_6H_{14}$; $45 \text{ mL} \times \dfrac{0.66 \text{ g}}{\text{mL}} \times \dfrac{1 \text{ mol}}{86.17 \text{ g}} = 0.34 \text{ mol } C_6H_{14}$

$\text{mass \% pentane} = \dfrac{\text{mass pentane}}{\text{total mass}} \times 100 = \dfrac{16 \text{ g}}{16 \text{ g} + 30. \text{ g}} \times 100 = 35\%$

$\chi_{\text{pentane}} = \dfrac{\text{mol pentane}}{\text{total mol}} = \dfrac{0.22 \text{ mol}}{0.22 \text{ mol} + 0.34 \text{ mol}} = 0.39$

$\text{molality} = \dfrac{\text{mol pentane}}{\text{kg hexane}} = \dfrac{0.22 \text{ mol}}{0.030 \text{ kg}} = 7.3 \text{ mol/kg}$

$\text{molarity} = \dfrac{\text{mol pentane}}{\text{L solution}} = \dfrac{0.22 \text{ mol}}{25 \text{ mL} + 45 \text{ mL}} \times \dfrac{1000 \text{ mL}}{1 \text{ L}} = 3.1 \text{ mol/L}$

31.   If we have 1.00 L of solution:

$1.37 \text{ mol citric acid} \times \dfrac{192.12 \text{ g}}{\text{mol}} = 263 \text{ g citric acid } (H_3C_6H_5O_7)$

$1.00 \times 10^3 \text{ mL solution} \times \dfrac{1.10 \text{ g}}{\text{mL}} = 1.10 \times 10^3 \text{ g solution}$

$\text{mass \% of citric acid} = \dfrac{263 \text{ g}}{1.10 \times 10^3 \text{ g}} \times 100 = 23.9\%$

In 1.00 L of solution, we have 263 g citric acid and $(1.10 \times 10^3 - 263) = 840$ g of H_2O.

$\text{molality} = \dfrac{1.37 \text{ mol citric acid}}{0.84 \text{ kg } H_2O} = 1.6 \text{ mol/kg}$

$840 \text{ g } H_2O \times \dfrac{1 \text{ mol}}{18.02 \text{ g}} = 47 \text{ mol } H_2O$; $\chi_{\text{citric acid}} = \dfrac{1.37}{47 + 1.37} = 0.028$

Since citric acid is a triprotic acid, the number of protons citric acid can provide is three times the molarity. Therefore, normality = $3 \times$ molarity:

$\text{normality} = 3 \times 1.37 \, M = 4.11 \, N$

Energetics of Solutions and Solubility

33.   Using Hess's law:

$$NaI(s) \rightarrow Na^+(g) + I^-(g) \qquad\qquad \Delta H = -\Delta H_{LE} = -(-686 \text{ kJ/mol})$$
$$Na^+(g) + I^-(g) \rightarrow Na^+(aq) + I^-(aq) \qquad \Delta H = \Delta H_{hyd} = -694 \text{ kJ/mol}$$

$$NaI(s) \rightarrow Na^+(aq) + I^-(aq) \qquad\qquad \Delta H_{soln} = -8 \text{ kJ/mol}$$

ΔH_{soln} refers to the heat released or gained when a solute dissolves in a solvent. Here, an ionic compound dissolves in water.

35.   Both $Al(OH)_3$ and NaOH are ionic compounds. Since the lattice energy is proportional to the charge of the ions, the lattice energy of aluminum hydroxide is greater than that of sodium hydroxide. The attraction of water molecules for Al^{3+} and OH^- cannot overcome the larger lattice energy and $Al(OH)_3$ is insoluble. For NaOH, the favorable hydration energy is large enough to overcome the smaller lattice energy and NaOH is soluble.

37.   Water is a polar solvent and dissolves polar solutes and ionic solutes. Carbon tetrachloride (CCl_4) is a nonpolar solvent and dissolves nonpolar solutes (like dissolves like).

   a. CCl_4; CO_2 is a nonpolar molecule.   b. Water; NH_4NO_3 is an ionic solid.

   c. Water; CH_3COCH_3 is polar molecule. d. Water; $HC_2H_3O_2$ (acetic acid) is a polar molecule.

   e. CCl_4; $CH_3CH_2CH_2CH_2CH_3$ is a nonpolar molecule.

39.   Water is a polar molecule capable of hydrogen bonding. Polar molecules, especially molecules capable of hydrogen bonding, and ions are all attracted to water. For covalent compounds, as polarity increases, the attraction to water increases. For ionic compounds, as the charge of the ions increases and/or the size of the ions decreases, the attraction to water increases.

   a. CH_3CH_2OH; CH_3CH_2OH is polar while $CH_3CH_2CH_3$ is nonpolar.

   b. $CHCl_3$; $CHCl_3$ is polar while CCl_4 is nonpolar.

   c. CH_3CH_2OH; CH_3CH_2OH is much more polar than $CH_3(CH_2)_{14}CH_2OH$.

41.   As the length of the hydrocarbon chain increases, the solubility decreases. The –OH end of the alcohols can hydrogen bond with water. The hydrocarbon chain, however, is basically nonpolar and interacts poorly with water. As the hydrocarbon chain gets longer, a greater portion of the molecule cannot interact with the water molecules and the solubility decreases, i.e., the effect of the –OH group decreases as the alcohols get larger.

43. $C = kP$, $\dfrac{8.21 \times 10^{-4}\text{ mol}}{L} = k \times 0.790$ atm, $k = 1.04 \times 10^{-3}$ mol/L•atm

$C = kP$, $C = \dfrac{1.04 \times 10^{-3}\text{ mol}}{L\,\text{atm}} \times 1.10$ atm $= 1.14 \times 10^{-3}$ mol/L

Vapor Pressures of Solutions

45. $P_{H_2O} = \chi_{H_2O}\, P^{\circ}_{H_2O}$; $\chi_{H_2O} = \dfrac{\text{mol }H_2O\text{ in solution}}{\text{total mol in solution}}$

50.0 g $C_6H_{12}O_6 \times \dfrac{1\text{ mol }C_6H_{12}O_6}{180.16\text{ g }C_6H_{12}O_6} = 0.278$ mol glucose

600.0 g $H_2O \times \dfrac{1\text{ mol}}{18.02\text{ g}} = 33.30$ mol H_2O; Total mol $= 0.278 + 33.30 = 33.58$ mol

$\chi_{H_2O} = \dfrac{33.30}{33.58} = 0.9917$; $P_{H_2O} = \chi_{H_2O}\, P^{\circ}_{H_2O} = 0.9917 \times 23.8$ torr $= 23.6$ torr

47. $P_B = \chi_B P^{\circ}_B$, $\chi_B = P_B/P^{\circ}_B = 0.900$ atm/0.930 atm $= 0.968$

$0.968 = \dfrac{\text{mol benzene}}{\text{total mol}}$; mol benzene $= 78.11$ g $C_6H_6 \times \dfrac{1\text{ mol}}{78.11\text{ g}} = 1.000$ mol

Let x = mol solute, then: $\chi_B = 0.968 = \dfrac{1.000\text{ mol}}{1.000 + x}$, $0.968 + 0.968\,x = 1.000$, $x = 0.033$ mol

molar mass $= \dfrac{10.0\text{ g}}{0.033\text{ mol}} = 303$ g/mol $\approx 3.0 \times 10^2$ g/mol

49.   a.   25 mL $C_5H_{12} \times \dfrac{0.63\text{ g}}{\text{mL}} \times \dfrac{1\text{ mol}}{72.15\text{ g}} = 0.22$ mol C_5H_{12}

45 mL $C_6H_{14} \times \dfrac{0.66\text{ g}}{\text{mL}} \times \dfrac{1\text{ mol}}{86.17\text{ g}} = 0.34$ mol C_6H_{14}; total mol $= 0.22 + 0.34 = 0.56$ mol

$\chi^L_{pen} = \dfrac{\text{mol pentane in solution}}{\text{total mol in solution}} = \dfrac{0.22\text{ mol}}{0.56\text{ mol}} = 0.39$, $\chi^L_{hex} = 1.00 - 0.39 = 0.61$

$P_{pen} = \chi^L_{pen} P^{\circ}_{pen} = 0.39(511\text{ torr}) = 2.0 \times 10^2$ torr; $P_{hex} = 0.61(150.\text{ torr}) = 92$ torr

$P_{total} = P_{pen} + P_{hex} = 2.0 \times 10^2 + 92 = 292$ torr $= 290$ torr

b.   From Chapter 5 on gases, the partial pressure of a gas is proportional to the number of moles of gas present. For the vapor phase:

$$\chi^V_{pen} = \frac{mol\ pentane\ in\ vapor}{total\ mol\ vapor} = \frac{P_{pen}}{P_{total}} = \frac{2.0 \times 10^2\ torr}{290\ torr} = 0.69$$

Note: In the Solutions Guide, we added V or L to the mole fraction symbol to emphasize which value we are solving. If the L or V is omitted, then the liquid phase is assumed.

51. $P_{total} = P_{meth} + P_{prop}$, 174 torr $= \chi^L_{meth}$ (303 torr) $+ \chi^L_{prop}$ (44.6 torr); $\chi^L_{prop} = 1.000 - \chi^L_{meth}$

174 $= 303\ \chi^L_{meth} + (1.000 - \chi^L_{meth})\ 44.6$, $\frac{129}{258} = \chi^L_{meth} = 0.500$; $\chi^L_{prop} = 1.000 - 0.500 = 0.500$

53.   Compared to H_2O, solution d (methanol/water) will have the highest vapor pressure because methanol is more volatile than water. Both solution b (glucose/water) and solution c (NaCl/water) will have a lower vapor pressure than water by Raoult's law. NaCl dissolves to give Na^+ ions and Cl^- ions; glucose is a nonelectrolyte. Since there are more solute particles in solution c, the vapor pressure of solution c will be the lowest.

55.   50.0 g $CH_3COCH_3 \times \frac{1\ mol}{58.08\ g} = 0.861$ mol acetone

50.0 g $CH_3OH \times \frac{1\ mol}{32.04\ g} = 1.56$ mol methanol

$\chi^L_{acetone} = \frac{0.861}{0.861 + 1.56} = 0.356$; $\chi^L_{methanol} = 1.000 - \chi^L_{acetone} = 0.644$

$P_{total} = P_{methanol} + P_{acetone} = 0.644(143\ torr) + 0.356(271\ torr) = 92.1\ torr + 96.5\ torr = 188.6\ torr$

Since partial pressures are proportional to the moles of gas present, then in the vapor phase:

$\chi^V_{acetone} = \frac{P_{acetone}}{P_{total}} = \frac{96.5\ torr}{188.6\ torr} = 0.512$; $\chi^V_{methanol} = 1.000 - 0.512 = 0.488$

The actual vapor pressure of the solution (161 torr) is less than the calculated pressure assuming ideal behavior (188.6 torr). Therefore, the solution exhibits negative deviations from Raoult's law. This occurs when the solute-solvent interactions are stronger than in pure solute and pure solvent.

Colligative Properties

57.   molality $= m = \frac{mol\ solute}{kg\ solvent} = \frac{4.9\ g\ C_{12}H_{22}O_{11}}{175\ g\ H_2O} \times \frac{1000\ g}{kg} \times \frac{1\ mol\ C_{12}H_{22}O_{11}}{342.30\ g\ C_{12}H_{22}O_{11}} = 0.082$ molal

$\Delta T_b = K_b m = \frac{0.51°C}{molal} \times 0.082$ molal $= 0.042°C$

The boiling point is raised from 100.000°C to 100.042°C. We assumed P = 1 atm and ample significant figures in the boiling point of pure water.

59. $\Delta T_f = K_f m$, $\Delta T_f = 1.50°C = \dfrac{1.86°C}{molal} \times m$, $m = 0.806$ mol/kg

0.200 kg $H_2O \times \dfrac{0.806 \text{ mol } C_3H_8O_3}{\text{kg } H_2O} \times \dfrac{92.09 \text{ g } C_3H_8O_3}{\text{mol } C_3H_8O_3} = 14.8$ g $C_3H_8O_3$

61. molality $= m = \dfrac{50.0 \text{ g } C_2H_6O_2}{50.0 \text{ g } H_2O} \times \dfrac{1000 \text{ g}}{\text{kg}} \times \dfrac{1 \text{ mol}}{62.07 \text{ g}} = 16.1$ mol/kg

$\Delta T_f = K_f m = 1.86°C/molal \times 16.1$ molal $= 29.9°C$; $T_f = 0.0°C - 29.9°C = -29.9°C$

$\Delta T_b = K_b m = 0.51°C/molal \times 16.1$ molal $= 8.2°C$; $T_b = 100.0°C + 8.2°C = 108.2°C$

63. $\Delta T_f = K_f m$, $m = \dfrac{\Delta T_f}{K_f} = \dfrac{0.240°C}{4.70°C \text{ kg/mol}} = \dfrac{5.11 \times 10^{-2} \text{ mol biomolecule}}{\text{kg solvent}}$

The mol of biomolecule present is:

0.0150 kg solvent $\times \dfrac{5.11 \times 10^{-2} \text{ mol biomolecule}}{\text{kg solvent}} = 7.67 \times 10^{-4}$ mol biomolecule

From the problem, 0.350 g biomolecule were used that must contain 7.67×10^{-4} mol biomolecule. The molar mass of the biomolecule is:

molar mass $= \dfrac{0.350 \text{ g}}{7.67 \times 10^{-4} \text{ mol}} = 456$ g/mol

65. a. $M = \dfrac{1.0 \text{ g protein}}{L} \times \dfrac{1 \text{ mol}}{9.0 \times 10^4 \text{ g}} = 1.1 \times 10^{-5}$ mol/L; $\pi = MRT$

At 298 K: $\pi = \dfrac{1.1 \times 10^{-5} \text{ mol}}{L} \times \dfrac{0.08206 \text{ L atm}}{\text{mol K}} \times 298 \text{ K} \times \dfrac{760 \text{ torr}}{\text{atm}}$, $\pi = 0.20$ torr

Since d = 1.0 g/cm³, 1.0 L solution has a mass of 1.0 kg. Since only 1.0 g of protein is present per liter of solution, 1.0 kg of H_2O is present and molality equals molarity.

$\Delta T_f = K_f m = \dfrac{1.86°C}{molal} \times 1.1 \times 10^{-5}$ molal $= 2.0 \times 10^{-5}°C$

b. Osmotic pressure is better for determining the molar mass of large molecules. A temperature change of $10^{-5}°C$ is very difficult to measure. A change in height of a column of mercury by 0.2 mm (0.2 torr) is not as hard to measure precisely.

67. $\pi = MRT$, $M = \dfrac{\pi}{RT} = \dfrac{8.00 \text{ atm}}{0.08206 \text{ L atm/mol·K} \times 298 \text{ K}} = 0.327$ mol/L

Properties of Electrolyte Solutions

69. $Na_3PO_4(s) \rightarrow 3\ Na^+(aq) + PO_4^{3-}(aq)$, $i = 4.0$; $CaBr_2(s) \rightarrow Ca^{2+}(aq) + 2\ Br^-(aq)$, $i = 3.0$

$KCl(s) \rightarrow K^+(aq) + Cl^-(aq)$, $i = 2.0$.

The effective particle concentrations of the solutions are:

4.0(0.010 molal) = 0.040 molal for Na_3PO_4 solution; 3.0(0.020 molal) = 0.060 molal for $CaBr_2$ solution; 2.0(0.020 molal) = 0.040 molal for KCl solution; slightly greater than 0.020 molal for HF solution since HF only partially dissociates in water (it is a weak acid).

a.   The 0.010 m Na_3PO_4 solution and the 0.020 m KCl solution both have effective particle concentrations of 0.040 m (assuming complete dissociation), so both of these solutions should have the same boiling point as the 0.040 m $C_6H_{12}O_6$ solution (a nonelectrolyte).

b. $P = \chi P°$; As the solute concentration decreases, the solvent's vapor pressure increases since χ increases. Therefore, the 0.020 m HF solution will have the highest vapor pressure since it has the smallest effective particle concentration.

c. $\Delta T = K_f m$; The 0.020 m $CaBr_2$ solution has the largest effective particle concentration so it will have the largest freezing point depression (largest ΔT).

71.   a. $MgCl_2(s) \rightarrow Mg^{2+}(aq) + 2\ Cl^-(aq)$, $i = 3.0$ mol ions/mol solute

$\Delta T_f = iK_f m = 3.0 \times 1.86°C/\text{molal} \times 0.050\ \text{molal} = 0.28°C$; $T_f = -0.28°C$ (Assuming water freezes at $0.00°C$.)

$\Delta T_b = iK_b m = 3.0 \times 0.51°C/\text{molal} \times 0.050\ \text{molal} = 0.077°C$; $T_b = 100.077°C$ (Assuming water boils at $100.000°C$.)

b. $FeCl_3(s) \rightarrow Fe^{3+}(aq) + 3\ Cl^-(aq)$, $i = 4.0$ mol ions/mol solute

$\Delta T_f = iK_f m = 4.0 \times 1.86°C/\text{molal} \times 0.050\ \text{molal} = 0.37°C$; $T_f = -0.37°C$

$\Delta T_b = iK_b m = 4.0 \times 0.51°C/\text{molal} \times 0.050\ \text{molal} = 0.10°C$; $T_b = 100.10°C$

73. $\Delta T_f = iK_f m$, $i = \dfrac{\Delta T_f}{K_f m} = \dfrac{0.110°C}{1.86°\ C/\text{molal} \times 0.0225\ \text{molal}} = 2.63$ for 0.0225 m $CaCl_2$

$i = \dfrac{0.440}{1.86 \times 0.0910} = 2.60$ for 0.0910 m $CaCl_2$; $i = \dfrac{1.330}{1.86 \times 0.278} = 2.57$ for 0.278 m $CaCl_2$

$i_{ave} = (2.63 + 2.60 + 2.57)/3 = 2.60$

Note that i is less than the ideal value of 3.0 for $CaCl_2$. This is due to ion pairing in solution. Also note that as molality increases, i decreases. More ion pairing occurs as concentration increases.

75. $\pi = iMRT = 3.0 \times 0.50$ mol/L $\times$ 0.08206 L atm/K•mol $\times$ 298 K = 37 atm

Because of ion pairing in solution, we would expect i to be less than 3.0, which results in fewer solute particles in solution, which results in a lower osmotic pressure than calculated above.

Additional Exercises

77.   a. $NH_4NO_3(s) \rightarrow NH_4^+(aq) + NO_3^-(aq)$ $\Delta H_{soln} = ?$

Heat gain by dissolution process = heat loss by solution; We will keep all quantities positive in order to avoid sign errors. Since the temperature of the water decreased, the dissolution of NH_4NO_3 is endothermic (ΔH is positive). Mass of solution = 1.60 + 75.0 = 76.6 g.

heat loss by solution $= \dfrac{4.18\ J}{g\ °C} \times 76.6$ g $\times$ (25.00°C - 23.34°C) = 532 J

$\Delta H_{soln} = \dfrac{532\ J}{1.60\ g\ NH_4NO_3} \times \dfrac{80.05\ g\ NH_4NO_3}{mol\ NH_4NO_3} = 2.66 \times 10^4$ J/mol = 26.6 kJ/mol

b.   We will use Hess's law to solve for the lattice energy. The lattice energy equation is:

$NH_4^+(g) + NO_3^-(g) \rightarrow NH_4NO_3(s)$ ΔH = lattice energy

$NH_4^+(g) + NO_3^-(g) \rightarrow NH_4^+(aq) + NO_3^-(aq)$ $\Delta H = \Delta H_{hyd}$ = -630. kJ/mol
$NH_4^+(aq) + NO_3^-(aq) \rightarrow NH_4NO_3(s)$ $\Delta H = -\Delta H_{soln}$ = -26.6 kJ/mol
_____

$NH_4^+(g) + NO_3^-(g) \rightarrow NH_4NO_3(s)$ $\Delta H = \Delta H_{hyd} - \Delta H_{soln}$ = -657 kJ/mol

79.   a.   Water boils when the vapor pressure equals the pressure above the water. In an open pan $P_{atm} \approx 1.0$ atm. In a pressure cooker, $P_{inside} > 1.0$ atm, and water boils at a higher temperature. The higher the cooking temperature, the faster the cooking time.

b.   Salt dissolves in water forming a solution with a melting point lower than that of pure water ($\Delta T_f = K_f m$). This happens in water on the surface of ice. If it is not too cold, the ice melts. This won't work if the ambient temperature is lower than the depressed freezing point of the salt solution.

c.   When water freezes from a solution, it freezes as pure water, leaving behind a more concentrated salt solution.   Therefore, the melt of frozen sea ice is pure water.

d.   On the CO_2 phase diagram in chapter 10, the triple point is above 1 atm, so $CO_2(g)$ is the stable phase at 1 atm and room temperature. $CO_2(l)$ can't exist at normal atmospheric pressures. Therefore, dry ice sublimes instead of boils. In a fire extinguisher, P > 1 atm and $CO_2(l)$ can exist. When CO_2 is released from the fire extinguisher, $CO_2(g)$ forms as predicted from the phase diagram.

e. Adding a solute to a solvent increases the boiling point and decreases the freezing point of the solvent. Thus, the solvent is a liquid over a wider range of temperatures when a solute is dissolved.

81. Since partial pressures are proportional to the moles of gas present, then $\chi_{CS_2}^{V} = P_{CS_2}/P_{tot}$.

$$P_{CS_2} = \chi_{CS_2}^{V} P_{tot} = 0.855 \ (263 \text{ torr}) = 225 \text{ torr}$$

$$P_{CS_2} = \chi_{CS_2}^{L} P_{CS_2}^{\circ}, \ \chi_{CS_2}^{L} = \frac{P_{CS_2}}{P_{CS_2}^{\circ}} = \frac{225 \text{ torr}}{375 \text{ torr}} = 0.600$$

83. Out of 100.00 g, there are:

$$31.57 \text{ g C} \times \frac{1 \text{ mol C}}{12.01 \text{ g}} = 2.629 \text{ mol C}; \quad \frac{2.629}{2.629} = 1.000$$

$$5.30 \text{ g H} \times \frac{1 \text{ mol H}}{1.008 \text{ g}} = 5.26 \text{ mol H}; \quad \frac{5.26}{2.629} = 2.00$$

$$63.13 \text{ g O} \times \frac{1 \text{ mol O}}{16.00 \text{ g}} = 3.946 \text{ mol O}; \quad \frac{3.946}{2.629} = 1.501$$

empirical formula: $C_2H_4O_3$; Use the freezing point data to determine the molar mass.

$$m = \frac{\Delta T_f}{K_f} = \frac{5.20°C}{1.86°C/\text{molal}} = 2.80 \text{ molal}$$

$$\text{mol solute} = 0.0250 \text{ kg} \times \frac{2.80 \text{ mol solute}}{\text{kg}} = 0.0700 \text{ mol solute}$$

$$\text{molar mass} = \frac{10.56 \text{ g}}{0.0700 \text{ mol}} = 151 \text{ g/mol}$$

The empirical formula mass of $C_2H_4O_3$ = 76.05 g/mol. Since the molar mass is about twice the empirical mass, the molecular formula is $C_4H_8O_6$, which has a molar mass of 152.10 g/mol.

Note: We use the experimental molar mass to determine the molecular formula. Knowing this, we calculate the molar mass precisely from the molecular formula using the periodic table.

85. If ideal, NaCl dissociates completely and i = 2.00. $\Delta T_f = iK_f m$; Assuming water freezes at 0.00°C:

$$1.28°C = 2 \times 1.86°C \text{ kg/mol} \times m, \ m = 0.344 \text{ mol NaCl/kg } H_2O$$

Assume an amount of solution which contains 1.00 kg of water (solvent).

$$0.344 \text{ mol NaCl} \times \frac{58.44 \text{ g}}{\text{mol}} = 20.1 \text{ g NaCl}; \quad \text{mass \% NaCl} = \frac{20.1 \text{ g}}{1.00 \times 10^3 \text{ g} + 20.1 \text{ g}} \times 100 = 1.97\%$$

Challenge Problems

87. $\chi_{pen}^V = 0.15 = \dfrac{P_{pen}}{P_{total}}$; $P_{pen} = \chi_{pen}^L P_{pen}^\circ = \chi_{pen}^L (511 \text{ torr})$; $P_{total} = P_{pen} + P_{hex} = \chi_{pen}^L(511) + \chi_{hex}^L(150.)$

Since $\chi_{hex}^L = 1.000 - \chi_{pen}^L$, then: $P_{total} = \chi_{pen}^L(511) + (1.000 - \chi_{pen}^L)(150.) = 150. + 361 \, \chi_{pen}^L$

$\chi_{pen}^V = \dfrac{P_{pen}}{P_{total}}$, $0.15 = \dfrac{\chi_{pen}^L (511)}{150. + 361 \, \chi_{pen}^L}$, $0.15 \,(150. + 361 \, \chi_{pen}^L) = 511 \chi_{pen}^L$

$23 + 54 \, \chi_{pen}^L = 511 \, \chi_{pen}^L$, $\chi_{pen}^L = \dfrac{23}{457} = 0.050$

89. $\Delta T_f = 5.51 - 2.81 = 2.70°C$; $m = \dfrac{\Delta T_f}{K_f} = \dfrac{2.7°C}{5.12°C/molal} = 0.527 \text{ molal}$

Let x = mass of naphthalene (molar mass: 128.2 g/mol). Then $1.60 - x$ = mass of anthracene (molar mass = 178.2 g/mol).

$\dfrac{x}{128.2}$ = moles naphthalene and $\dfrac{1.60 - x}{178.2}$ = moles anthracene

$\dfrac{0.527 \text{ mol solute}}{\text{kg solvent}} = \dfrac{\dfrac{x}{128.2} + \dfrac{1.60 - x}{178.2}}{0.0200 \text{ kg solvent}}$, $1.05 \times 10^{-2} = \dfrac{178.2\,x + 1.60\,(128.2) - 128.2\,x}{128.2\,(178.2)}$

$50.0\,x + 205 = 240.$, $50.0\,x = 35$, $x = 0.70 \text{ g naphthalene}$

So mixture is: $\dfrac{0.70 \text{ g}}{1.60 \text{ g}} \times 100 = 44\%$ naphthalene by mass and 56% anthracene by mass

91. $HCO_2H \rightarrow H^+ + HCO_2^-$; Only 4.2% of HCO_2H ionizes. The amount of H^+ or HCO_2^- produced is:

$0.042 \times 0.10 \, M = 0.0042 \, M$

The amount of HCO_2H remaining in solution after ionization is:

$0.10 \, M - 0.0042 \, M = 0.10 \, M$

The total molarity of species present $= M_{HCO_2H} + M_{H^+} + M_{HCO_2^-} = 0.10 + 0.0042 + 0.0042 = 0.11 \, M$.

Assuming $0.11 \, M = 0.11$ molal and assuming ample significant figures in the freezing point and boiling point of water at P = 1 atm:

$\Delta T = K_f m = 1.86°C/molal \times 0.11 \text{ molal} = 0.20°C$; freezing point = -0.20°C

$\Delta T = K_b m = 0.51°C/molal \times 0.11 \text{ molal} = 0.056°C$; boiling point = 100.056°C

93.   a.   Assuming $MgCO_3(s)$ does not dissociate, the solute concentration in water is:

$$\frac{560\ \mu g\ MgCO_3(s)}{mL} = \frac{560\ mg}{L} = \frac{560 \times 10^{-3}\ g}{L} \times \frac{1\ mol\ MgCO_3}{84.32\ g} = 6.6 \times 10^{-3}\ mol\ MgCO_3/L$$

An applied pressure of 8.0 atm will purify water up to a solute concentration of:

$$M = \frac{\pi}{RT} = \frac{8.0\ atm}{0.08206\ L\ atm/mol{\cdot}K \times 300.\ K} = \frac{0.32\ mol}{L}$$

When the concentration of $MgCO_3(s)$ reaches 0.32 mol/L, the reverse osmosis unit can no longer purify the water. Let V = volume (L) of water remaining after purifying 45 L of H_2O. When V + 45 L of water have been processed, the moles of solute particles will equal:

6.6×10^{-3} mol/L $\times$ (45 L + V) = 0.32 mol/L $\times$ V

Solving: 0.30 = (0.32 - 0.0066) $\times$ V, V = 0.96 L

The minimum total volume of water that must be processed is 45 L + 0.96 L = 46 L.

Note: If $MgCO_3$ does dissociate into Mg^{2+} and CO_3^{2-} ions, the solute concentration will increase to 1.3×10^{-2} *M* and at least 47 L of water must be processed.

   b.   No; A reverse osmosis system that applies 8.0 atm can only purify water with a solute concentration less than 0.32 mol/L. Salt water has a solute concentration of 2(0.60 *M*) = 1.20 *M* ions. The solute concentration of salt water is much too high for this reverse osmosis unit to work.

CHAPTER TWELVE

CHEMICAL KINETICS

Questions

9.   The rate of a chemical reaction varies with time. Consider the general reaction:

$$A \rightarrow \text{Products where rate} = \frac{-\Delta[A]}{\Delta t}$$

If we graph [A] vs. t, it would roughly look like the dark line in the following plot.

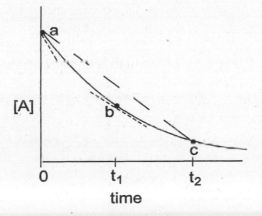

An instantaneous rate is the slope of a tangent line to the graph of [A] vs. t. We can determine the instantaneous rate at any time during the reaction. On the plot, tangent lines at $t = 0$ and $t = t_1$ are drawn. The slope of these tangent lines would be the instantaneous rates at $t \approx 0$ and $t = t_1$. We call the instantaneous rate at $t \approx 0$ the initial rate. The average rate is measured over a period of time. For example, the slope of the line connecting points a and c is the average rate of the reaction over the entire length of time 0 to t_2 (average rate $= \Delta[A]/\Delta t$). An average rate is determined over some time period while an instantaneous rate is determined at one specific time. The rate which is largest is generally the initial rate. At $t \approx 0$, the slope of the tangent line is greatest, which means the rate is largest at $t \approx 0$.

11.   a.   The greater the frequency of collisions, the greater the opportunities for molecules to react, and, hence, the greater the rate.

157

b. Chemical reactions involve the making and breaking of chemical bonds. The kinetic energy of the collisions can be used to break bonds. So, as the kinetic energy of the collisions increases, the rate increases.

c. For a reaction to occur, it is the reactive portion of each molecule that must be involved in a collision. Only some of all the possible collisions have the correct orientation to convert reactants to products.

13.   A catalyst increases the rate of a reaction by providing reactants with an alternate pathway (mechanism) to convert to products. This alternate pathway has a lower activation energy, thus increasing the rate of the reaction.

A heterogeneous catalyst is in a different phase than the reactants. The catalyst is usually a solid, although a catalyst in a liquid phase can act as a heterogeneous catalyst for some gas phase reactions. Since the catalyzed reaction has a different mechanism than the uncatalyzed reaction, the catalyzed reaction most likely will have a different rate law.

15.   a. Activation energy and ΔE are independent of each other. Activation energy depends on the path reactants to take to convert to products. The overall energy change, ΔE, depends only on the initial and final energy states of the reactants and products. ΔE is path independent.

b. The rate law can be determined only from experiment, not from the overall balanced reaction.

c. Most reactions occur by a series of steps. The rate of the reaction is determined by the rate of the slowest step in the mechanism.

Exercises

Reaction Rates

17.   The coefficients in the balanced reaction relate the rate of disappearance of reactants to the rate of production of products. From the balanced reaction, the rate of production of P_4 will be 1/4 the rate of disappearance of PH_3, and the rate of production of H_2 will be 6/4 the rate of disappearance of PH_3. By convention, all rates are given as positive values.

$$\text{Rate} = -\frac{\Delta[PH_3]}{\Delta t} = \frac{(0.0048 \text{ mol}/2.0 \text{ L})}{s} = 2.4 \times 10^{-3} \text{ mol/L} \cdot s$$

$$\frac{\Delta[P_4]}{\Delta t} = -\frac{1}{4}\frac{\Delta[PH_3]}{\Delta t} = 2.4 \times 10^{-3}/4 = 6.0 \times 10^{-4} \text{ mol/L} \cdot s$$

$$\frac{\Delta[H_2]}{\Delta t} = -\frac{6}{4}\frac{\Delta[PH_3]}{\Delta t} = 6(2.4 \times 10^{-3})/4 = 3.6 \times 10^{-3} \text{ mol/L} \cdot s$$

19.   a.   The units for rate are always mol/L•s.   b.   Rate = k; k must have units of mol/L•s.

       c.   Rate = k[A], $\dfrac{mol}{L\,s} = k\left(\dfrac{mol}{L}\right)$ d.   Rate = k[A]^2, $\dfrac{mol}{L\,s} = k\left(\dfrac{mol}{L}\right)^2$

             k must have units of s^{-1}.   k must have units of L/mol•s.

       e.   L^2/mol^2•s

Rate Laws from Experimental Data: Initial Rates Method

21.   a.   In the first two experiments, [NO] is held constant and [Cl$_2$] is doubled. The rate also doubled.
             Thus, the reaction is first order with respect to Cl$_2$. Or mathematically: Rate = k[NO]^x[Cl$_2$]^y

             $$\dfrac{0.36}{0.18} = \dfrac{k(0.10)^x(0.20)^y}{k(0.10)^x(0.10)^y} = \dfrac{(0.20)^y}{(0.10)^y},\ 2.0 = 2.0^y,\ y = 1$$

             We can get the dependence on NO from the second and third experiments. Here, as the NO
             concentration doubles (Cl$_2$ concentration is constant), the rate increases by a factor of four.
             Thus, the reaction is second order with respect to NO. Or mathematically:

             $$\dfrac{1.45}{0.36} = \dfrac{k(0.20)^x(0.20)}{k(0.10)^x(0.20)} = \dfrac{(0.20)^x}{(0.10)^x},\ 4.0 = 2.0^x,\ x = 2;\ \text{So, Rate} = k[NO]^2[Cl_2]$$

             Try to examine experiments where only one concentration changes at a time. The more
             variables that change, the harder it is to determine the orders. Also, these types of problems can
             usually be solved by inspection. In general, we will solve using a mathematical approach, but
             keep in mind you probably can solve for the orders by simple inspection of the data.

       b.   The rate constant k can be determined from the experiments. From experiment 1:

             $$\dfrac{0.18\ mol}{L\ min} = k\left(\dfrac{0.10\ mol}{L}\right)^2\left(\dfrac{0.10\ mol}{L}\right),\ k = 180\ L^2/mol^2\text{•min}$$

             From the other experiments:

             k = 180 L^2/mol^2•min (2nd exp.); k = 180 L^2/mol^2•min (3rd exp.)

             The average rate constant is k$_{mean}$ = 1.8 × 10^2 L^2/mol^2•min.

23.   a.   Rate = k[NOCl]^n; Using experiments two and three:

             $$\dfrac{2.66 \times 10^4}{6.64 \times 10^3} = \dfrac{k(2.0 \times 10^{16})^n}{k(1.0 \times 10^{16})^n},\ 4.01 = 2.0^n,\ n = 2;\ \text{Rate} = k[NOCl]^2$$

b. $\dfrac{5.98 \times 10^4 \text{ molecules}}{cm^3 \text{ s}} = k \left(\dfrac{3.0 \times 10^{16} \text{ molecules}}{cm^3} \right)^2$, $k = 6.6 \times 10^{-29}$ cm³/molecules•s

The other three experiments give $(6.7, 6.6 \text{ and } 6.6) \times 10^{-29}$ cm³/molecules•s, respectively.

The mean value for k is 6.6×10^{-29} cm³/molecules•s.

c. $\dfrac{6.6 \times 10^{-29} \text{ cm}^3}{\text{molecules s}} \times \dfrac{1 \text{ L}}{1000 \text{ cm}^3} \times \dfrac{6.022 \times 10^{23} \text{molecules}}{\text{mol}} = \dfrac{4.0 \times 10^{-8} \text{ L}}{\text{mol s}}$

25.   a. Rate = $k[Hb]^x[CO]^y$; Comparing the first two experiments, [CO] is unchanged, [Hb] doubles, and the rate doubles. Therefore, the reaction is first order in Hb. Comparing the second and third experiments, [Hb] is unchanged, [CO] triples. and the rate triples. Therefore, $y = 1$ and the reaction is first order in CO.

b. Rate = k[Hb][CO]

c. From the first experiment:

0.619 μmol/L•s = k (2.21 μmol/L)(1.00 μmol/L), k = 0.280 L/μmol•s

The second and third experiments give similar k values, so $k_{mean} = 0.280$ L/μmol•s.

d. Rate = k[Hb][CO] = $\dfrac{0.280 \text{ L}}{\mu\text{mol s}} \times \dfrac{3.36 \,\mu\text{mol}}{\text{L}} \times \dfrac{2.40 \,\mu\text{mol}}{\text{L}} = 2.26$ μmol/L•s

Integrated Rate Laws

27.   The first assumption to make is that the reaction is first order. For a first-order reaction, a graph of ln $[H_2O_2]$ vs time will yield a straight line. If this plot is not linear, then the reaction is not first order and we make another assumption. The data and plot for the first-order assumption is below.

Time (s)	$[H_2O_2]$ (mol/L)	ln $[H_2O_2]$
0	1.00	0.000
120.	0.91	-0.094
300.	0.78	-0.25
600.	0.59	-0.53
1200.	0.37	-0.99
1800.	0.22	-1.51
2400.	0.13	-2.04
3000.	0.082	-2.50
3600.	0.050	-3.00

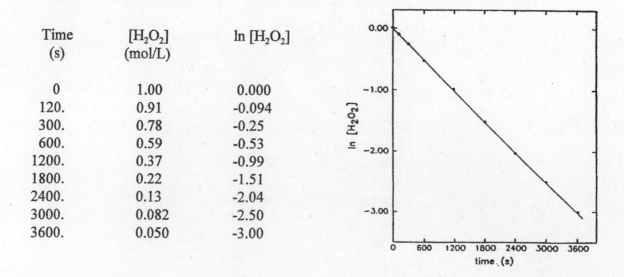

Note: We carried extra significant figures in some of the ln values in order to reduce round-off error. For the plots, we will do this most of the time when the ln function is involved.

The plot of ln $[H_2O_2]$ vs. time is linear. Thus, the reaction is first order. The rate law and integrated rate law are: Rate = $k[H_2O_2]$ and ln $[H_2O_2]$ = -kt + ln $[H_2O_2]_o$.

We determine the rate constant k by determining the slope of the ln $[H_2O_2]$ vs time plot (slope = -k). Using two points on the curve gives:

$$\text{slope} = -k = \frac{\Delta y}{\Delta x} = \frac{0 - (3.00)}{0 - 3600.} = -8.3 \times 10^{-4} \text{ s}^{-1}, \quad k = 8.3 \times 10^{-4} \text{ s}^{-1}$$

To determine $[H_2O_2]$ at 4000. s, use the integrated rate law where at t = 0, $[H_2O_2]_o$ = 1.00 M.

$$\ln [H_2O_2] = -kt + \ln [H_2O_2]_o \quad \text{or} \quad \ln\left(\frac{[H_2O_2]}{[H_2O_2]_o}\right) = -kt$$

$$\ln\left(\frac{[H_2O_2]}{1.00}\right) = -8.3 \times 10^{-4} \text{ s}^{-1} \times 4000. \text{ s}, \quad \ln [H_2O_2] = -3.3, \quad [H_2O_2] = e^{-3.3} = 0.037 \ M$$

29.   Assume the reaction is first order and see if the plot of ln $[NO_2]$ vs. time is linear. If this isn't linear, try the second-order plot of $1/[NO_2]$ vs. time. The data and plots follow.

Time (s)	$[NO_2]$ (M)	ln $[NO_2]$	$1/[NO_2]$ (M^{-1})
0	0.500	-0.693	2.00
1.20×10^3	0.444	-0.812	2.25
3.00×10^3	0.381	-0.965	2.62
4.50×10^3	0.340	-1.079	2.94
9.00×10^3	0.250	-1.386	4.00
1.80×10^4	0.174	-1.749	5.75

The plot of $1/[NO_2]$ vs. time is linear. The reaction is second order in NO_2. The rate law and integrated rate law are: Rate = $k[NO_2]^2$ and $\dfrac{1}{[NO_2]} = kt + \dfrac{1}{[NO_2]_o}$.

The slope of the plot $1/[NO_2]$ vs. t gives the value of k. Using a couple of points on the plot:

$$\text{slope} = k = \frac{\Delta y}{\Delta x} = \frac{(5.75 - 2.00)\, M^{-1}}{(1.80 \times 10^4 - 0)\, s} = 2.08 \times 10^{-4}\ L/mol\bullet s$$

To determine $[NO_2]$ at 2.70×10^4 s, use the integrated rate law where $1/[NO_2]_o = 1/0.500\ M = 2.00\ M^{-1}$.

$$\frac{1}{[NO_2]} = kt + \frac{1}{[NO_2]_o},\quad \frac{1}{[NO_2]} = \frac{2.08 \times 10^{-4}\ L}{mol\ s} \times 2.70 \times 10^4\ s + 2.00\ M^{-1}$$

$$\frac{1}{[NO_2]} = 7.62,\quad [NO_2] = 0.131\ M$$

31.   a. Since the $[C_2H_5OH]$ vs. time plot was linear, the reaction is zero order in C_2H_5OH. The slope of the $[C_2H_5OH]$ vs. time plot equals -k. Therefore, the rate law, the integrated rate law and the rate constant value are: Rate $= k[C_2H_5OH]^0 = k$; $[C_2H_5OH] = -kt + [C_2H_5OH]_o$; $k = 4.00 \times 10^{-5}$ mol/L$\bullet$s.

      b. The half-life expression for a zero-order reaction is: $t_{1/2} = [A]_o/2k$.

$$t_{1/2} = \frac{[C_2H_5OH]_o}{2k} = \frac{1.25 \times 10^{-2}\ mol/L}{2 \times 4.00 \times 10^{-5}\ mol/L\bullet s} = 156\ s$$

      Note: we could have used the integrated rate law to solve for $t_{1/2}$ where $[C_2H_5OH] = (1.25 \times 10^{-2}/2)$ mol/L.

      c. $[C_2H_5OH] = -kt + [C_2H_5OH]_o$, 0 mol/L $= -(4.00 \times 10^{-5}$ mol/L$\bullet$s) t $+ 1.25 \times 10^{-2}$ mol/L

$$t = \frac{1.25 \times 10^{-2}\ mol/L}{4.00 \times 10^{-5}\ mol/L\bullet s} = 313\ s$$

33.   The first assumption to make is that the reaction is first order. For a first-order reaction, a graph of ln $[C_4H_6]$ vs. t should yield a straight line. If this isn't linear, then try the second-order plot of $1/[C_4H_6]$ vs. t. The data and the plots follow.

Time	195	604	1246	2180	6210 s
$[C_4H_6]$	1.6×10^{-2}	1.5×10^{-2}	1.3×10^{-2}	1.1×10^{-2}	$0.68 \times 10^{-2}\ M$
ln $[C_4H_6]$	-4.14	-4.20	-4.34	-4.51	-4.99
$1/[C_4H_6]$	62.5	66.7	76.9	90.9	$147\ M^{-1}$

Note: To reduce round-off error, we carried extra sig. figs. in the data points.

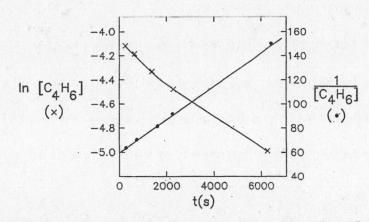

The natural log plot is not linear, so the reaction is not first order. Since the second-order plot of $1/[C_4H_6]$ vs. t is linear, we can conclude that the reaction is second order in butadiene. The rate law is:

$$Rate = k[C_4H_6]^2$$

For a second order reaction, the integrated rate law is: $\dfrac{1}{[C_4H_6]} = kt + \dfrac{1}{[C_4H_6]_o}$

The slope of the straight line equals the value of the rate constant. Using the points on the line at 1000. and 6000. s:

$$k = slope = \frac{144\ L/mol - 73\ L/mol}{6000.\ s - 1000.\ s} = 1.4 \times 10^{-2}\ L/mol \cdot s$$

35.   Since the $1/[A]$ vs. time plot is linear with a positive slope, the reaction is second order with respect to A. The y-intercept in the plot will equal $1/[A]_o$. Extending the plot, the y-intercept will be about 10, so $1/10 = 0.1\ M = [A]_o$.

37.   a. $[A] = -kt + [A]_o$, $[A] = -(5.0 \times 10^{-2}\ mol/L \cdot s)\ t + 1.0 \times 10^{-3}\ mol/L$

b.   The half-life expression for a zero-order reaction is: $t_{1/2} = \dfrac{[A]_o}{2\ k}$

$$t_{1/2} = \frac{1.0 \times 10^{-3}\ mol/L}{2 \times 5.0 \times 10^{-2}\ mol/L \cdot s} = 1.0 \times 10^{-2}\ s$$

c. $[A] = -5.0 \times 10^{-2}\ mol/L \cdot s \times 5.0 \times 10^{-3}\ s + 1.0 \times 10^{-3}\ mol/L = 7.5 \times 10^{-4}\ mol/L$

Since $7.5 \times 10^{-4}\ M$ A remains, $2.5 \times 10^{-4}\ M$ A reacted, which means that $2.5 \times 10^{-4}\ M$ B has been produced.

39.   a.   If the reaction is 38.5% complete, then 38.5% of the original concentration is consumed, leaving 61.5%.

$[A] = 61.5\%$ of $[A]_o$ or $[A] = 0.615\ [A]_o$; $\ln\left(\dfrac{[A]}{[A]_o}\right) = -kt$, $\ln\left(\dfrac{0.615\ [A]_o}{[A]_o}\right) = -k(480.\ s)$

$\ln(0.615) = -k(480.\ s)$, $-0.486 = -k(480.\ s)$, $k = 1.01 \times 10^{-3}\ s^{-1}$

b. $t_{1/2} = (\ln 2)/k = 0.6931/1.01 \times 10^{-3}\ s^{-1} = 686\ s$

c. 25% complete: $[A] = 0.75\ [A]_o$; $\ln(0.75) = -1.01 \times 10^{-3}\ (t)$, $t = 280\ s$

75% complete: $[A] = 0.25\ [A]_o$; $\ln(0.25) = -1.01 \times 10^{-3}\ (t)$, $t = 1.4 \times 10^3\ s$

Or, we know it takes $2 \times t_{1/2}$ for reaction to be 75% complete. $t = 2 \times 686\ s = 1370\ s$

95% complete: $[A] = 0.05\ [A]_o$; $\ln(0.05) = -1.01 \times 10^{-3}\ (t)$, $t = 3 \times 10^3\ s$

41.   For a second-order reaction: $t_{1/2} = \dfrac{1}{k[A]_o}$ or $k = \dfrac{1}{t_{1/2}[A]_o}$

$$k = \dfrac{1}{143\ s(0.060\ mol/L)} = 0.12\ L/mol \cdot s$$

43.   Successive half-lives increase in time for a second-order reaction. Therefore, assume the reaction is second order in A.

$$t_{1/2} = \dfrac{1}{k[A]_o},\ k = \dfrac{1}{t_{1/2}[A]_o} = \dfrac{1}{10.0\ min\ (0.10\ M)} = 1.0\ L/mol \cdot min$$

a. $\dfrac{1}{[A]} = kt + \dfrac{1}{[A]_o} = \dfrac{1.0\ L}{mol\ min} \times 80.0\ min + \dfrac{1}{0.10\ M} = 90.\ M^{-1}$, $[A] = 1.1 \times 10^{-2}\ M$

b.   30.0 min = 2 half-lives, so 25% of original A is remaining.

$[A] = 0.25(0.10\ M) = 0.025\ M$

Reaction Mechanisms

45.   For elementary reactions, the rate law can be written using the coefficients in the balanced equation to determine orders.

a. Rate = $k[CH_3NC]$ b. Rate = $k[O_3][NO]$

c. Rate = $k[O_3]$ d. Rate = $k[O_3][O]$

47.   A mechanism consists of a series of elementary reactions where the rate law for each step can be determined using the coefficients in the balanced equations. For a plausible mechanism, the rate law derived from a mechanism must agree with the rate law determined from experiment. To derive the rate law from the mechanism, the rate of the reaction is assumed to equal the rate of the slowest step in the mechanism.

Since step 1 is the rate-determining step, the rate law for this mechanism is: Rate = k[C₄H₉Br]. To get the overall reaction, we sum all the individual steps of the mechanism.

Summing all steps gives:

$$C_4H_9Br \rightarrow C_4H_9^+ + Br^-$$
$$C_4H_9^+ + H_2O \rightarrow C_4H_9OH_2^+$$
$$C_4H_9OH_2^+ + H_2O \rightarrow C_4H_9OH + H_3O^+$$

$$\overline{C_4H_9Br + 2\ H_2O \rightarrow C_4H_9OH + Br^- + H_3O^+}$$

Intermediates in a mechanism are species that are neither reactants nor products, but that are formed and consumed during the reaction sequence. The intermediates for this mechanism are $C_4H_9^+$ and $C_4H_9OH_2^+$.

Temperature Dependence of Rate Constants and the Collision Model

49.   In the following plot, R = reactants, P = products, E_a = activation energy and RC = reaction coordinate which is the same as reaction progress. Note for this reaction that ΔE is positive since the products are at a higher energy than the reactants.

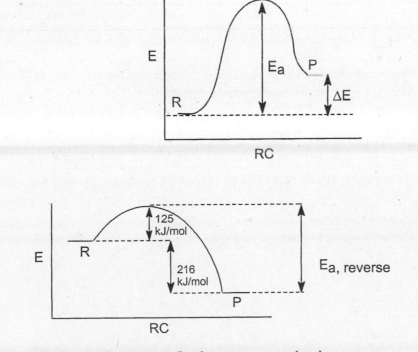

51.

The activation energy for the reverse reaction is:

$$E_{a,\ reverse} = 216\ kJ/mol + 125\ kJ/mol = 341\ kJ/mol$$

53.   The Arrhenius equation is: $k = A \exp(-E_a/RT)$ or in logarithmic form, $\ln k = -E_a/RT + \ln A$.
      Hence, a graph of $\ln k$ vs. $1/T$ should yield a straight line with a slope equal to $-E_a/R$ since the
      logarithmic form of the Arrhenius equation is in the form of a straight line equation, $y = mx + b$.
      Note: We carried extra significant figures in the following $\ln k$ values in order to reduce round off
      error.

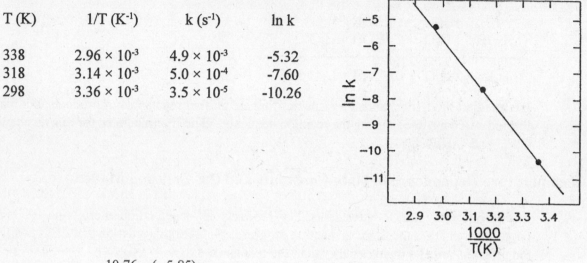

T (K)	1/T (K^{-1})	k (s^{-1})	ln k
338	2.96×10^{-3}	4.9×10^{-3}	-5.32
318	3.14×10^{-3}	5.0×10^{-4}	-7.60
298	3.36×10^{-3}	3.5×10^{-5}	-10.26

$$\text{Slope} = \frac{-10.76 - (-5.85)}{(3.40 \times 10^{-3} - 3.00 \times 10^{-3}) \text{ K}^{-1}} = -1.2 \times 10^4 \text{ K} = -E_a/R$$

$$E_a = -\text{slope} \times R = 1.2 \times 10^4 \text{ K} \times \frac{8.3145 \text{ J}}{\text{K mol}}, \quad E_a = 1.0 \times 10^5 \text{ J/mol} = 1.0 \times 10^2 \text{ kJ/mol}$$

55. $k = A \exp(-E_a/RT)$ or $\ln k = \dfrac{-E_a}{RT} + \ln A$ (the Arrhenius equation)

For two conditions: $\ln\left(\dfrac{k_2}{k_1}\right) = \dfrac{E_a}{R}\left(\dfrac{1}{T_1} - \dfrac{1}{T_2}\right)$ (Assuming A is temperature independent.)

Let $k_1 = 2.0 \times 10^3 \text{ s}^{-1}$, $T_1 = 298 \text{ K}$; $k_2 = ?$, $T_2 = 348 \text{ K}$; $E_a = 15.0 \times 10^3 \text{ J/mol}$

$$\ln\left(\frac{k_2}{2.0 \times 10^3 \text{ s}^{-1}}\right) = \frac{15.0 \times 10^3 \text{ J/mol}}{8.3145 \text{ J/mol} \cdot \text{K}}\left(\frac{1}{298 \text{ K}} - \frac{1}{348 \text{ K}}\right) = 0.87$$

$$\ln\left(\frac{k_2}{2.0 \times 10^3}\right) = 0.87, \quad \frac{k_2}{2.0 \times 10^3} = e^{0.87} = 2.4, \quad k_2 = 2.4(2.0 \times 10^3) = 4.8 \times 10^3 \text{ s}^{-1}$$

57. $\ln\left(\dfrac{k_2}{k_1}\right) = \dfrac{E_a}{R}\left(\dfrac{1}{T_1} - \dfrac{1}{T_2}\right); \quad \dfrac{k_2}{k_1} = 7.00,\ T_1 = 295\ K,\ E_a = 54.0 \times 10^3\ J/mol$

$\ln(7.00) = \dfrac{54.0 \times 10^3\ J/mol}{8.3145\ J/mol \cdot K}\left(\dfrac{1}{295\ K} - \dfrac{1}{T_2}\right),\ \dfrac{1}{295} - \dfrac{1}{T_2} = 3.00 \times 10^{-4}$

$\dfrac{1}{T_2} = 3.09 \times 10^{-3},\ T_2 = 324\ K = 51\,°C$

59. $H_3O^+(aq) + OH^-(aq) \rightarrow 2\ H_2O(l)$ should have the faster rate. H_3O^+ and OH^- will be electrostatically attracted to each other; Ce^{4+} and Hg_2^{2+} will repel each other (so E_a is much larger).

Catalysts

61. a. NO is the catalyst. NO is present in the first step of the mechanism on the reactant side, but it is not a reactant since it is regenerated in the second step.

   b. NO_2 is an intermediate. Intermediates also never appear in the overall balanced equation. In a mechanism, intermediates always appear first on the product side while catalysts always appear first on the reactant side.

   c. $k = A\ \exp(-E_a/RT); \quad \dfrac{k_{cat}}{k_{un}} = \dfrac{A\ \exp[-E_a(cat)/RT]}{A\ \exp[-E_a(un)/RT]} = \exp\left(\dfrac{E_a(un) - E_a(cat)}{RT}\right)$

   $\dfrac{k_{cat}}{k_{un}} = \exp\left(\dfrac{2100\ J/mol}{8.3145\ J/mol \cdot K \times 298\ K}\right) = e^{0.85} = 2.3$

   The catalyzed reaction is 2.3 times faster than the uncatalyzed reaction at 25°C.

63. The reaction at the surface of the catalyst is assumed to follow the steps:

Thus, CH_2D-CH_2D should be the product. If the mechanism is possible, then the reaction must be:

$$C_2H_4 + D_2 \rightarrow CH_2DCH_2D$$

If we got this product, then we could conclude that this is a possible mechanism. If we got some other product, e.g., CH_3CHD_2, then we would conclude that the mechanism is wrong. Even though this mechanism correctly predicts the products of the reaction, we cannot say conclusively that this is the correct mechanism; we might be able to conceive of other mechanisms that would give the same products as our proposed one.

Additional Exercises

65.   Rate = $k[NO]^x[O_2]^y$; comparing the first two experiments, $[O_2]$ is unchanged, $[NO]$ is tripled, and the rate increases by a factor of nine. Therefore, the reaction is second order in NO ($3^2 = 9$). The order of O_2 is more difficult to determine. Comparing the second and third experiments;

$$\frac{3.13 \times 10^{17}}{1.80 \times 10^{17}} = \frac{k\,(2.50 \times 10^{18})^2(2.50 \times 10^{18})^y}{k\,(3.00 \times 10^{18})^2(1.00 \times 10^{18})^y}, \; 1.74 = 0.694\,(2.50)^y, \; 2.51 = 2.50^y, \; y = 1$$

Rate = $k[NO]^2[O_2]$; From experiment 1:

$$2.00 \times 10^{16} \text{ molecules/cm}^3 \text{•s} = k\,(1.00 \times 10^{18} \text{ molecules/cm}^3)^2\,(1.00 \times 10^{18} \text{ molecules/cm}^3)$$

$$k = 2.00 \times 10^{-38} \text{ cm}^6/\text{molecules}^2\text{•s} = k_{mean}$$

$$\text{Rate} = \frac{2.00 \times 10^{-38} \text{ cm}^6}{\text{molecules}^2\text{•s}} \times \left(\frac{6.21 \times 10^{18} \text{ molecules}}{\text{cm}^3} \right)^2 \times \frac{7.36 \times 10^{18} \text{ molecules}}{\text{cm}^3}$$

$$= 5.68 \times 10^{18} \text{ molecules/cm}^3\text{•s}$$

67.   From 338 K data, a plot of $\ln[N_2O_5]$ vs. t is linear and the slope = -4.86×10^{-3} (plot not included). This tells us the reaction is first order in N_2O_5 with $k = 4.86 \times 10^{-3}$ at 338 K.

From 318 K data, the slope of $\ln[N_2O_5]$ vs t plot is equal to -4.98×10^{-4}, so $k = 4.98 \times 10^{-4}$ at 318 K. We now have two values of k at two temperatures, so we can solve for E_a.

$$\ln\left(\frac{k_2}{k_1}\right) = \frac{E_a}{R}\left(\frac{1}{T_1} - \frac{1}{T_2}\right), \; \ln\left(\frac{4.86 \times 10^{-3}}{4.98 \times 10^{-4}}\right) = \frac{E_a}{8.3145 \text{ J/K • mol}}\left(\frac{1}{318 \text{ K}} - \frac{1}{338 \text{ K}}\right)$$

$$E_a = 1.0 \times 10^5 \text{ J/mol} = 1.0 \times 10^2 \text{ kJ/mol}$$

69.   At high [S], the enzyme is completely saturated with substrate. Once the enzyme is completely saturated, the rate of decomposition of ES can no longer increase, and the overall rate remains constant.

Challenge Problems

71.   Rate = $k[I^-]^x[OCl^-]^y[OH^-]^z$; Comparing the first and second experiments:

$$\frac{18.7 \times 10^{-3}}{9.4 \times 10^{-3}} = \frac{k(0.0026)^x\,(0.012)^y\,(0.10)^z}{k(0.0013)^x\,(0.012)^y\,(0.10)^z}, \; 2.0 = 2.0^x, \; x = 1$$

Comparing the first and third experiments:

$$\frac{9.4 \times 10^{-3}}{4.7 \times 10^{-3}} = \frac{k(0.0013)\,(0.012)^y\,(0.10)^z}{k(0.0013)\,(0.0060)^y\,(0.10)^z}, \; 2.0 = 2.0^y, \; y = 1$$

Comparing the first and sixth experiments:

$$\frac{4.7 \times 10^{-3}}{9.4 \times 10^{-3}} = \frac{k(0.0013)\,(0.012)\,(0.20)^z}{k(0.0013)\,(0.012)\,(0.10)^z}, \quad 1/2 = 2.0^z, \quad z = -1$$

Rate $= \dfrac{k[I^-][OCl^-]}{[OII^-]}$; The presence of OH^- decreases the rate of the reaction.

For the first experiment:

$$\frac{9.4 \times 10^{-3}\ \text{mol}}{L\ s} = k\,\frac{(0.0013\ \text{mol/L})\,(0.012\ \text{mol/L})}{(0.10\ \text{mol/L})}, \quad k = 60.3\ s^{-1} = 60.\ s^{-1}$$

For all experiments, $k_{mean} = 60.\ s^{-1}$.

73.   a.   We check for first-order dependence by graphing ln [concentration] vs. time for each set of data. The rate dependence on NO is determined from the first set of data since the ozone concentration is relatively large compared to the NO concentration, so $[O_3]$ is effectively constant.

Time (ms)	[NO] (molecules/cm³)	ln [NO]
0	6.0×10^8	20.21
100.	5.0×10^8	20.03
500.	2.4×10^8	19.30
700.	1.7×10^8	18.95
1000.	9.9×10^7	18.41

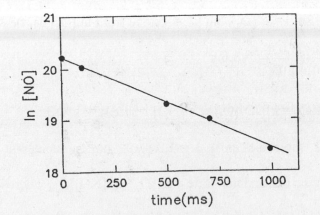

Since ln [NO] vs. t is linear, the reaction is first order with respect to NO.

We follow the same procedure for ozone using the second set of data. The data and plot are:

Time (ms)	$[O_3]$ (molecules/cm³)	$\ln [O_3]$
0	1.0×10^{10}	23.03
50.	8.4×10^9	22.85
100.	7.0×10^9	22.67
200.	4.9×10^9	22.31
300.	3.4×10^9	21.95

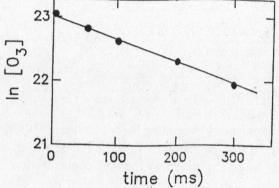

The plot of $\ln [O_3]$ vs. t is linear. Hence, the reaction is first order with respect to ozone.

b.   Rate = $k[NO][O_3]$ is the overall rate law.

c.   For NO experiment, Rate = $k'[NO]$ and k' = -(slope from graph of $\ln [NO]$ vs. t).

$$k' = \text{-slope} = -\frac{18.41 - 20.21}{(1000. - 0) \times 10^{-3} \text{ s}} = 1.8 \text{ s}^{-1}$$

For ozone experiment, Rate = $k''[O_3]$ and k'' = -(slope from $\ln [O_3]$ vs. t).

$$k'' = \text{-slope} = -\frac{(21.95 - 23.03)}{(300. - 0) \times 10^{-3} \text{ s}} = 3.6 \text{ s}^{-1}$$

d.   From NO experiment, Rate = $k[NO][O_3] = k'[NO]$ where $k' = k[O_3]$.

$k' = 1.8 \text{ s}^{-1} = k(1.0 \times 10^{14} \text{ molecules/cm}^3)$, $k = 1.8 \times 10^{-14} \text{ cm}^3/\text{molecules}\bullet\text{s}$

We can check this from the ozone data. Rate = $k''[O_3] = k[NO][O_3]$ where $k'' = k[NO]$.

$k'' = 3.6 \text{ s}^{-1} = k(2.0 \times 10^{14} \text{ molecules/cm}^3)$, $k = 1.8 \times 10^{-14} \text{ cm}^3/\text{molecules}\bullet\text{s}$

Both values of k agree.

75.   a.   If the interval between flashes is 16.3 sec, then the rate is:

1 flash/16.3 s = 6.13×10^{-2} s^{-1} = k

Interval	k	T
16.3 s	6.13×10^{-2} s^{-1}	21.0°C (294.2 K)
13.0 s	7.69×10^{-2} s^{-1}	27.8°C (301.0 K)

$\ln\left(\dfrac{k_2}{k_1}\right) = \dfrac{E_a}{R}\left(\dfrac{1}{T_1} - \dfrac{1}{T_2}\right)$; Solving: $E_a = 2.5 \times 10^4$ J/mol = 25 kJ/mol

b. $\ln\left(\dfrac{k}{6.13 \times 10^{-2}}\right) = \dfrac{2.5 \times 10^4 \text{ J/mol}}{8.3145 \text{ J/K} \bullet \text{mol}}\left(\dfrac{1}{294.2 \text{ K}} - \dfrac{1}{303.2 \text{ K}}\right) = 0.30$

$k = e^{0.30} \times 6.13 \times 10^{-2} = 8.3 \times 10^{-2}$ s^{-1}; Interval = 1/k = 12 seconds

c.

T	Interval	54-2(Intervals)
21.0 °C	16.3 s	21 °C
27.8 °C	13.0 s	28 °C
30.0 °C	12 s	30. °C

This rule of thumb gives excellent agreement to two significant figures.

CHAPTER THIRTEEN

CHEMICAL EQUILIBRIUM

Questions

9.   a.   The rates of the forward and reverse reactions are equal at equilibrium.

 b.   There is no net change in the composition (as long as temperature is constant).

11.   The equilibrium constant is a number that tells us the relative concentrations (pressures) of reactants and products at equilibrium. An equilibrium position is a set of concentrations that satisfy the equilibrium constant expression. More than one equilibrium position can satisfy the same equilibrium constant expression.

 Table 13.1 of the text illustrates this nicely. Each of the three experiments in Table 13.1 have different equilibrium positions; that is, each experiment has different equilibrium concentrations. However, when these equilibrium concentrations are inserted into the equilibrium constant expression, each experiment gives the same value for K. The equilibrium position depends on the initial concentrations one starts with. Since there are an infinite number of initial conditions, there are an infinite number of equilibrium positions. However, each of these infinite equilibrium positions will always give the same value for the equilibrium constant (assuming temperature is constant).

13.   For the gas phase reaction $a\,A + b\,B \rightleftharpoons c\,C + d\,D$

 the equilibrium constant expression is: $K = \dfrac{[C]^c\,[D]^d}{[A]^a\,[B]^b}$

 and the reaction quotient has the same form: $Q = \dfrac{[C]^c\,[D]^d}{[A]^a\,[B]^b}$

The difference is that in the expression for K we use equilibrium concentrations, i.e., [A], [B], [C], and [D] are all in equilibrium with each other. Any set of concentrations can be plugged into the reaction quotient expression. Typically, we plug initial concentrations into the Q expression and then compare the value of Q to K to see how far we are from equilibrium. If Q = K, the reaction is at equilibrium with these concentrations. If Q ≠ K, then the reaction will have to shift either to products or to reactants to reach equilibrium.

Exercises

Characteristics of Chemical Equilibrium

15.   No, equilibrium is a dynamic process. Both reactions:

$$H_2O + CO \rightarrow H_2 + CO_2 \text{ and } H_2 + CO_2 \rightarrow H_2O + CO$$

are occurring, but at equal rates. Thus, ^{14}C atoms will be distributed between CO and CO_2.

17. $H_2O(g) + CO(g) \rightleftharpoons H_2(g) + CO_2(g) \quad K = \dfrac{[H_2][CO_2]}{[H_2O][CO]} = 2.0$

K is a unitless number since there is an equal number of moles of product gases as compared to moles of reactant gases in the balanced equation. Therefore, we can use units of molecules per liter instead of moles per liter to determine K.

By trial and error, if 3 molecules of CO react, then 3 molecules of H_2O must react, and 3 molecules each of H_2 and CO_2 are formed. We would have 6 - 3 = 3 molecules CO, 8 - 3 = 5 molecules H_2O, 0 + 3 = 3 molecules H_2, and 0 + 3 = 3 molecules CO_2 present. This will be an equilibrium mixture if K = 2.0:

$$K = \dfrac{\left(\dfrac{3 \text{ molecules } H_2}{L}\right)\left(\dfrac{3 \text{ molecules } CO_2}{L}\right)}{\left(\dfrac{5 \text{ molecules } H_2O}{L}\right)\left(\dfrac{3 \text{ molecules } CO}{L}\right)} = \dfrac{3}{5}$$

Since this mixture does not give a value of K = 2.0, this is not an equilibrium mixture. Let's try 4 molecules of CO reacting to reach equilibrium.

molecules CO remaining = 6 - 4 = 2 molecules CO;
molecules H_2O remaining = 8 - 4 = 4 molecules H_2O;
molecules H_2 present = 0 + 4 = 4 molecules H_2;
molecules CO_2 present = 0 + 4 = 4 molecules CO_2

$$K = \dfrac{\left(\dfrac{4 \text{ molecules } H_2}{L}\right)\left(\dfrac{4 \text{ molecules } CO_2}{L}\right)}{\left(\dfrac{4 \text{ molecules } H_2O}{L}\right)\left(\dfrac{2 \text{ molecules } CO}{L}\right)} = 2.0$$

Since K = 2.0 for this reaction mixture, we are at equilibrium.

The Equilibrium Constant

19.   a. $K = \dfrac{[NO]^2}{[N_2][O_2]}$ b. $K = \dfrac{[NO_2]^2}{[N_2O_4]}$

c. $K = \dfrac{[SiCl_4][H_2]^2}{[SiH_4][Cl_2]^2}$ d. $K = \dfrac{[PCl_3]^2[Br_2]^3}{[PBr_3]^2[Cl_2]^3}$

21. $K = 1.3 \times 10^{-2} = \dfrac{[NH_3]^2}{[N_2][H_2]^3}$ for $N_2(g) + 3\ H_2(g) \rightleftharpoons 2\ NH_3(g)$

When a reaction is reversed, then $K_{new} = 1/K_{original}$. When a reaction is multiplied through by a value of n, then $K_{new} = (K_{original})^n$.

a. $1/2\ N_2(g) + 3/2\ H_2(g) \rightleftharpoons NH_3(g)$ $K' = \dfrac{[NH_3]}{[N_2]^{1/2}[H_2]^{3/2}} = K^{1/2} = (1.3 \times 10^{-2})^{1/2} = 0.11$

b. $2\ NH_3(g) \rightleftharpoons N_2(g) + 3\ H_2(g)$ $K'' = \dfrac{[N_2][H_2]^3}{[NH_3]^2} = \dfrac{1}{K} = \dfrac{1}{1.3 \times 10^{-2}} = 77$

c. $NH_3(g) \rightleftharpoons 1/2\ N_2(g) + 3/2\ H_2(g)$ $K''' = \dfrac{[N_2]^{1/2}[H_2]^{3/2}}{[NH_3]} = \left(\dfrac{1}{K}\right)^{1/2} = \left(\dfrac{1}{1.3 \times 10^{-2}}\right)^{1/2} = 8.8$

d. $2\ N_2(g) + 6\ H_2(g) \rightleftharpoons 4\ NH_3(g)$ $K = \dfrac{[NH_3]^4}{[N_2]^2[H_2]^6} = (K)^2 = (1.3 \times 10^{-2})^2 = 1.7 \times 10^{-4}$

23. $K = \dfrac{[NO]^2}{[N_2][O_2]} = \dfrac{(4.7 \times 10^{-4})^2}{(0.041)(0.0078)} = 6.9 \times 10^{-4}$

25. $[NO] = \dfrac{4.5 \times 10^{-3}\ mol}{3.0\ L} = 1.5 \times 10^{-3}\ M$; $[Cl_2] = \dfrac{2.4\ mol}{3.0\ L} = 0.80\ M$

$[NOCl] = \dfrac{1.0\ mol}{3.0\ L} = 0.33\ M$; $K = \dfrac{[NO]^2[Cl_2]}{[NOCl]^2} = \dfrac{(1.5 \times 10^{-3})^2(0.80)}{(0.33)^2} = 1.7 \times 10^{-5}$

27. $K_p = \dfrac{P_{NO}^2 \times P_{O_2}}{P_{NO_2}^2} = \dfrac{(6.5 \times 10^{-5})^2(4.5 \times 10^{-5})}{(0.55)^2} = 6.3 \times 10^{-13}$

29. $K_p = K(RT)^{\Delta n}$ where Δn = sum of gaseous product coefficients - sum of gaseous reactant coefficients. For this reaction, $\Delta n = 1 - 2 = -1$.

$$K_p = \frac{3.7 \times 10^9}{(0.08206 \times 298)} = 1.5 \times 10^8$$

31.   Solids and liquids do not appear in the equilibrium expression. Only gases and dissolved solutes appear in the equilibrium expression.

a. $K = \dfrac{1}{[O_2]^5}$; $K_p = \dfrac{1}{P_{O_2}^5}$

b. $K = [N_2O][H_2O]^2$; $K_p = P_{N_2O} \times P_{H_2O}^2$

c. $K = \dfrac{1}{[CO_2]}$; $K_p = \dfrac{1}{P_{CO_2}}$

d. $K = \dfrac{[SO_2]^8}{[O_2]^8}$; $K_p = \dfrac{P_{SO_2}^8}{P_{O_2}^8}$

33.   Since solids do not appear in the equilibrium constant expression, $K = 1/[O_2]^3$.

$$[O_2] = \frac{1.0 \times 10^{-3} \text{ mol}}{2.0 \text{ L}}; \quad K = \frac{1}{[O_2]^3} = \frac{1}{\left(\dfrac{1.0 \times 10^{-3}}{2.0}\right)^3} = \frac{1}{(5.0 \times 10^{-4})^3} = 8.0 \times 10^9$$

Equilibrium Calculations

35. $2 \text{ NO(g)} \rightleftharpoons \text{N}_2\text{(g)} + \text{O}_2\text{(g)}$ $K = \dfrac{[N_2][O_2]}{[NO]^2} = 2.4 \times 10^3$

Use the reaction quotient Q to determine which way the reaction shifts to reach equilibrium. For the reaction quotient, initial concentrations given in a problem are used to calculate the value for Q. If $Q < K$, then the reaction shifts right to reach equilibrium. If $Q > K$, then the reaction shifts left to reach equilibrium. If $Q = K$, then the reaction does not shift in either direction since the reaction is at equilibrium.

a. $[N_2] = \dfrac{2.0 \text{ mol}}{1.0 \text{ L}} = 2.0 \, M$; $[O_2] = \dfrac{2.6 \text{ mol}}{1.0 \text{ L}} = 2.6 \, M$; $[NO] = \dfrac{0.024 \text{ mol}}{1.0 \text{ L}} = 0.024 \, M$

$$Q = \frac{[N_2]_0[O_2]_0}{[NO]_0^2} = \frac{(2.0)(2.6)}{(0.024)^2} = 9.0 \times 10^3$$

$Q > K$ so the reaction shifts left to produce more reactants in order to reach equilibrium.

b. $[N_2] = \dfrac{0.62 \text{ mol}}{2.0 \text{ L}} = 0.31 \, M$; $[O_2] = \dfrac{4.0 \text{ mol}}{2.0 \text{ L}} = 2.0 \, M$; $[NO] = \dfrac{0.032 \text{ mol}}{2.0 \text{ L}} = 0.016 \, M$

$$Q = \frac{(0.31)(2.0)}{(0.016)^2} = 2.4 \times 10^3 = K; \text{ at equilibrium}$$

c. $[N_2] = \dfrac{2.4 \text{ mol}}{3.0 \text{ L}} = 0.80\,M;\quad [O_2] = \dfrac{1.7 \text{ mol}}{3.0 \text{ L}} = 0.57\,M;\quad [NO] = \dfrac{0.060 \text{ mol}}{3.0 \text{ L}} = 0.020\,M$

$Q = \dfrac{(0.80)(0.57)}{(0.020)^2} = 1.1 \times 10^3 < K;$ Reaction shifts right to reach equilibrium.

37. $CaCO_3(s) \rightleftharpoons CaO(s) + CO_2(g)\quad K_p = P_{CO_2} = 1.04$

a. $Q = P_{CO_2};$ We only need the partial pressure of CO_2 to determine Q since solids do not appear in equilibrium expressions (or Q expressions). At this temperature all CO_2 will be in the gas phase. $Q = 2.55$ so $Q > K_p;$ Reaction will shift to the left to reach equilibrium; the mass of CaO will decrease.

b. $Q = 1.04 = K_p$ so the reaction is at equilibrium; mass of CaO will not change.

c. $Q = 1.04 = K_p$ so the reaction is at equilibrium; mass of CaO will not change.

d. $Q = 0.211 < K_p;$ The reaction will shift to the right to reach equilibrium; the mass of CaO will increase.

39. $K = \dfrac{[HF]^2}{[H_2][F_2]} = 2.1 \times 10^3;\quad 2.1 \times 10^3 = \dfrac{[HF]^2}{(0.0021)(0.0021)},\quad [HF] = 0.096\,M$

41. $SO_2(g) + NO_2(g) \rightleftharpoons SO_3(g) + NO(g)\quad K = \dfrac{[SO_3][NO]}{[SO_2][NO_2]}$

To determine K, we must calculate the equilibrium concentrations. The initial concentrations are:

$[SO_3]_o = [NO]_o = 0;\quad [SO_2]_o = [NO_2]_o = \dfrac{2.00 \text{ mol}}{1.00 \text{ L}} = 2.00\,M$

Next, we determine the change required to reach equilibrium. At equilibrium, $[NO] = 1.30$ mol/1.00 L = 1.30 M. Since there was zero NO present initially, 1.30 M of SO_2 and 1.30 M NO_2 must have reacted to produce 1.30 M NO as well as 1.30 M SO_3, all required by the balanced reaction. The equilibrium concentration for each substance is the sum of the initial concentration plus the change in concentration necessary to reach equilibrium. The equilibriumoncentrations are:

$[SO_3] = [NO] = 0 + 1.30\,M = 1.30\,M;\quad [SO_2] = [NO_2] = 2.00\,M - 1.30\,M = 0.70\,M$

We now use these equilibrium concentrations to calculate K:

$K = \dfrac{[SO_3][NO]}{[SO_2][NO_2]} = \dfrac{(1.30)(1.30)}{(0.70)(0.70)} = 3.4$

43. When solving equilibrium problems, a common method to summarize all the information in the problem is to set up a table. We call this table the ICE table since it summarizes initial concentrations, changes that must occur to reach equilibrium and equilibrium concentrations (the sum of the initial and change columns). For the change column, we will generally use the variable x, which will be defined as the amount of reactant (or product) that must react to reach equilibrium. In this problem, the reaction must shift right to reach equilibrium since there are no products present initially. Therefore, x is defined as the amount of reactant SO_3 that reacts to reach equilibrium, and we use the coefficients in the balanced equation to relate the net change in SO_3 to the net change in SO_2 and O_2. The general ICE table for this problem is:

$$2\,SO_3(g) \;\rightleftharpoons\; 2\,SO_2(g) \;+\; O_2(g) \qquad K = \frac{[SO_2]^2[O_2]}{[SO_3]^2}$$

Initial   12.0 mol/3.0 L   0   0
            Let x mol/L of SO_3 react to reach equilibrium
Change $-x$ $\rightarrow$ $+x$ $+x/2$
Equil. $4.0 - x$ x $x/2$

From the problem, we are told that the equilibrium SO_2 concentration is 3.0 mol/3.0 L = 1.0 M ($[SO_2]_e = 1.0\ M$). From the ICE table, $[SO_2]_e = x$ so $x = 1.0$. Solving for the other equilibrium concentrations: $[SO_3]_e = 4.0 - x = 4.0 - 1.0 = 3.0\ M$; $[O_2] = x/2 = 1.0/2 = 0.50\ M$.

$$K = \frac{[SO_2]^2[O_2]}{[SO_3]^2} = \frac{(1.0)^2(0.50)}{(3.0)^2} = 0.056$$

Alternate Method: Fractions in the change column can be avoided (if you want) by defining x differently. If we were to let $2x$ mol/L of SO_3 react to reach equilibrium, then the ICE table is:

$$2\,SO_3(g) \;\rightleftharpoons\; 2\,SO_2(g) \;+\; O_2(g) \qquad K = \frac{[SO_2]^2[O_2]}{[SO_3]^2}$$

Initial   4.0 M 0   0
            Let $2x$ mol/L of SO_3 react to reach equilibrium
Change $-2x$ $\rightarrow$ $+2x$ $+x$
Equil. $4.0 - 2x$ $2x$ x

Solving: $2x = [SO_2]_e = 1.0\ M$, $x = 0.50\ M$; $[SO_3]_e = 4.0 - 2(0.50) = 3.0\ M$; $[O_2]_e = x = 0.50\ M$

These are exactly the same equilibrium concentrations as solved for previously, thus K will be the same (as it must be). The moral of the story is to define x in a manner that is most comfortable for you. Your final answer is independent of how you define x initially.

45. Q = 1.00, which is less than K. Reaction shifts to the right to reach equilibrium. Summarizing the equilibrium problem in a table:

$$SO_2(g) \quad + \quad NO_2(g) \quad \rightleftharpoons \quad SO_3(g) \quad + \quad NO(g) \quad K = 3.75$$

Initial	0.800 M	0.800 M	0.800 M	0.800 M

x mol/L of SO_2 reacts to reach equilibrium

Change	$-x$	$-x \rightarrow$	$+x$	$+x$
Equil.	$0.800 - x$	$0.800 - x$	$0.800 + x$	$0.800 + x$

Plug the equilibrium concentrations into the equilibrium constant expression:

$$K = \frac{[SO_3][NO]}{[SO_2][NO_2]} = 3.75 = \frac{(0.800 + x)^2}{(0.800 - x)^2}; \text{ Take the square root of both sides and solve for } x:$$

$$\frac{0.800 + x}{0.800 - x} = 1.94, \ 0.800 + x = 1.55 - 1.94\,x, \ 2.94\,x = 0.75, \ x = 0.26\ M$$

The equilibrium concentrations are:

$$[SO_3] = [NO] = 0.800 + x = 0.800 + 0.26 = 1.06\ M; \ [SO_2] = [NO_2] = 0.800 - x = 0.54\ M$$

47. Since only reactants are present initially, the reaction must proceed to the right to reach equilibrium. Summarizing the problem in a table:

$$N_2(g) \quad + \quad O_2(g) \quad \rightleftharpoons \quad 2\ NO(g) \quad K_p = 0.050$$

Initial	0.80 atm	0.20 atm	0

x atm of N_2 reacts to reach equilibrium

Change	$-x$	$-x \rightarrow$	$+2x$
Equil.	$0.80 - x$	$0.20 - x$	$2x$

$$K_p = 0.050 = \frac{P_{NO}^2}{P_{N_2} \times P_{O_2}} = \frac{(2x)^2}{(0.80 - x)(0.20 - x)}, \ 0.050(0.16 - 1.00\,x + x^2) = 4\,x^2$$

$$4\,x^2 = 8.0 \times 10^{-3} - 0.050\,x + 0.050\,x^2, \ 3.95\,x^2 + 0.050\,x - 8.0 \times 10^{-3} = 0$$

Solving using the quadratic formula (see Appendix 1.4 of the text):

$$x = \frac{-b \pm (b^2 - 4ac)^{1/2}}{2a} = \frac{-0.050 \pm [(0.050)^2 - 4(3.95)(-8.0 \times 10^{-3})]^{1/2}}{2(3.95)}$$

$x = 3.9 \times 10^{-2}$ atm or $x = -5.2 \times 10^{-2}$ atm; Only $x = 3.9 \times 10^{-2}$ atm makes sense (x cannot be negative), so the equilibrium NO partial pressure is:

$$P_{NO} = 2x = 2(3.9 \times 10^{-2}\text{ atm}) = 7.8 \times 10^{-2}\text{ atm}$$

49. $2 SO_2(g)$ + $O_2(g)$ $\rightleftharpoons$ $2 SO_3(g)$ $K_p = 0.25$

Initial   0.50 atm   0.50 atm   0
               $2x$ atm of SO_2 reacts to reach equilibrium
Change   -2x   -x $\rightarrow$ +2x
Equil.   0.50 - 2x   0.50 - x   2x

$-0.00763 \leq x$

$$K_p = 0.25 = \frac{P_{SO_3}^2}{P_{SO_2}^2 \times P_{O_2}} = \frac{(2x)^2}{(0.50 - 2x)^2(0.50 - x)}$$

This will give a cubic equation. Graphing calculators can be used to solve this expression. If you don't have a graphing calculator, an alternative method for solving a cubic equation is to use the method of successive approximations (see Appendix 1.4 of the text). The first step is to guess a value for x. Since the value of K is small (K < 1), then not much of the forward reaction will occur to reach equilibrium. This tells us that x is small. Lets guess that $x = 0.050$ atm. Now we take this estimated value for x and substitute it into the equation everywhere that x appears except for one. For equilibrium problems, we will substitute the estimated value for x into the denominator, then solve for the numerator value of x. We continue this process until the estimated value of x and the calculated value of x converge on the same number. This is the same answer we would get if we were to solve the cubic equation exactly. Applying the method of successive approximations and carrying extra significant figures:

$$\frac{4x^2}{[0.50 - 2(0.050)]^2 [0.50 - (0.050)]} = \frac{4x^2}{(0.40)^2(0.45)} = 0.25, \; x = 0.067$$

$$\frac{4x^2}{[0.50 - 2(0.067)]^2 [0.50 - (0.067)]} = \frac{4x^2}{(0.366)^2(0.433)} = 0.25, \; x = 0.060$$

$$\frac{4x^2}{(0.38)^2(0.44)} = 0.25, \; x = 0.063; \quad \frac{4x^2}{(0.374)^2(0.437)} = 0.25, \; x = 0.062$$

The next trial gives the same value for $x = 0.062$ atm. We are done except for determining the equilibrium concentrations. They are:

$P_{SO_2} = 0.50 - 2x = 0.50 - 2(0.062) = 0.376 = 0.38$ atm

$P_{O_2} = 0.50 - x = 0.438 = 0.44$ atm; $P_{SO_3} = 2x = 0.124 = 0.12$ atm

51.   a.   The reaction must proceed to products to reach equilibrium since only reactants are present initially. Summarizing the problem in a table:

	$2\ NOCl(g)$	$\rightleftharpoons$	$2\ NO(g)$	$+$	$Cl_2(g)$	$K = 1.6 \times 10^{-5}$
Initial	$\dfrac{2.0\ \text{mol}}{2.0\ \text{L}}$		0		0	

$2x$ mol/L of NOCl reacts to reach equilibrium

Change	$-2x$	$\rightarrow$	$+2x$	$+x$	
Equil.	$1.0 - 2x$		$2x$	x	

$$K = 1.6 \times 10^{-5} = \frac{[NO]^2[Cl_2]}{[NOCl]^2} = \frac{(2x)^2\,(x)}{(1.0 - 2x)^2}$$

If we assume that $1.0 - 2x \approx 1.0$ (from the size of K, we know that not much reaction will occur so x is small), then:

$$1.6 \times 10^{-5} = \frac{4x^3}{1.0^2}, \quad x = 1.6 \times 10^{-2}; \text{ Now we must check the assumption.}$$

$$1.0 - 2x = 1.0 - 2(0.016) = 0.97 = 1.0 \text{ (to proper significant figures)}$$

Our error is about 3%, i.e., $2x$ is 3.2% of 1.0 M. Generally, if the error we introduce by making simplifying assumptions is less than 5%, we go no further and the assumption is said to be valid. We call this the 5% rule. Solving for the equilibrium concentrations:

$$[NO] = 2x = 0.032\ M; \ \ [Cl_2] = x = 0.016\ M; \ \ [NOCl] = 1.0 - 2x = 0.97\ M \approx 1.0\ M$$

Note: If we were to solve this cubic equation exactly (a long and tedious process), we would get $x = 0.016$. This is the exact same answer we determined by making a simplifying assumption. We saved time and energy. Whenever K is a very small value, always make the assumption that x is small. If the assumption introduces an error of less than 5%, then the answer you calculated making the assumption will be considered the correct answer.

   b.

	$2\ NOCl(g)$	$\rightleftharpoons$	$2\ NO(g)$	$+$	$Cl_2(g)$
Initial	$1.0\ M$		$1.0\ M$	0	

$2x$ mol/L of NOCl reacts to reach equilibrium

Change	$-2x$	$\rightarrow$	$+2x$	$+x$	
Equil.	$1.0 - 2x$		$1.0 + 2x$	x	

$$1.6 \times 10^{-5} = \frac{(1.0 + 2x)^2(x)}{(1.0 - 2x)^2} \approx \frac{(1.0)^2(x)}{(1.0)^2} \quad \text{(assuming } 2x \ll 1.0)$$

$x = 1.6 \times 10^{-5}$; Assumptions are great ($2x$ is 3.2×10^{-3}% of 1.0).

$[Cl_2] = 1.6 \times 10^{-5}\ M$ and $[NOCl] = [NO] = 1.0\ M$

c. $2\ NOCl(g)$ $\rightleftharpoons$ $2\ NO(g)$ + $Cl_2(g)$

Initial   2.0 M 0   1.0 M
           $2x$ mol/L of NOCl reacts to reach equilibrium
Change $-2x$ $\rightarrow$ $+2x$ $+x$
Equil. $2.0 - 2x$ $2x$ $1.0 + x$

$$1.6 \times 10^{-5} = \frac{(2x)^2(1.0 + x)}{(2.0 - 2x)^2} \approx \frac{4x^2}{4.0}\quad \text{(assuming } x \ll 1.0)$$

Solving: $x = 4.0 \times 10^{-3}$; Assumptions good (x is 0.4% of 1.0 and $2x$ is 0.4% of 2.0).

$[Cl_2] = 1.0 + x = 1.0\ M$; $[NO] = 2(4.0 \times 10^{-3}) = 8.0 \times 10^{-3}$; $[NOCl] = 2.0\ M$

53. $2\ CO_2(g)$ $\rightleftharpoons$ $2\ CO(g)$ + $O_2(g)$ $K = \dfrac{[CO]^2[O_2]}{[CO_2]^2} = 2.0 \times 10^{-6}$

Initial   2.0 mol/5.0 L   0   0
           $2x$ mol/L of CO_2 reacts to reach equilibrium
Change $-2x$ $\rightarrow$ $+2x$ $+x$
Equil. $0.40 - 2x$ $2x$ x

$$K - 2.0 \times 10^{-6} = \frac{[CO]^2[O_2]}{[CO_2]^2} = \frac{(2x)^2(x)}{(0.40 - 2x)^2};\ \text{Assuming } 2x \ll 0.40:$$

$$2.0 \times 10^{-6} \approx \frac{4x^3}{(0.40)^2},\ 2.0 \times 10^{-6} = \frac{4x^3}{0.16},\ x = 4.3 \times 10^{-3}\ M$$

Checking assumption: $\dfrac{2(4.3 \times 10^{-3})}{0.40} \times 100 = 2.2\%$; Assumption valid by the 5% rule.

$[CO_2] = 0.40 - 2x = 0.40 - 2(4.3 \times 10^{-3}) = 0.39\ M$

$[CO] = 2x = 2(4.3 \times 10^{-3}) = 8.6 \times 10^{-3}\ M$; $[O_2] = x = 4.3 \times 10^{-3}\ M$

55.   This is a typical equilibrium problem except that the reaction contains a solid. Whenever solids and liquids are present, we basically ignore them in the equilibrium problem.

$NH_4OCONH_2(s)$ $\rightleftharpoons$ $2\ NH_3(g)$ + $CO_2(g)$ $K_p = 2.9 \times 10^{-3}$

Initial   -   0   0
           Some NH_4OCONH_2 decomposes to produce $2x$ atm of NH_3 and x atm of CO_2.
Change   - $\rightarrow$ $+2x$ $+x$
Equil.   - $2x$ x

$K_p = 2.9 \times 10^{-3} = P_{NH_3}^2 \times P_{CO_2} = (2x)^2(x) = 4x^3$

$$x = \left(\frac{2.9 \times 10^{-3}}{4} \right)^{1/3} = 9.0 \times 10^{-2} \text{ atm}; \quad P_{NH_3} = 2x = 0.18 \text{ atm}; \quad P_{CO_2} = x = 9.0 \times 10^{-2} \text{ atm}$$

$$P_{total} = P_{NH_3} + P_{CO_2} = 0.18 \text{ atm} + 0.090 \text{ atm} = 0.27 \text{ atm}$$

Le Chatelier's Principle

57. a. No effect; Adding more of a pure solid or pure liquid has no effect on the equilibrium position.

  b. Shifts left; HF(g) will be removed by reaction with the glass. As HF(g) is removed, the reaction will shift left to produce more HF(g).

  c. Shifts right; As $H_2O(g)$ is removed, the reaction will shift right to produce more $H_2O(g)$.

59. a. right   b. right   c. no effect; He(g) is neither a reactant nor a product.

  d. left; Since the reaction is exothermic, heat is a product:

$$CO(g) + H_2O(g) \rightarrow H_2(g) + CO_2(g) + \text{Heat}$$

  Increasing T will add heat. The equilibrium shifts to the left to use up the added heat.

  e. no effect; Since there are equal numbers of gas molecules on both sides of the reaction, a change in volume has no effect on the equilibrium position.

61. a. left   b. right   c. left

  d. no effect (reactant and product concentrations are unchanged)

  e. no effect; Since there are equal numbers of product and reactant gas molecules, a change in volume has no effect on the equilibrium position.

  f. right; A decrease in temperature will shift the equilibrium to the right since heat is a product in this reaction (as is true in all exothermic reactions).

63. In an exothermic reaction, heat is a product. To maximize product yield , one would want as low a temperature as possible since high temperatures would shift the reaction left (away from products). Since temperature changes also change the value of K, then at low temperatures the value of K will be largest, which maximizes yield of products.

Additional Exercises

65.
$$
\begin{array}{ll}
O(g) + NO(g) \rightleftharpoons NO_2(g) & K = 1/6.8 \times 10^{-49} = 1.5 \times 10^{48} \\
NO_2(g) + O_2(g) \rightleftharpoons NO(g) + O_3(g) & K = 1/5.8 \times 10^{-34} = 1.7 \times 10^{33}
\end{array}
$$

$$O_2(g) + O(g) \rightleftharpoons O_3(g) \qquad\qquad K = (1.5 \times 10^{48})(1.7 \times 10^{33}) = 2.6 \times 10^{81}$$

67. $3 H_2(g)$ + $N_2(g)$ $\rightleftharpoons$ $2 NH_3(g)$

Initial $[H_2]_o$ $[N_2]_o$ 0
               x mol/L of N_2 reacts to reach equilibrium
Change $-3x$ $-x$ $\rightarrow$ $+2x$
Equil $[H_2]_o - 3x$ $[N_2]_o - x$ $2x$

From the problem:

$[NH_3]_e = 4.0\ M = 2x$, $x = 2.0\ M$; $[H_2]_e = 5.0\ M = [H_2]_o - 3x$; $[N_2]_e = 8.0\ M = [N_2]_o - x$

$5.0\ M = [H_2]_o - 3(2.0\ M)$, $[H_2]_o = 11.0\ M$; $8.0\ M = [N_2]_o - 2.0\ M$, $[N_2]_o = 10.0\ M$

69.   a. $P_{PCl_5} = \dfrac{nRT}{V} = \dfrac{\dfrac{2.450\ g\ PCl_5}{208.22\ g/mol} \times \dfrac{0.08206\ L\ atm}{mol\ K} \times 600.\ K}{0.500\ L} = 1.16$ atm

       b. $PCl_5(g)$ $\rightleftharpoons$ $PCl_3(g)$ + $Cl_2(g)$ $K_p = \dfrac{P_{PCl_3} \times P_{Cl_2}}{P_{PCl_5}} = 11.5$

Initial   1.16 atm   0   0
           x atm of PCl_5 reacts to reach equilibrium
Change $-x$ $\rightarrow$ $+x$ $+x$
Equil. $1.16 - x$ x x

$K_p = \dfrac{x^2}{1.16 - x} = 11.5$, $x^2 + 11.5\,x - 13.3 = 0$;   Using the quadratic formula: $x = 1.06$ atm

$P_{PCl_5} = 1.16 - 1.06 = 0.10$ atm

       c. $P_{PCl_3} = P_{Cl_2} = 1.06$ atm; $P_{PCl_5} = 0.10$ atm

$P_{tot} = P_{PCl_5} + P_{PCl_3} + P_{Cl_2} = 0.10 + 1.06 + 1.06 = 2.22$ atm

       d.   Percent dissociation $= \dfrac{x}{1.16} \times 100 = \dfrac{1.06}{1.16} \times 100 = 91.4\%$

71. $K = \dfrac{[HF]^2}{[H_2][F_2]} = \dfrac{(0.400)^2}{(0.0500)\,(0.0100)} = 320.$; 0.200 mol F_2/5.00 L = 0.0400 $M\ F_2$ added

From LeChatelier's principle, added F_2 causes the reaction to shift right to reestablish equilibrium.

                   $H_2(g)$ + $F_2(g)$ $\rightleftharpoons$ $2 HF(g)$

Initial   0.0500 M 0.0500 M 0.400 M
           x mol/L of F_2 reacts to reach equilibrium
Change $-x$ $-x$ $\rightarrow$ $+2x$
Equil. $0.0500 - x$ $0.0500 - x$ $0.400 + 2x$

$$K = 320. = \frac{(0.400 + 2x)^2}{(0.0500 - x)^2};$$ Taking the square root of the equation:

$$17.9 = \frac{0.400 + 2x}{0.0500 - x}, \quad 0.895 - 17.9\,x = 0.400 + 2x, \quad 19.9\,x = 0.495, \quad x = 0.0249 \text{ mol/L}$$

[HF] = 0.400 + 2(0.0249) = 0.450 M; [H$_2$] = [F$_2$] = 0.0500 - 0.0249 = 0.0251 M

73.   H$^+$ + OH$^-$ → H$_2$O; Sodium hydroxide (NaOH) will react with the H$^+$ on the product side of the reaction. This effectively removes H$^+$ from the equilibrium, which will shift the reaction to the right to produce more H$^+$ and CrO$_4^{2-}$. Since more CrO$_4^{2-}$ is produced, the solution turns yellow.

Challenge Problems

75.   There is a little trick we can use to solve this problem in order to avoid solving a cubic equation. Since K for this reaction is very small, the dominant reaction is the reverse reaction. We will let the products react to completion by the reverse reaction, then we will solve the forward equilibrium problem to determine the equilibrium concentrations. Summarizing these steps to solve in a table:

	2 NOCl(g) ⇌	2 NO(g) +	Cl$_2$(g)	K = 1.6 × 10^{-5}
Before	0	2.0 M	1.0 M	
	Let 1.0 mol/L Cl$_2$ react completely.			(K is small, reactants dominate.)
Change	+2.0 ←	-2.0	-1.0	React completely
After	2.0	0	0	New initial conditions
	2x mol/L of NOCl reacts to reach equilibrium			
Change	-2x →	+2x	+x	
Equil.	2.0 - 2x	2x	x	

$$K = 1.6 \times 10^{-5} = \frac{(2x)^2 (x)}{(2.0 - 2x)^2} \approx \frac{4x^3}{2.0^2} \quad \text{(assuming } 2.0 - 2x \approx 2.0\text{)}$$

$x^3 = 1.6 \times 10^{-5}$, $x = 2.5 \times 10^{-2}$ Assumption good by the 5% rule (2x is 2.5% of 2.0).

[NOCl] = 2.0 - 0.050 = 1.95 M = 2.0 M; [NO] = 0.050 M; [Cl$_2$] = 0.025 M

Note: If we do not break this problem into two parts (a stoichiometric part and an equilibrium part), we are faced with solving a cubic equation. The set-up would be:

	2 NOCl ⇌	2 NO +	Cl$_2$
Initial	0	2.0 M	1.0 M
Change	+2y ←	-2y	-y
Equil.	2y	2.0 - 2y	1.0 - y

$$1.6 \times 10^{-5} = \frac{(2.0 - 2y)^2 (1.0 - y)}{(2y)^2};$$ If we say that y is small to simplify the problem, then:

$$1.6 \times 10^{-5} = \frac{2.0^2}{4y^2};$$ We get $y = 250$. This is impossible!

To solve this equation, we cannot make any simplifying assumptions; we have to solve a cubic equation. If you don't have a graphing calculator, this is difficult. Alternatively, we can use some chemical common sense and solve the problem as illustrated above.

77. $N_2(g)$ + $3 H_2(g)$ $\rightleftharpoons$ $2 NH_3(g)$

Initial   0   0 P_0 P_0 = initial pressure of NH_3 in atm
                 $2x$ atm of NH_3 reacts to reach equilibrium
Change $+x$ $+3x$ $\leftarrow$ $-2x$
Equil. x $3x$ $P_0 - 2x$

From problem, $P_0 - 2x = \dfrac{P_0}{2.00}$, so $P_0 = 4.00\,x$

$$K_p = \frac{(4.00\,x - 2x)^2}{(x)(3x)^3} = \frac{(2.00\,x)^2}{(x)(3x)^3} = \frac{4.00\,x^2}{27x^4} = \frac{4.00}{27x^2} = 5.3 \times 10^5, \quad x = 5.3 \times 10^{-4} \text{ atm}$$

$P_0 = 4.00\,x = 4.00 \times (5.3 \times 10^{-4}) \text{ atm} = 2.1 \times 10^{-3} \text{ atm}$

79. $N_2O_4(g) \rightleftharpoons 2 NO_2(g)$ $K_p = \dfrac{P_{NO_2}^2}{P_{N_2O_4}} = \dfrac{(1.20)^2}{0.34} = 4.2$

Doubling the volume decreases each partial pressure by a factor of 2 (P = nRT/V).

P_{NO_2} = 0.600 atm and $P_{N_2O_4}$ = 0.17 atm are the new partial pressures.

$Q = \dfrac{(0.600)^2}{0.17} = 2.1$, so Q < K;   Equilibrium will shift to the right.

                   $N_2O_4(g)$ $\rightleftharpoons$ $2 NO_2(g)$

Initial   0.17 atm   0.600 atm
                 x atm of N_2O_4 reacts to reach equilibrium
Change $-x$ $\rightarrow$ $+2x$
Equil. $0.17 - x$ $0.600 + 2x$

$K_p = 4.2 = \dfrac{(0.600 + 2x)^2}{(0.17 - x)}$, $4x^2 + 6.6\,x - 0.354 = 0$ (carrying extra sig. figs.)

Solving using the quadratic formula: $x = 0.052$

P_{NO_2} = 0.600 + 2(0.052) = 0.704 atm; $P_{N_2O_4}$ = 0.17 - 0.052 = 0.12 atm

81. $SO_3(g)$ $\rightleftharpoons$ $SO_2(g)$ + $1/2\ O_2(g)$

Initial	P_0		0	0
Change	$-x$	$\rightarrow$	$+x$	$+x/2$
Equil.	$P_0 - x$		x	$x/2$

P_0 = initial pressure of SO_3

The average molar mass of the mixture is:

$$\text{average molar mass} = \frac{dRT}{P} = \frac{(1.60\ \text{g/L})\ (0.08206\ \text{L atm mol}^{-1}\ \text{K}^{-1})\ (873\ \text{K})}{1.80\ \text{atm}} = 63.7\ \text{g/mol}$$

The average molar mass is determined by:

$$\text{average molar mass} = \frac{n_{SO_3}\ (80.07\ \text{g/mol}) + n_{SO_2}\ (64.07\ \text{g/mol}) + n_{O_2}\ (32.00\ \text{g/mol})}{n_{total}}$$

Since χ_A = mol fraction of component $A = n_A/n_{total} = P_A/P_{total}$, then:

$$63.7\ \text{g/mol} = \frac{P_{SO_3}\ (80.07) + P_{SO_2}\ (64.07) + P_{O_2}\ (32.00)}{P_{total}}$$

$P_{total} = P_0 - x + x + x/2 = P_0 + x/2 = 1.80\ \text{atm},\ \ P_0 = 1.80 - x/2$

$$63.7 = \frac{(P_0 - x)\ (80.07) + x(64.07) + \frac{x}{2}(32.00)}{1.80}$$

$$63.7 = \frac{(1.80 - 3/2x)\ (80.07) + x(64.07) + \frac{x}{2}(32.00)}{1.80}$$

$115 = 144 - 120.1\ x + 64.07\ x + 16.00\ x,\ \ 40.0\ x = 29,\ \ x = 0.73\ \text{atm}$

$P_{SO_3} = P_0 - x = 1.80 - 3/2\ x = 0.71\ \text{atm};\ \ P_{SO_2} = 0.73\ \text{atm};\ \ P_{O_2} = x/2 = 0.37\ \text{atm}$

$$K_p = \frac{P_{SO_2} \times P_{O_2}^{1/2}}{P_{SO_3}} = \frac{(0.73)\ (0.37)^{1/2}}{(0.71)} = 0.63$$

CHAPTER FOURTEEN

ACIDS AND BASES

Questions

17. a. Arrhenius acid: produce H^+ in water
     b. Brønsted-Lowry acid: proton (H^+) donor
     c. Lewis acid: electron pair acceptor

The Lewis definition is most general. The Lewis definition can apply to all Arrhenius and Brønsted-Lowry acids; H^+ has an empty 1s orbital and forms bonds to all bases by accepting a pair of electrons from the base. In addition, the Lewis definition incorporates other reactions not typically considered acid-base reactions, e.g., $BF_3(g) + NH_3(g) \rightarrow F_3B-NH_3(s)$. NH_3 is something we usually consider a base and it is a base in this reaction using the Lewis definition; NH_3 donates a pair of electrons to form the N–B bond.

19. $H_2O \rightleftharpoons H^+ + OH^-$ $\qquad K_w = [H^+][OH^-] = 1.0 \times 10^{-14}$

Neutral solution: $[H^+] = [OH^-]$; $[H^+] = 1.0 \times 10^{-7}\ M$; $pH = -\log(1.0 \times 10^{-7}) = 7.00$

21. Neutrally charged organic compounds containing at least one nitrogen atom generally behave as weak bases. The nitrogen atom has an unshared pair of electrons around it. This lone pair of electrons is used to form a bond to H^+.

23. In general, the weaker the acid, the stronger the conjugate base and vice versa.

     a. Since acid strength increases as the X – H bond strength decreases, conjugate base strength will increase as the strength of the X – H bond increases.
     b. Since acid strength increases as the electronegativity of neighboring atoms increases, conjugate base strength will decrease as the electronegativity of neighboring atoms increases.
     c. Since acid strength increases as the number of oxygen atoms increases, conjugate base strength decreases as the number of oxygen atoms increase.

25. a. H_2O and $CH_3CO_2^-$

     b. An acid-base reaction can be thought of as a competition between two opposing bases. Since this equilibrium lies far to the left ($K_a < 1$), $CH_3CO_2^-$ is a stronger base than H_2O.

     c. The acetate ion is a better base than water and produces basic solutions in water. When we put acetate ion into solution as the only major basic species, the reaction is:

$$CH_3CO_2^- + H_2O \rightleftharpoons CH_3CO_2H + OH^-$$

Now the competition is between $CH_3CO_2^-$ and OH^- for the proton. Hydroxide ion is the strongest base possible in water. The equilibrium above lies far to the left, resulting in a K_b value less than one. Those species we specifically call weak bases ($10^{-14} < K_b < 1$) lie between H_2O and OH^- in base strength. Weak bases are stronger bases than water but are weaker bases than OH^-.

27.   When an acid dissociates, ions are produced. The conductivity of the solution is a measure of the number of ions. In addition, the colligative properties, which are discussed in Chapter 11, depend on the number of particles present. So measurements of osmotic pressure, vapor pressure lowering, freezing point depression or boiling point elevation will also allow us to determine the extent of ionization.

Exercises

Nature of Acids and Bases

29.   a. $HClO_4(aq) + H_2O(l) \rightarrow H_3O^+(aq) + ClO_4^-(aq)$. Only the forward reaction is indicated since $HClO_4$ is a strong acid and is basically 100% dissociated in water. For acids, the dissociation reaction is commonly written without water as a reactant. The common abbreviation for this reaction is: $HClO_4(aq) \rightarrow H^+(aq) + ClO_4^-(aq)$. This reaction is also called the K_a reaction as the equilibrium constant for this reaction is called K_a.

   b.   Propanoic acid is a weak acid, so it is only partially dissociated in water. The dissociation reaction is: $CH_3CH_2CO_2H(aq) + H_2O(l) \rightleftharpoons H_3O^+(aq) + CH_3CH_2CO_2^-(aq)$ or $CH_3CH_2CO_2H(aq) \rightleftharpoons H^+(aq) + CH_3CH_2CO_2^-(aq)$.

   c. NH_4^+ is a weak acid. Similar to propanoic acid, the dissociation reaction is:

   $$NH_4^+(aq) + H_2O(l) \rightleftharpoons H_3O^+(aq) + NH_3(aq) \text{ or } NH_4^+(aq) \rightleftharpoons H^+(aq) + NH_3(aq)$$

31.   An acid is a proton (H^+) donor and a base is a proton acceptor. A conjugate acid-base pair differs by only a proton (H^+).

	Acid	Base	Conjugate Base of Acid	Conjugate Acid of Base
a.	HF	H_2O	F^-	H_3O^+
b.	H_2SO_4	H_2O	HSO_4^-	H_3O^+
c.	HSO_4^-	H_2O	SO_4^{2-}	H_3O^+

33.   Strong acids have a $K_a \gg 1$ and weak acids have $K_a < 1$. Table 14.2 in the text lists some K_a values for weak acids. K_a values for strong acids are hard to determine so they are not listed in the text. However, there are only a few common strong acids so if you memorize the strong acids, then all other acids will be weak acids. The strong acids to memorize are HCl, HBr, HI, HNO_3, $HClO_4$ and H_2SO_4.

a. $HClO_4$ is a strong acid.
b. HOCl is a weak acid ($K_a = 3.5 \times 10^{-8}$).
c. H_2SO_4 is a strong acid.
d. H_2SO_3 is a weak diprotic acid since the K_{a1} and K_{a2} values are less than one.

35.   The K_a value is directly related to acid strength. As K_a increases, acid strength increases. For water, use K_w when comparing the acid strength of water to other species. The K_a values are:

HNO_3: strong acid ($K_a \gg 1$); HOCl: $K_a = 3.5 \times 10^{-8}$

NH_4^+: $K_a = 5.6 \times 10^{-10}$; H_2O: $K_a = K_w = 1.0 \times 10^{-14}$

From the K_a values, the ordering is: $HNO_3 > HOCl > NH_4^+ > H_2O$.

37.   a. HCl is a strong acid and water is a very weak acid with $K_a = K_w = 1.0 \times 10^{-14}$. HCl is a much stronger acid than H_2O.

        b. H_2O, $K_a = K_w = 1.0 \times 10^{-14}$; HNO_2, $K_a = 4.0 \times 10^{-4}$; HNO_2 is a stronger acid than H_2O since K_a for $HNO_2 > K_a$ for H_2O.

        c. HOC_6H_5, $K_a = 1.6 \times 10^{-10}$; HCN, $K_a = 6.2 \times 10^{-10}$; HCN is a stronger acid than HOC_6H_5 since K_a for HCN $> K_a$ for HOC_6H_5.

Autoionization of Water and the pH Scale

39.   At 25 °C, the relationship: $[H^+][OH^-] = K_w = 1.0 \times 10^{-14}$ always holds for aqueous solutions. When $[H^+]$ is greater than 1.0×10^{-7} M, the solution is acidic; when $[H^+]$ is less than 1.0×10^{-7} M, the solution is basic; when $[H^+] = 1.0 \times 10^{-7}$ M, the solution is neutral. In terms of $[OH^-]$, an acidic solution has $[OH^-] < 1.0 \times 10^{-7}$ M, a basic solution has $[OH^-] > 1.0 \times 10^{-7}$ M, and a neutral solution has $[OH^-] = 1.0 \times 10^{-7}$ M.

        a. $[OH^-] = \dfrac{K_w}{[H^+]} = \dfrac{1.0 \times 10^{-14}}{1.0 \times 10^{-7}} = 1.0 \times 10^{-7}$ M; The solution is neutral.

        b. $[OH^-] = \dfrac{1.0 \times 10^{-14}}{6.7 \times 10^{-4}} = 1.5 \times 10^{-11}$ M; The solution is acidic.

        c. $[OH^-] = \dfrac{1.0 \times 10^{-14}}{1.9 \times 10^{-11}} = 5.3 \times 10^{-4}$ M; The solution is basic.

        d. $[OH^-] = \dfrac{1.0 \times 10^{-14}}{2.3} = 4.3 \times 10^{-15}$ M; The solution is acidic.

41.   a. Since the value of the equilibrium constant increases as the temperature increases, the reaction is endothermic. In endothermic reactions, heat is a reactant so an increase in temperature (heat) shifts the reaction to produce more products and increases K in the process.

b. $H_2O(l) \rightleftharpoons H^+(aq) + OH^-(aq)$ $K_w = 5.47 \times 10^{-14} = [H^+][OH^-]$ at $50\,^\circ C$

In pure water $[H^+] = [OH^-]$, so $5.47 \times 10^{-14} = [H^+]^2$, $[H^+] = 2.34 \times 10^{-7}\ M = [OH^-]$

43. pH = -log $[H^+]$; pOH = -log $[OH^-]$; At $25\,^\circ C$, pH + pOH = 14.00; For Exercise 14.39:

a. pH = -log $[H^+]$ = -log (1.0×10^{-7}) = 7.00; pOH = 14.00 - pH = 14.00 - 7.00 = 7.00

b. pH = -log (6.7×10^{-4}) = 3.17; pOH = 14.00 - 3.17 = 10.83

c. pH = -log (1.9×10^{-11}) = 10.72; pOH = 14.00 - 10.72 = 3.28

d. pH = -log (2.3) = -0.36; pOH = 14.00 - (-0.36) = 14.36

Note that pH is less than zero when $[H^+]$ is greater than 1.0 M (an extremely acidic solution).

For Exercise 14.40:

a. pOH = -log $[OH^-]$ = -log (3.6) = -0.56; pH = 14.00 - pOH = 14.00 - (-0.56) = 14.56

b. pOH = -log (9.7×10^{-9}) = 8.01; pH = 14.00 - 8.01 = 5.99

c. pOH = -log (2.2×10^{-3}) = 2.66; pH = 14.00 - 2.66 = 11.34

d. pOH = -log (1.0×10^{-7}) = 7.00; pH = 14.00 - 7.00 = 7.00

Note that pH is greater than 14.00 when $[OH^-]$ is greater than 1.0 M (an extremely basic solution).

45. pOH = 14.00 - pH = 14.00 - 6.77 = 7.23; $[H^+] = 10^{-pH} = 10^{-6.77} = 1.7 \times 10^{-7}\ M$

$$[OH^-] = \frac{K_w}{[H^+]} = \frac{1.0 \times 10^{-14}}{1.7 \times 10^{-7}} = 5.9 \times 10^{-8}\ M \text{ or } [OH^-] = 10^{-pOH} = 10^{-7.23} = 5.9 \times 10^{-8}\ M$$

The sample of milk is slightly acidic since the pH is less than 7.00 at $25\,^\circ C$.

Solutions of Acids

47. All the acids in this problem are strong acids that are always assumed to completely dissociate in water. The general dissociation reaction for a strong acid is: $HA(aq) \rightarrow H^+(aq) + A^-(aq)$ where A^- is the conjugate base of the strong acid HA. For 0.250 M solutions of these strong acids, 0.250 M H^+ and 0.250 M A^- are present when the acids completely dissociate. The amount of H^+ donated from water will be insignificant in this problem since H_2O is a very weak acid.

a. Major species present after dissociation = H^+, ClO_4^- and H_2O;

pH = -log $[H^+]$ = -log (0.250) = 0.602

b.   Major species = H^+, NO_3^- and H_2O; pH = 0.602

49.   Both are strong acids.

$0.0500 \text{ L} \times 0.050 \text{ mol/L} = 2.5 \times 10^{-3}$ mol HCl $= 2.5 \times 10^{-3}$ mol $H^+ + 2.5 \times 10^{-3}$ mol Cl^-

$0.1500 \text{ L} \times 0.10 \text{ mol/L} = 1.5 \times 10^{-2}$ mol $HNO_3 = 1.5 \times 10^{-2}$ mol $H^+ + 1.5 \times 10^{-2}$ mol NO_3^-

$$[H^+] = \frac{(2.5 \times 10^{-3} + 1.5 \times 10^{-2}) \text{ mol}}{0.2000 \text{ L}} = 0.088 \ M; \ [OH^-] = \frac{K_w}{[H^+]} = 1.1 \times 10^{-13} \ M$$

$$[Cl^-] = \frac{2.5 \times 10^{-3} \text{ mol}}{0.2000 \text{ L}} = 0.013 \ M; \ [NO_3^-] = \frac{1.5 \times 10^{-2} \text{ mol}}{0.2000 \text{ L}} = 0.075 \ M$$

51. $[H^+] = 10^{-pH} = 10^{-2.50} = 3.2 \times 10^{-3} \ M$. Since HCl is a strong acid, a $3.2 \times 10^{-3} \ M$ HCl solution will produce $3.2 \times 10^{-3} \ M \ H^+$ giving a pH = 2.50.

53.   a. HNO_2 ($K_a = 4.0 \times 10^{-4}$) and H_2O ($K_a = K_w = 1.0 \times 10^{-14}$) are the major species. HNO_2 is a much stronger acid than H_2O so it is the major source of H^+. However, HNO_2 is a weak acid ($K_a < 1$) so it only partially dissociates in water. We must solve an equilibrium problem to determine $[H^+]$. In the Solutions Guide, we will summarize the initial, change and equilibrium concentrations into one table called the ICE table. Solving the weak acid problem:

$$HNO_2 \rightleftharpoons H^+ + NO_2^-$$

Initial   0.250 M ~0   0
              x mol/L HNO_2 dissociates to reach equilibrium
Change $-x$ $\rightarrow$ $+x$ $+x$
Equil.   0.250 $-x$ x x

$$K_a = \frac{[H^+][NO_2^-]}{[HNO_2]} = 4.0 \times 10^{-4} = \frac{x^2}{0.250 - x}; \ \text{If we assume } x \ll 0.250, \text{ then:}$$

$$4.0 \times 10^{-4} \approx \frac{x^2}{0.250}, x = \sqrt{4.0 \times 10^{-4}(0.250)} = 0.010 \ M$$

We must check the assumption: $\dfrac{x}{0.250} \times 100 = \dfrac{0.010}{0.250} \times 100 = 4.0\%$

All the assumptions are good. The H^+ contribution from water ($10^{-7} \ M$) is negligible, and x is small compared to 0.250 (percent error = 4.0%). If the percent error is less than 5% for an assumption, we will consider it a valid assumption (called the 5% rule). Finishing the problem: $x = 0.010 \ M = [H^+]$; pH = $-\log(0.010) = 2.00$

b. CH_3CO_2H ($K_a = 1.8 \times 10^{-5}$) and H_2O ($K_a = K_w = 1.0 \times 10^{-14}$) are the major species. CH_3CO_2H is the major source of H^+. Solving the weak acid problem:

$$CH_3CO_2H \;\rightleftharpoons\; H^+ \;+\; CH_3CO_2^-$$

Initial	0.250 M		~0	0

x mol/L CH$_3$CO$_2$H dissociates to reach equilibrium

Change	-x	$\rightarrow$	+x	+x
Equil.	0.250 - x		x	x

$$K_a = \frac{[H^+][CH_3CO_2^-]}{[CH_3CO_2H]} = 1.8 \times 10^{-5} = \frac{x^2}{0.250-x} \approx \frac{x^2}{0.250} \quad \text{(assuming } x \ll 0.250)$$

$x = 2.1 \times 10^{-3}\ M$; Checking assumption: $\dfrac{2.1 \times 10^{-3}}{0.250} \times 100 = 0.84\%$. Assumptions good.

$[H^+] = x = 2.1 \times 10^{-3}\ M$; pH = $-\log(2.1 \times 10^{-3}) = 2.68$

55.   20.0 mL glacial acetic acid $\times \dfrac{1.05\ g}{mL} \times \dfrac{1\ mol}{60.05\ g} = 0.350$ mol HC$_2$H$_3$O$_2$

Initial concentration of HC$_2$H$_3$O$_2$ = $\dfrac{0.350\ mol}{0.2500\ L} = 1.40\ M$

$$HC_2H_3O_2 \;\rightleftharpoons\; H^+ \;+\; C_2H_3O_2^- \qquad K_a = 1.8 \times 10^{-5}$$

Initial	1.40 M		~0	0

x mol/L HC$_2$H$_3$O$_2$ dissociates to reach equilibrium

Change	-x	$\rightarrow$	+x	+x
Equil.	1.40 - x		x	x

$$K_a = 1.8 \times 10^{-5} = \frac{[H^+][C_2H_3O_2^-]}{[HC_2H_3O_2]} = \frac{x^2}{1.40-x} \approx \frac{x^2}{1.40}$$

$x = [H^+] = 5.0 \times 10^{-3}\ M$; pH = 2.30   Assumptions good (x is 0.36% of 1.40).

57.   This is a weak acid in water. Solving the weak acid problem:

$$HF \;\rightleftharpoons\; H^+ \;+\; F^- \qquad K_a = 7.2 \times 10^{-4}$$

Initial	0.020 M		~0	0

x mol/L HF dissociates to reach equilibrium

Change	-x	$\rightarrow$	+x	+x
Equil.	0.020 - x		x	x

$$K_a = 7.2 \times 10^{-4} = \frac{[H^+][F^-]}{[HF]} = \frac{x^2}{0.020-x} \approx \frac{x^2}{0.020} \quad \text{(assuming } x \ll 0.020)$$

$x = [H^+] = 3.8 \times 10^{-3}\ M$; Check assumptions: $\dfrac{x}{0.020} \times 100 = \dfrac{3.8 \times 10^{-3}}{0.020} \times 100 = 19\%$

The assumption $x \ll 0.020$ is not good (x is more than 5% of 0.020). We must solve $x^2/(0.020 - x) = 7.2 \times 10^{-4}$ exactly by using either the quadratic formula or by the method of successive approximations (see Appendix 1.4 of text). Using successive approximations, we let $0.016\ M$ be a new approximation for [HF]. That is, in the denominator, try $x = 0.0038$ (the value of x we calculated making the normal assumption), so $0.020 - 0.0038 = 0.016$, then solve for a new value of x in the numerator.

$$\frac{x^2}{0.020 - x} \approx \frac{x^2}{0.016} = 7.2 \times 10^{-4},\ \ x = 3.4 \times 10^{-3}$$

We use this new value of x to further refine our estimate of [HF], i.e., $0.020 - x = 0.020 - 0.0034 - 0.0166$ (carry extra significant figure).

$$\frac{x^2}{0.020 - x} \approx \frac{x^2}{0.0166} = 7.2 \times 10^{-4},\ \ x = 3.5 \times 10^{-3}$$

We repeat until we get an answer that repeats itself. This would be the same answer we would get solving exactly using the quadratic equation. In this case it is: $x = 3.5 \times 10^{-3}$

So: $[H^+] = [F^-] = x = 3.5 \times 10^{-3}\ M$; $[OH^-] = K_w/[H^+] = 2.9 \times 10^{-12}\ M$

$[HF] = 0.020 - x = 0.020 - 0.0035 = 0.017\ M$; pH $= 2.46$

Note: When the 5% assumption fails, use whichever method you are most comfortable with to solve exactly. The method of successive approximations is probably fastest when the percent error is less than ~25% (unless you have a calculator that can solve quadratic equations).

59.   Major species: $HC_2H_2ClO_2$ ($K_a = 1.35 \times 10^{-3}$) and H_2O; Major source of H^+: $HC_2H_2ClO_2$

$$HC_2H_2ClO_2 \quad \rightleftharpoons \quad H^+ \;+\; C_2H_2ClO_2^-$$

Initial	0.10 M	~0	0
	x mol/L $HC_2H_2ClO_2$ dissociates to reach equilibrium		
Change	$-x$ $\rightarrow$	$+x$	$+x$
Equil.	$0.10 - x$	x	x

$K_a = 1.35 \times 10^{-3} = \dfrac{x^2}{0.10 - x} \approx \dfrac{x^2}{0.10}$, $x = 1.2 \times 10^{-2}\ M$

Checking the assumptions finds that x is 12% of 0.10 which fails the 5% rule. We must solve $1.35 \times 10^{-3} = x^2/(0.10 - x)$ exactly using either the method of successive approximations or the quadratic equation. Using either method gives $x = [H^+] = 1.1 \times 10^{-2}\ M$. pH $= -\log [H^+] = -\log (1.1 \times 10^{-2}) = 1.96$.

61.   a. HCl is a strong acid. It will produce 0.10 M H^+. HOCl is a weak acid. Let's consider the equilibrium:

$$HOCl \rightleftharpoons H^+ + OCl^- \qquad K_a = 3.5 \times 10^{-8}$$

Initial	0.10 M	0.10 M	0

x mol/L HOCl dissociates to reach equilibrium

Change	$-x$	$\rightarrow$ $+x$	$+x$
Equil.	0.10 - x	0.10 + x	x

$$K_a = 3.5 \times 10^{-8} = \frac{[H^+][OCl^-]}{[HOCl]} = \frac{(0.10 + x)(x)}{0.10 - x} \approx x, \; x = 3.5 \times 10^{-8} M$$

Assumptions are great (x is 3.5×10^{-5}% of 0.10). We are really assuming that HCl is the only important source of H^+, which it is. The [H^+] contribution from HOCl, x, is negligible. Therefore, [H^+] = 0.10 M; pH = 1.00

b. HNO_3 is a strong acid, giving an initial concentration of H^+ equal to 0.050 M. Consider the equilibrium:

$$HC_2H_3O_2 \rightleftharpoons H^+ + C_2H_3O_2^- \qquad K_a = 1.8 \times 10^{-5}$$

Initial	0.50 M	0.050 M	0

x mol/L $HC_2H_3O_2$ dissociates to reach equilibrium

Change	$-x$	$\rightarrow$ $+x$	$+x$
Equil.	0.50 - x	0.050 + x	x

$$K_a = 1.8 \times 10^{-5} = \frac{[H^+][C_2H_3O_2^-]}{[HC_2H_3O_2]} = \frac{(0.050 + x)x}{(0.50 - x)} \approx \frac{0.050\,x}{0.50}$$

$x = 1.8 \times 10^{-4}$; Assumptions are good (well within the 5% rule).

[H^+] = 0.050 + x = 0.050 M and pH = 1.30

63.   In all parts of this problem, acetic acid ($HC_2H_3O_2$) is the best weak acid present. We must solve a weak acid problem.

a. $$HC_2H_3O_2 \rightleftharpoons H^+ + C_2H_3O_2^-$$

Initial	0.50 M	~0	0

x mol/L $HC_2H_3O_2$ dissociates to reach equilibrium

Change	$-x$	$\rightarrow$ $+x$	$+x$
Equil.	0.50 - x	x	x

$$K_a = 1.8 \times 10^{-5} = \frac{[H^+][C_2H_3O_2^-]}{[HC_2H_3O_2]} = \frac{x^2}{0.50 - x} \approx \frac{x^2}{0.50}$$

$x = [H^+] = [C_2H_3O_2^-] = 3.0 \times 10^{-3} M$ Assumptions good.

$$\text{Percent dissociation} = \frac{[H^+]}{[HC_2H_3O_2]_o} \times 100 = \frac{3.0 \times 10^{-3}}{0.50} \times 100 = 0.60\%$$

b. The setups for solutions b and c are similar to solution a except the final equation is slightly different, reflecting the new concentration of $HC_2H_3O_2$.

$$K_a = 1.8 \times 10^{-5} = \frac{x^2}{0.050 - x} \approx \frac{x^2}{0.050}$$

$x = [H^+] = [C_2H_3O_2^-] = 9.5 \times 10^{-4}\ M$ Assumptions good.

% dissociation $= \dfrac{9.5 \times 10^{-4}}{0.050} \times 100 = 1.9\%$

c. $K_a = 1.8 \times 10^{-5} = \dfrac{x^2}{0.0050 - x} \approx \dfrac{x^2}{0.0050}$

$x = [H^+] = [C_2H_3O_2^-] = 3.0 \times 10^{-4}\ M;$ Check assumptions.

Assumption that x is negligible is borderline (6.0% error). We should solve exactly. Using the method of successive approximations (see Appendix 1.4 of text):

$$1.8 \times 10^{-5} = \frac{x^2}{0.0050 - 3.0 \times 10^{-4}} = \frac{x^2}{0.0047}, \quad x = 2.9 \times 10^{-4}$$

Next trial also gives $x = 2.9 \times 10^{-4}$.

% dissociation $= \dfrac{2.9 \times 10^{-4}}{5.0 \times 10^{-3}} \times 100 = 5.8\%$

d. As we dilute a solution, all concentrations decrease. Dilution will shift the equilibrium to the side with the greater number of particles. For example, suppose we double the volume of an equilibrium mixture of a weak acid by adding water, then:

$$Q = \frac{\left(\dfrac{[H^+]_{eq}}{2}\right)\left(\dfrac{[X^-]_{eq}}{2}\right)}{\left(\dfrac{[HX]_{eq}}{2}\right)} = \frac{1}{2}K_a$$

$Q < K_a$, so the equilibrium shifts to the right or towards a greater percent dissociation.

e. $[H^+]$ depends on the initial concentration of weak acid and on how much weak acid dissociates. For solutions a-c the initial concentration of acid decreases more rapidly than the percent dissociation increases. Thus, $[H^+]$ decreases.

65.   Let HX symbolize the weak acid. Setup the problem like a typical weak acid equilibrium problem.

$$HX \quad \rightleftharpoons \quad H^+ \quad + \quad X^-$$

Initial   0.15 M ~0   0
              x mol/L HX dissociates to reach equilibrium
Change   -x $\rightarrow$ +x +x
Equil.   0.15 - x x x

If the acid is 3.0% dissociated, then $x = [H^+]$ is 3.0% of 0.15: $x = 0.030 \times (0.15\ M) = 4.5 \times 10^{-3}\ M$. Now that we know the value of x, we can solve for K_a.

$$K_a = \frac{[H^+][X^-]}{[HX]} = \frac{x^2}{0.15 - x} = \frac{(4.5 \times 10^{-3})^2}{0.15 - 4.5 \times 10^{-3}} = 1.4 \times 10^{-4}$$

67.   Setup the problem using the K_a equilibrium reaction for HOCN.

$$HOCN \quad \rightleftharpoons \quad H^+ \quad + \quad OCN^-$$

Initial   0.0100 M ~0   0
              x mol/L HOBr dissociates to reach equilibrium
Change   -x $\rightarrow$ +x +x
Equil.   0.0100 - x x x

$$K_a = \frac{[H^+][OCN^-]}{[HOCN]} = \frac{x^2}{0.0100 - x};\ \text{Since pH} = 2.77,\ \text{then: } x = [H^+] = 10^{-pH} = 10^{-2.77} = 1.7 \times 10^{-3}\ M$$

$$K_a = \frac{(1.7 \times 10^{-3})^2}{0.0100 - 1.7 \times 10^{-3}} = 3.5 \times 10^{-4}$$

69.   Major species: HCOOH and H_2O; Major source of H^+: HCOOH

$$HCOOH \quad \rightleftharpoons \quad H^+ \quad + \quad HCOO^-$$

Initial   C   ~0   0   where C = [HCOOH]$_o$
              x mol/L HCOOH dissociates to reach equilibrium
Change   -x $\rightarrow$ +x +x
Equil.   C - x x x

$$K_a = 1.8 \times 10^{-4} = \frac{[H^+][HCOO^-]}{[HCOOH]} = \frac{x^2}{C - x}\ \text{where } x = [H^+]$$

$$1.8 \times 10^{-4} = \frac{[H^+]^2}{C - [H^+]};\ \text{Since pH} = 2.70,\ \text{then: } [H^+] = 10^{-2.70} = 2.0 \times 10^{-3}\ M$$

$$1.8 \times 10^{-4} = \frac{(2.0 \times 10^{-3})^2}{C - (2.0 \times 10^{-3})},\ C - (2.0 \times 10^{-3}) = \frac{4.0 \times 10^{-6}}{1.8 \times 10^{-4}},\ C = 2.4 \times 10^{-2}\ M$$

A 0.024 M formic acid solution will have pH = 2.70.

Solutions of Bases

71.   a. $NH_3(aq) + H_2O(l) \rightleftharpoons NH_4^+(aq) + OH^-(aq)$ $K_b = \dfrac{[NH_4^+][OH^-]}{[NH_3]}$

      b. $C_5H_5N(aq) + H_2O(l) \rightleftharpoons C_5H_5NH^+(aq) + OH^-(aq)$ $K_b = \dfrac{[C_5H_5NH^+][OH^-]}{[C_5H_5N]}$

73. NO_3^-: $K_b \ll K_w$ since HNO_3 is a strong acid. All conjugate bases of strong acids have no base strength. H_2O: $K_b = K_w = 1.0 \times 10^{-14}$; NH_3: $K_b = 1.8 \times 10^{-5}$; C_5H_5N: $K_b = 1.7 \times 10^{-9}$

      $NH_3 > C_5H_5N > H_2O > NO_3^-$ (As K_b increases, base strength increases.)

75.   a. NH_3 b. NH_3 c. OH^- d. CH_3NH_2

      The base with the largest K_b value is the strongest base. OH^- is the strongest base possible in water.

77. $NaOH(aq) \rightarrow Na^+(aq) + OH^-(aq)$; NaOH is a strong base which completely dissociates into Na^+ and OH^-. The initial concentration of NaOH will equal the concentration of OH^- donated by NaOH.

      a. $[OH^-] = 0.10\ M$; $pOH = -\log[OH^-] = -\log(0.10) = 1.00$

           $pH = 14.00 - pOH = 14.00 - 1.00 = 13.00$

           Note that H_2O is also present, but the amount of OH^- produced by H_2O will be insignificant compared to the $0.10\ M$ OH^- produced from the NaOH.

      b.   The $[OH^-]$ concentration donated by the NaOH is $1.0 \times 10^{-10}\ M$. Water by itself donates $1.0 \times 10^{-7}\ M$. In this problem, water is the major OH^- contributor and $[OH^-] = 1.0 \times 10^{-7}\ M$.

           $pOH = -\log(1.0 \times 10^{-7}) = 7.00$; $pH = 14.00 - 7.00 = 7.00$

      c. $[OH^-] = 2.0\ M$; $pOH = -\log(2.0) = -0.30$; $pH = 14.00 - (-0.30) = 14.30$

79.   a.   Major species: K^+, OH^-, H_2O (KOH is a strong base.)

           $[OH^-] = 0.015\ M$, $pOH = -\log(0.015) = 1.82$; $pH = 14.00 - pOH = 12.18$

      b.   Major species: Ba^{2+}, OH^-, H_2O; $Ba(OH)_2(aq) \rightarrow Ba^{2+}(aq) + 2\ OH^-(aq)$; Since each mol of the strong base $Ba(OH)_2$ dissolves in water to produce two mol OH^-, then $[OH^-] = 2(0.015\ M) = 0.030\ M$.

           $pOH = -\log(0.030) = 1.52$; $pH = 14.00 - 1.52 = 12.48$

81. $pH = 10.50$; $pOH = 14.00 - 10.50 = 3.50$; $[OH^-] = 10^{-3.50} = 3.2 \times 10^{-4}\ M$

      $KOH(aq) \rightleftharpoons K^+(aq) + OH^-(aq)$; Since KOH is a strong base, a $3.2 \times 10^{-4}\ M$ KOH solution will produce a $pH = 10.50$ solution.

83. NH_3 is a weak base with $K_b = 1.8 \times 10^{-5}$. The major species present will be NH_3 and H_2O ($K_b = K_w = 1.0 \times 10^{-14}$). Since NH_3 has a much larger K_b value compared to H_2O, NH_3 is the stronger base present and will be the major producer of OH^-. To determine the amount of OH^- produced from NH_3, we must perform an equilibrium calculation.

$$NH_3(aq) \; + \; H_2O(l) \; \rightleftharpoons \; NH_4^+(aq) \; + \; OH^-(aq)$$

Initial	0.150 M	0	~0

x mol/L NH_3 reacts with H_2O to reach equilibrium

Change	$-x$	$\rightarrow$	$+x$	$+x$
Equil.	0.150 - x		x	x

$$K_b = 1.8 \times 10^{-5} = \frac{[NH_4^+][OH^-]}{[NH_3]} = \frac{x^2}{0.150 - x} \approx \frac{x^2}{0.150} \quad \text{(assuming } x << 0.150\text{)}$$

$x = [OH^-] = 1.6 \times 10^{-3} \, M$; Check assumptions: x is 1.1% of 0.150 so the assumption 0.150 - $x \approx$ 0.150 is valid by the 5% rule. Also, the contribution of OH^- from water will be insignificant (which will usually be the case). Finishing the problem, $pOH = -\log[OH^-] = -\log(1.6 \times 10^{-3} \, M) = 2.80$; $pH = 14.00 - pOH = 14.00 - 2.80 = 11.20$.

85.   These are solutions of weak bases in water. We must solve the equilibrium weak base problem.

a. $$(C_2H_5)_3N + H_2O \; \rightleftharpoons \; (C_2H_5)_3NH^+ + OH^- \qquad K_b = 4.0 \times 10^{-4}$$

Initial	0.20 M	0	~0

x mol/L of $(C_2H_5)_3N$ reacts with H_2O to reach equilibrium

Change	$-x$	$\rightarrow$	$+x$	$+x$
Equil.	0.20 - x		x	x

$$K_b = 4.0 \times 10^{-4} = \frac{[(C_2H_5)_3NH^+][OH^-]}{[(C_2H_5)_3N]} = \frac{x^2}{0.20 - x} \approx \frac{x^2}{0.20}, \; x = [OH^-] = 8.9 \times 10^{-3} \, M$$

Assumptions good (x is 4.5% of 0.20). $[OH^-] = 8.9 \times 10^{-3} \, M$

$$[H^+] = \frac{K_w}{[OH^-]} = \frac{1.0 \times 10^{-14}}{8.9 \times 10^{-3}} = 1.1 \times 10^{-12} \, M; \; pH = 11.96$$

b. $$HONH_2 + H_2O \; \rightleftharpoons \; HONH_3^+ + OH^- \qquad K_b = 1.1 \times 10^{-8}$$

Initial	0.20 M	0	~0
Equil.	0.20 - x	x	x

$$K_b = 1.1 \times 10^{-8} = \frac{x^2}{0.20 - x} \approx \frac{x^2}{0.20}, \; x = [OH^-] = 4.7 \times 10^{-5} \, M; \; \text{Assumptions good.}$$

$[H^+] = 2.1 \times 10^{-10} \, M; \; pH = 9.68$

87.   This is a solution of a weak base in water. We must solve the weak base equilibrium problem.

$$C_2H_5NH_2 \; + \; H_2O \; \rightleftharpoons \; C_2H_5NH_3^+ \; + \; OH^- \qquad K_b = 5.6 \times 10^{-4}$$

Initial   0.20 M 0   ~0

x mol/L $C_2H_5NH_2$ reacts with H_2O to reach equilibrium

Change   -x $\rightarrow$ +x +x
Equil.   0.20 - x x x

$$K_b = \frac{[C_2H_5NH_3^+][OH^-]}{[C_2H_5NH_2]} = \frac{x^2}{0.20 - x} \approx \frac{x^2}{0.20} \quad \text{(assuming } x \ll 0.20\text{)}$$

$x = 1.1 \times 10^{-2}$;   Checking assumption: $\dfrac{1.1 \times 10^{-2}}{0.20} \times 100 = 5.5\%$

Assumption fails the 5% rule. We must solve exactly using either the quadratic equation or the method of successive approximations (see Appendix 1.4 of the text). Using successive approximations and carrying extra significant figures:

$$\frac{x^2}{0.20 - 0.011} = \frac{x^2}{0.189} = 5.6 \times 10^{-4}, \; x = 1.0 \times 10^{-2} \, M \quad \text{(consistent answer)}$$

$$x = [OH^-] = 1.0 \times 10^{-2} \, M; \; [H^+] = \frac{K_w}{[OH^-]} = \frac{1.0 \times 10^{-14}}{1.0 \times 10^{-2}} = 1.0 \times 10^{-12} \, M; \; pH = 12.00$$

89.   To solve for percent ionization, just solve the weak base equilibrium problem.

a. $$NH_3 \; + \; H_2O \; \rightleftharpoons \; NH_4^+ \; + \; OH^- \qquad K_b = 1.8 \times 10^{-5}$$

Initial   0.10 M 0   ~0
Equil.   0.10 - x x x

$$K_b = 1.8 \times 10^{-5} = \frac{x^2}{0.10 - x} \approx \frac{x^2}{0.10}, \; x = [OH^-] = 1.3 \times 10^{-3} \, M; \; \text{Assumptions good.}$$

$$\text{Percent ionization} = \frac{[OH^-]}{[NH_3]_o} \times 100 = \frac{1.3 \times 10^{-3} \, M}{0.10 \, M} \times 100 = 1.3\%$$

b. $$NH_3 \; + \; H_2O \; \rightleftharpoons \; NH_4^+ \; + \; OH^-$$

Initial   0.010 M 0   ~0
Equil.   0.010 - x x x

$$1.8 \times 10^{-5} = \frac{x^2}{0.010 - x} \approx \frac{x^2}{0.010}, \; x = [OH^-] = 4.2 \times 10^{-4} \, M; \; \text{Assumptions good.}$$

$$\text{Percent ionization} = \frac{4.2 \times 10^{-4}}{0.010} \times 100 = 4.2\%$$

Note: For the same base, the percent ionization increases as the initial concentration of base decreases.

91.   Let cod = codeine, $C_{18}H_{21}NO_3$; using the K_b reaction to solve:

$$cod \; + \; H_2O \; \rightleftharpoons \; codH^+ \; + \; OH^-$$

Initial	$1.7 \times 10^{-3} \, M$	0	~0
	x mol/L codeine reacts with H_2O to reach equilibrium		
Change	$-x$	$\rightarrow$	$+x$	$+x$
Equil.	$1.7 \times 10^{-3} - x$	x	x

$K_b = \dfrac{x^2}{1.7 \times 10^{-3} - x}$; Since pH = 9.59, then pOH = 14.00 - 9.59 = 4.41.

$[OH^-] = x = 10^{-4.41} = 3.9 \times 10^{-5} \, M$; $K_b = \dfrac{(3.9 \times 10^{-5})^2}{1.7 \times 10^{-3} - 3.9 \times 10^{-5}} = 9.2 \times 10^{-7}$

Polyprotic Acids

93. $H_2SO_3(aq) \rightleftharpoons HSO_3^-(aq) + H^+(aq)$ $K_{a_1} = \dfrac{[HSO_3^-][H^+]}{[H_2SO_3]}$

$HSO_3^-(aq) \rightleftharpoons SO_3^{2-}(aq) + H^+(aq)$ $K_{a_2} = \dfrac{[SO_3^{2-}][H^+]}{[HSO_3^-]}$

95.   In both these polyprotic acid problems, the dominate equilibrium is the K_{a_1} reaction. The amount of H^+ produced from the subsequent K_a reactions will be minimal since they are all have much smaller K_a values.

a. $H_3PO_4 \; \rightleftharpoons \; H^+ \; + \; H_2PO_4^-$ $K_{a_1} = 7.5 \times 10^{-3}$

Initial	0.10 M	~0	0
	x mol/L H_3PO_4 dissociates to reach equilibrium		
Change	$-x$	$\rightarrow$	$+x$	$+x$
Equil.	0.10 - x	x	x

$K_{a_1} = 7.5 \times 10^{-3} = \dfrac{[H^+][H_2PO_4^-]}{[H_3PO_4]} = \dfrac{x^2}{0.10 - x} \approx \dfrac{x^2}{0.10}$, $x = 2.7 \times 10^{-2}$

Assumption is bad (x is 27% of 0.10). Using successive approximations:

$\dfrac{x^2}{0.10 - 0.027} = 7.5 \times 10^{-3}$, $x = 2.3 \times 10^{-2}$; $\dfrac{x^2}{0.10 - 0.023} = 7.5 \times 10^{-3}$, $x = 2.4 \times 10^{-2}$
                                                                                    (consistent answer)

$x = [H^+] = 2.4 \times 10^{-2} \, M$; pH = -log $(2.4 \times 10^{-2}) = 1.62$

b. H_2CO_3 $\rightleftharpoons$ H^+ + HCO_3^- $K_{a_1} = 4.3 \times 10^{-7}$

	H_2CO_3	H^+	HCO_3^-
Initial	0.10 M	~0	0
Equil.	0.10 - x	x	x

$$K_{a_1} = 4.3 \times 10^{-7} = \frac{[H^+][HCO_3^-]}{[H_2CO_3]} = \frac{x^2}{(0.10 - x)} \approx \frac{x^2}{0.10}$$

$x = [H^+] = 2.1 \times 10^{-4}\ M$; pH = 3.68; Assumptions good.

97.   The dominant H^+ producer is the strong acid H_2SO_4. A 2.0 M H_2SO_4 solution produces 2.0 M HSO_4^- and 2.0 M H^+. However, HSO_4^- is a weak acid which could also add H^+ to the solution.

HSO_4^- $\rightleftharpoons$ H^+ + SO_4^{2-}

	HSO_4^-	H^+	SO_4^{2-}
Initial	2.0 M	2.0 M	0

x mol/L HSO_4^- dissociates to reach equilibrium

	HSO_4^-		H^+	SO_4^{2-}
Change	-x	$\rightarrow$	+x	+x
Equil.	2.0 - x		2.0 + x	x

$$K_{a_2} = 1.2 \times 10^{-2} = \frac{[H^+][SO_4^{2-}]}{[HSO_4^-]} = \frac{(2.0 + x)(x)}{2.0 - x} \approx \frac{2.0\,(x)}{2.0}, \quad x = 1.2 \times 10^{-2}$$

Since x is 0.60% of 2.0, the assumption is valid by the 5% rule. The amount of additional H^+ from HSO_4^- is 1.2×10^{-2}. The total amount of H^+ present is:

$[H^+] = 2.0 + 1.2 \times 10^{-2} = 2.0\ M$; pH = -log (2.0) = -0.30

Note: In this problem, H^+ from HSO_4^- could have been ignored. However, this is not usually the case, especially in more dilute solutions of H_2SO_4.

Acid-Base Properties of Salts

99.   One difficult aspect of acid-base chemistry is recognizing what types of species are present in solution, i.e., whether a species is a strong acid, strong base, weak acid, weak base or a neutral species. Below are some ideas and generalizations to keep in mind that will help in recognizing types of species present.

a.   Memorize the following strong acids: HCl, HBr, HI, HNO_3, $HClO_4$ and H_2SO_4

b.   Memorize the following strong bases: LiOH, NaOH, KOH, RbOH, $Ca(OH)_2$, $Sr(OH)_2$ and $Ba(OH)_2$

c.   All weak acids have a K_a value less than 1 but greater than K_w. Some weak acids are in Table 14.2 of the text. All weak bases have a K_b value less than 1 but greater than K_w. Some weak bases are in Table 14.3 of the text.

d.   All conjugate bases of weak acids are weak bases, i.e., all have a K_b value less than 1 but greater than K_w. Some examples of these are the conjugate bases of the weak acids in Table 14.2 of the text.

e. All conjugate acids of weak bases are weak acids, i.e., all have a K_a value less than 1 but greater than K_w. Some examples of these are the conjugate acids of the weak bases in Table 14.3 of the text.

f. Alkali metal ions (Li^+, Na^+, K^+, Rb^+, Cs^+) and heavier alkaline earth metal ions (Ca^{2+}, Sr^{2+}, Ba^{2+}) have no acidic or basic properties in water.

g. All conjugate bases of strong acids (Cl^-, Br^-, I^-, NO_3^-, ClO_4^-, HSO_4^-) have no basic properties in water ($K_b \ll K_w$) and only HSO_4^- has any acidic properties in water.

Lets apply these ideas to this problem to see what type of species are present. The letters in parenthesis is/are the generalization(s) above which identifies the species.

KOH: strong base (b)

KCl: neutral; K^+ and Cl^- have no acidic/basic properties (f and g).

KCN: CN^- is a weak base, $K_b = 1.0 \times 10^{-14}/6.2 \times 10^{-10} = 1.6 \times 10^{-5}$ (c and d). Ignore K^+(f).

NH₄Cl: NH_4^+ is a weak acid, $K_a = 5.6 \times 10^{-10}$ (c and e). Ignore Cl^-(g).

HCl: strong acid (a)

The most acidic solution will be the strong acid followed by the weak acid. The most basic solution will be the strong base followed by the weak base. The KCl solution will be between the acidic and basic solutions at pH = 7.00.

Most acidic → most basic; $HCl > NH_4Cl > KCl > KCN > KOH$

101. From the K_a values, acetic acid is a stronger acid than hypochlorous acid. Conversely, the conjugate base of acetic acid, $C_2H_3O_2^-$, will be a weaker base than the conjugate base of hypochlorous acid, OCl^-. Thus, the hypochlorite ion, OCl^-, is a stronger base than the acetate ion, $C_2H_3O_2^-$. In general, the stronger the acid, the weaker the conjugate base. This statement comes from the relationship $K_w = K_a \times K_b$, which holds for all conjugate acid-base pairs.

103. $NaN_3 \rightarrow Na^+ + N_3^-$; Azide, N_3^-, is a weak base since it is the conjugate base of a weak acid. All conjugate bases of weak acids are weak bases ($K_w < K_b < 1$). Ignore Na^+.

$$N_3^- + H_2O \rightleftharpoons HN_3 + OH^- \quad K_b = \frac{K_w}{K_a} = \frac{1.0 \times 10^{-14}}{1.9 \times 10^{-5}} = 5.3 \times 10^{-10}$$

Initial	0.010 M	0	~0
	x mol/L of N_3^- reacts with H_2O to reach equilibrium		
Change	$-x$ →	$+x$	$+x$
Equil.	0.010 - x	x	x

$$K_b = \frac{[HN_3][OH^-]}{[N_3^-]} = 5.3 \times 10^{-10} = \frac{x^2}{0.010 - x} \approx \frac{x^2}{0.010} \quad \text{(assuming } x \ll 0.010\text{)}$$

$x = [OH^-] = 2.3 \times 10^{-6} M$; $[H^+] = \dfrac{1.0 \times 10^{-14}}{2.3 \times 10^{-6}} = 4.3 \times 10^{-9} M$ Assumptions good.

$[HN_3] = [OH^-] = 2.3 \times 10^{-6} M$; $[Na^+] = 0.010 M$; $[N_3^-] = 0.010 - 2.3 \times 10^{-6} = 0.010 M$

105. a. $CH_3NH_3Cl \rightarrow CH_3NH_3^+ + Cl^-$: $CH_3NH_3^+$ is a weak acid. Cl^- is the conjugate base of a strong acid. Cl^- has no basic (or acidic) properties.

$$CH_3NH_3^+ \rightleftharpoons CH_3NH_2 + H^+ \quad K_a = \frac{[CH_3NH_2][H^+]}{[CH_3NH_3^+]} = \frac{K_w}{K_b} = \frac{1.00 \times 10^{-14}}{4.38 \times 10^{-4}} = 2.28 \times 10^{-11}$$

	$CH_3NH_3^+$	$\rightleftharpoons$	CH_3NH_2	+	H^+
Initial	0.10 M		0		~0

x mol/L $CH_3NH_3^+$ dissociates to reach equilibrium

| Change | $-x$ | $\rightarrow$ | $+x$ | | $+x$ |
| Equil. | 0.10 - x | | x | | x |

$$K_a = 2.28 \times 10^{-11} = \frac{x^2}{0.10 - x} \approx \frac{x^2}{0.10} \quad \text{(assuming } x \ll 0.10\text{)}$$

$x = [H^+] = 1.5 \times 10^{-6}\ M;\ \ pH = 5.82$ Assumptions good.

b. $NaCN \rightarrow Na^+ + CN^-$: CN^- is a weak base. Na^+ has no acidic (or basic) properties.

$$CN^- + H_2O \rightleftharpoons HCN + OH^- \quad K_b = \frac{K_w}{K_a} = \frac{1.0 \times 10^{-14}}{6.2 \times 10^{-10}} = 1.6 \times 10^{-5}$$

	CN^-	+	H_2O	$\rightleftharpoons$	HCN	+	OH^-
Initial	0.050 M				0		~0

x mol/L CN^- reacts with H_2O to reach equilibrium

| Change | $-x$ | | | $\rightarrow$ | $+x$ | | $+x$ |
| Equil. | 0.050 - x | | | | x | | x |

$$K_b = 1.6 \times 10^{-5} = \frac{[HCN][OH^-]}{[CN^-]} = \frac{x^2}{0.050 - x} \approx \frac{x^2}{0.050}$$

$x = [OH^-] = 8.9 \times 10^{-4}\ M;\ pOH = 3.05;\ pH = 10.95$ Assumptions good.

107. All these salts contain Na^+, which has no acidic/basic properties, and a conjugate base of a weak acid (except for NaCl where Cl^- is a neutral species.). All conjugate bases of weak acids are weak bases since K_b for these species is between K_w and 1. To identify the species, we will use the data given to determine the K_b value for the weak conjugate base. From the K_b value and data in Table 14.2 of the text, we can identify the conjugate base present by calculating the K_a value for the weak acid. We will use A^- as an abbreviation for the weak conjugate base.

	A^-	+	H_2O	$\rightleftharpoons$	HA	+	OH^-
Initial	0.100 mol/1.00 L				0		~0

x mol/L A^- reacts with H_2O to reach equilibrium

| Change | $-x$ | | | $\rightarrow$ | $+x$ | | $+x$ |
| Equil. | 0.100 - x | | | | x | | x |

$$K_b = \frac{[HA][OH^-]}{[A^-]} = \frac{x^2}{0.100 - x}; \text{ From the problem, pH = 8.07:}$$

pOH = 14.00 - 8.07 = 5.93; $[OH^-] = x = 10^{-5.93} = 1.2 \times 10^{-6} M$

$$K_b = \frac{(1.2 \times 10^{-6})^2}{0.100 - 1.2 \times 10^{-6}} = 1.4 \times 10^{-11} = K_b \text{ value for the conjugate base of a weak acid.}$$

The K_a value for the weak acid equals K_w/K_b: $K_a = \dfrac{1.0 \times 10^{-14}}{1.4 \times 10^{-11}} = 7.1 \times 10^{-4}$

From Table 14.2 of the text, this K_a value is closest to HF. Therefore, the unknown salt is NaF.

109. Major species present: $Al(H_2O)_6^{3+}$ ($K_a = 1.4 \times 10^{-5}$), NO_3^- (neutral) and H_2O ($K_w = 1.0 \times 10^{-14}$); $Al(H_2O)_6^{3+}$ is a stronger acid than water so it will be the dominant H^+ producer.

$$Al(H_2O)_6^{3+} \rightleftharpoons Al(H_2O)_5(OH)^{2+} + H^+$$

Initial	0.050 M	0	~0
	x mol/L $Al(H_2O)_6^{3+}$ dissociates to reach equilibrium		
Change	$-x$ $\rightarrow$	$+x$	$+x$
Equil.	0.050 - x	x	x

$$K_a = 1.4 \times 10^{-5} = \frac{[Al(H_2O)_5(OH)^{2+}][H^+]}{[Al(H_2O)_6^{3+}]} = \frac{x^2}{0.050 - x} \approx \frac{x^2}{0.050}$$

$x = 8.4 \times 10^{-4} M = [H^+]$; pH = -log (8.4×10^{-4}) = 3.08; Assumptions good.

111. Reference Table 14.6 of the text and the solution to Exercise 14.99 for some generalizations on acid-base properties of salts.

a. $NaNO_3 \rightarrow Na^+ + NO_3^-$ neutral; Neither species has any acidic/basic properties.

b. $NaNO_2 \rightarrow Na^+ + NO_2^-$ basic; NO_2^- is a weak base and Na^+ has no effect on pH.

$$NO_2^- + H_2O \rightleftharpoons HNO_2 + OH^- \quad K_b = \frac{K_w}{K_{a, HNO_2}} = \frac{1.0 \times 10^{-14}}{4.0 \times 10^{-4}} = 2.5 \times 10^{-11}$$

c. $C_5H_5NHClO_4 \rightarrow C_5H_5NH^+ + ClO_4^-$ acidic; $C_5H_5NH^+$ is a weak acid and ClO_4^- has no effect on pH.

$$C_5H_5NH^+ \rightleftharpoons H^+ + C_5H_5N \quad K_a = \frac{K_w}{K_{b, C_5H_5N}} = \frac{1.0 \times 10^{-14}}{1.7 \times 10^{-9}} = 5.9 \times 10^{-6}$$

d. $NH_4NO_2 \rightarrow NH_4^+ + NO_2^-$ acidic; NH_4^+ is a weak acid ($K_a = 5.6 \times 10^{-10}$) and NO_2^- is a weak base ($K_b = 2.5 \times 10^{-11}$). Since $K_{a, NH_4^+} > K_{b, NO_2^-}$, then the solution is acidic.

$NH_4^+ \rightleftharpoons H^+ + NH_3$ $K_a = 5.6 \times 10^{-10}$; $NO_2^- + H_2O \rightleftharpoons HNO_2 + OH^-$ $K_b = 2.5 \times 10^{-11}$

e. $KOCl \rightarrow K^+ + OCl^-$ basic; OCl^- is a weak base and K^+ has no effect on pH.

$$OCl^- + H_2O \rightleftharpoons HOCl + OH^- \quad K_b = \frac{K_w}{K_{a, HOCl}} = \frac{1.0 \times 10^{-14}}{3.5 \times 10^{-8}} = 2.9 \times 10^{-7}$$

f. $NH_4OCl \rightarrow NH_4^+ + OCl^-$ basic; NH_4^+ is a weak acid and OCl^- is a weak base. Since $K_{b, OCl^-} > K_{a, NH_4^+}$, then the solution is basic.

$NH_4^+ \rightleftharpoons NH_3 + H^+$ $K_a = 5.6 \times 10^{-10}$; $OCl^- + H_2O \rightleftharpoons HOCl + OH^-$ $K_b = 2.9 \times 10^{-7}$

Relationships Between Structure and Strengths of Acids and Bases

113.   a. $HIO_3 < HBrO_3$; As the electronegativity of the central atom increases, acid strength increases.

b. $HNO_2 < HNO_3$; As the number of oxygen atoms attached to the central nitrogen atom increases, acid strength increases.

c. $HOI < HOCl$; Same reasoning as in a.

d. $H_3PO_3 < H_3PO_4$; Same reasoning as in b.

115.   a. $H_2O < H_2S < H_2Se$; As the strength of the $H - X$ bond decreases, acid strength increases.

b. $CH_3CO_2H < FCH_2CO_2H < F_2CHCO_2H < F_3CCO_2H$; As the electronegativity of neighboring atoms increases, acid strength increases.

c. $NH_4^+ < HONH_3^+$; Same reason as in b.

d. $NH_4^+ < PH_4^+$; Same reason as in a.

117.   In general, metal oxides form basic solutions in water and nonmetal oxides form acidic solutions in water.

a. basic; $CaO(s) + H_2O(l) \rightarrow Ca(OH)_2(aq)$, $Ca(OH)_2$ is a strong base.

b. acidic; $SO_2(g) + H_2O(l) \rightarrow H_2SO_3(aq)$, H_2SO_3 is a weak diprotic acid.

c. acidic; $Cl_2O(g) + H_2O(l) \rightarrow 2\ HOCl(aq)$, HOCl is a weak acid.

Lewis Acids and Bases

119.   A Lewis base is an electron pair donor, and a Lewis acid is an electron pair acceptor.

   a. $B(OH)_3$, acid; H_2O, base   b. Ag^+, acid; NH_3, base   c. BF_3, acid; F^-, base

121. $Al(OH)_3(s) + 3 H^+(aq) \rightarrow Al^{3+}(aq) + 3 H_2O(l)$ (Brønsted-Lowry base, H^+ acceptor)

   $Al(OH)_3(s) + OH^-(aq) \rightarrow Al(OH)_4^-(aq)$ (Lewis acid, electron pair acceptor)

123. Fe^{3+} should be the stronger Lewis acid. Fe^{3+} is smaller and has a greater positive charge. Because of this, Fe^{3+} will be more strongly attracted to lone pairs of electrons as compared to Fe^{2+}.

Additional Exercises

125.   At pH = 2.000, $[H^+] = 10^{-2.000} = 1.00 \times 10^{-2}$ M; At pH = 4.000, $[H^+] = 10^{-4.000} = 1.00 \times 10^{-4}$ M

   $$\text{mol } H^+ \text{ present} = 0.0100 \text{ L} \times \frac{0.0100 \text{ mol } H^+}{\text{L}} = 1.00 \times 10^{-4} \text{ mol } H^+$$

   Let V = total volume of solution at pH = 4.000: 1.00×10^{-4} mol/L $= \dfrac{1.00 \times 10^{-4} \text{ mol } H^+}{V}$, V = 1.00 L

   Volume of water added = 1.00 L - 0.0100 L = 0.99 L = 990 mL

127.   The light bulb is bright because a strong electrolyte is present, i.e., a solute is present that dissolves to produce a lot of ions in solution. The pH meter value of 4.6 indicates that a weak acid is present. (If a strong acid were present, the pH would be close to zero.) Of the possible substances, only HCl (strong acid), NaOH (strong base) and NH_4Cl are strong electrolytes. Of these three substances, only NH_4Cl contains a weak acid (the HCl solution would have a pH close to zero and the NaOH solution would have a pH close to 14.0). NH_4Cl dissociates into NH_4^+ and Cl^- ions when dissolved in water. Cl^- is the conjugate base of a strong acid, so it has no basic (or acidic properties) in water. NH_4^+, however, is the conjugate acid of the weak base NH_3, so NH_4^+ is a weak acid and would produce a solution with a pH = 4.6 when the concentration is ~1 M.

129.   For $H_2C_6H_6O_6$. $K_{a_1} = 7.9 \times 10^{-5}$ and $K_{a_2} = 1.6 \times 10^{-12}$. Since $K_{a_1} >> K_{a_2}$, then the amount of H^+ produced by the K_{a_2} reaction will be negligible.

$$[H_2C_6H_6O_6]_o = \frac{0.500 \text{ g} \times \dfrac{1 \text{ mol } H_2C_6H_6O_6}{176.12 \text{ g}}}{0.2000 \text{ L}} = 0.0142 \text{ } M$$

	$H_2C_6H_6O_6(aq)$	$\rightleftharpoons$	$HC_6H_6O_6^-(aq)$	+	$H^+(aq)$	$K_{a_1} = 7.9 \times 10^{-5}$
Initial	0.0142 M		0		~0	
Equil.	0.0142 - x		x		x	

$$K_{a_1} = 7.9 \times 10^{-5} = \frac{x^2}{0.0142 - x} \approx \frac{x^2}{0.0142}, \quad x = 1.1 \times 10^{-3}; \quad \text{Assumption fails the 5\% rule.}$$

Solving by the method of successive approximations:

$$7.9 \times 10^{-5} = \frac{x^2}{0.0142 - 1.1 \times 10^{-3}}, \quad x = 1.0 \times 10^{-3}\ M \text{ (consistent answer)}$$

Since H^+ produced by the K_{a_2} reaction will be negligible, $[H^+] = 1.0 \times 10^{-3}$ and pH = 3.00.

131.   For this problem we will abbreviate $CH_2{=}CHCO_2H$ as Hacr and $CH_2{=}CHCO_2^-$ as acr$^-$.

a.   Solving the weak acid problem:

$$\text{Hacr} \quad \rightleftharpoons \quad H^+ \quad + \quad \text{acr}^- \qquad K_a = 5.6 \times 10^{-5}$$

	Hacr	H$^+$	acr$^-$
Initial	0.10 M	~0	0
Equil.	0.10 - x	x	x

$$\frac{x^2}{0.10 - x} = 5.6 \times 10^{-5} \approx \frac{x^2}{0.10}, \quad x = [H^+] = 2.4 \times 10^{-3}\ M;\ \text{pH} = 2.62;\ \text{Assumptions good.}$$

b.   % dissociation $= \dfrac{[H^+]}{[\text{Hacr}]_o} \times 100 = \dfrac{2.4 \times 10^{-3}}{0.10} \times 100 = 2.4\%$

c.   acr$^-$ is a weak base and the major source of OH$^-$ in this solution.

$$\text{acr}^- \ + \ H_2O \ \rightleftharpoons \ \text{Hacr} \ + \ OH^- \qquad K_b = \frac{K_w}{K_a} = \frac{1.0 \times 10^{-14}}{5.6 \times 10^{-5}}$$

	acr$^-$	Hacr	OH$^-$	
Initial	0.050 M	0	~0	$K_b = 1.8 \times 10^{-10}$
Equil.	0.050 - x	x	x	

$$K_b = \frac{[\text{Hacr}][OH^-]}{[\text{acr}^-]} = 1.8 \times 10^{-10} = \frac{x^2}{0.050 - x} \approx \frac{x^2}{0.050}$$

$x = [OH^-] = 3.0 \times 10^{-6}\ M;\ \text{pOH} = 5.52;\ \text{pH} = 8.48$ Assumptions good.

133.   a.

$$Fe(H_2O)_6^{3+} + H_2O \ \rightleftharpoons \ Fe(H_2O)_5(OH)^{2+} \ + \ H_3O^+$$

	Fe(H$_2$O)$_6^{3+}$	Fe(H$_2$O)$_5$(OH)$^{2+}$	H$_3$O$^+$
Initial	0.10 M	0	~0
Equil.	0.10 - x	x	x

$$K_a = \frac{[Fe(H_2O)_5(OH)^{2+}][H_3O^+]}{[Fe(H_2O)_6^{3+}]} = 6.0 \times 10^{-3} = \frac{x^2}{0.10 - x} \approx \frac{x^2}{0.10}$$

$x = 2.4 \times 10^{-2}$; Assumption is poor (x is 24% of 0.10). Using successive approximations:

$$\frac{x^2}{0.10 - 0.024} = 6.0 \times 10^{-3}, \quad x = 0.021$$

$$\frac{x^2}{0.10 - 0.021} = 6.0 \times 10^{-3}, \quad x = 0.022; \qquad \frac{x^2}{0.10 - 0.022} = 6.0 \times 10^{-3}, \quad x = 0.022$$

$x = [H^+] = 0.022\ M;\ \text{pH} = 1.66$

b. Because of the lower charge, Fe^{2+}(aq) will not be as strong an acid as Fe^{3+}(aq). A solution of iron(II) nitrate will be less acidic (have a higher pH) than a solution with the same concentration of iron(III) nitrate.

135.   Solution is acidic from $HSO_4^- \rightleftharpoons H^+ + SO_4^{2-}$. Solving the weak acid problem:

$$HSO_4^- \quad \rightleftharpoons \quad H^+ \quad + \quad SO_4^{2-} \quad\quad K_a = 1.2 \times 10^{-2}$$

Initial	0.10 M	~0	0
Equil.	0.10 - x	x	x

$$1.2 \times 10^{-2} = \frac{[H^+][SO_4^{2-}]}{[HSO_4^-]} = \frac{x^2}{0.10 - x} \approx \frac{x^2}{0.10}, \quad x = 0.035$$

Assumption is not good (x is 35% of 0.10). Using successive approximations:

$$\frac{x^2}{0.10 - x} \approx \frac{x^2}{0.10 - 0.035} = 1.2 \times 10^{-2}, \quad x = 0.028$$

$$\frac{x^2}{0.10 - 0.028} = 1.2 \times 10^{-2}, \quad x = 0.029; \quad\quad \frac{x^2}{0.10 - 0.029} = 1.2 \times 10^{-2}, \quad x = 0.029$$

$$x = [H^+] = 0.029 \ M; \quad pH = 1.54$$

137.   a. In the lungs, there is a lot of O_2 and the equilibrium favors $Hb(O_2)_4$. In the cells, there is a deficiency of O_2, and the equilibrium favors HbH_4^{4+}.

b. CO_2 is a weak acid, $CO_2 + H_2O \rightleftharpoons HCO_3^- + H^+$. Removing CO_2 essentially decreases H^+. $Hb(O_2)_4$ is then favored and O_2 is not released by hemoglobin in the cells. Breathing into a paper bag increases CO_2 in the blood, thus increasing H^+, which shifts the reaction left.

c. CO_2 builds up in the blood and it becomes too acidic, driving the equilibrium to the left. Hemoglobin can't bind O_2 as strongly in the lungs. Bicarbonate ion acts as a base in water and neutralizes the excess acidity.

139.   a. H_2SO_3 b. $HClO_3$ c. H_3PO_3

NaOH and KOH are soluble ionic compounds composed of Na^+ and K^+ cations and OH^- anions. All soluble ionic compounds dissolve to form the ions from which they are formed. In oxyacids, the compounds are all covalent compounds in which electrons are shared to form bonds (unlike ionic compounds). When these compounds are dissolved in water, the covalent bond between oxygen and hydrogen breaks to form H^+ ions.

Challenge Problems

141.   Since this is a very dilute solution of NaOH, we must worry about the amount of OH^- donated from the autoionization of water.

$$NaOH \rightarrow Na^+ + OH^-$$

$$H_2O \rightleftharpoons H^+ + OH^- \quad K_w = [H^+][OH^-] = 1.0 \times 10^{-14}$$

This solution, like all solutions, must be charge balanced, that is [positive charge] = [negative charge]. For this problem, the charge balance equation is:

$$[Na^+] + [H^+] = [OH^-], \text{ where } [Na^+] = 1.0 \times 10^{-7} \, M \text{ and } [H^+] = \frac{K_w}{[OH^-]}$$

Substituting into the charge balance equation:

$$1.0 \times 10^{-7} + \frac{1.0 \times 10^{-14}}{[OH^-]} = [OH^-], \quad [OH^-]^2 - 1.0 \times 10^{-7} \, [OH^-] - 1.0 \times 10^{-14} = 0$$

Using the quadratic formula to solve:

$$[OH^-] = \frac{-(-1.0 \times 10^{-7}) \pm [(-1.0 \times 10^{-7})^2 - 4(1)(-1.0 \times 10^{-14})]^{1/2}}{2(1)}$$

$$[OH^-] = 1.6 \times 10^{-7} \, M; \quad pOH = -\log(1.6 \times 10^{-7}) = 6.80; \quad pH = 7.20$$

143.

	HBrO	$\rightleftharpoons$	H^+	+	BrO^-	$K_a = 2 \times 10^{-9}$
Initial	$1.0 \times 10^{-6} \, M$		~0		0	
	x mol/L HBrO dissociates to reach equilibrium					
Change	$-x$	$\rightarrow$	$+x$		$+x$	
Equil.	$1.0 \times 10^{-6} - x$		x		x	

$$K_a = 2 \times 10^{-9} = \frac{x^2}{1.0 \times 10^{-6} - x} \approx \frac{x^2}{1.0 \times 10^{-6}}; \quad x = [H^+] = 4 \times 10^{-8} \, M; \quad pH = 7.4$$

Let's check the assumptions. This answer is impossible! We can't add a small amount of an acid to a neutral solution and get a basic solution. The highest possible pH for an acid in water is 7.0. In the correct solution, we would have to take into account the autoionization of water.

145.   Since NH_3 is so concentrated, we need to calculate the OH^- contribution from the weak base NH_3.

	NH_3	+	H_2O	$\rightleftharpoons$	NH_4^+	+	OH^-	$K_b = 1.8 \times 10^{-5}$
Initial	15.0 M				0		0.0100 M	(Assume no volume change.)
Equil.	15.0 - x				x		0.0100 + x	

$$K_b = 1.8 \times 10^{-5} = \frac{x(0.0100 + x)}{15.0 - x} \approx \frac{x(0.0100)}{15.0}, \ x = 0.027; \ \text{Assumption is horrible} \ (x \text{ is } 270\% \text{ of } 0.0100).$$

Using the quadratic formula:

$$1.8 \times 10^{-5}(15.0 - x) = 0.0100 \, x + x^2, \ x^2 + 0.0100 \, x - 2.7 \times 10^{-4} = 0$$

$$x = 1.2 \times 10^{-2}, \ [OH^-] = 1.2 \times 10^{-2} + 0.0100 = 0.022 \, M$$

147. PO_4^{3-} is the conjugate base of HPO_4^{2-}. The K_a value for HPO_4^{2-} is $K_{a_3} = 4.8 \times 10^{-13}$.

$$PO_4^{3-}(aq) + H_2O(l) \ \rightleftharpoons \ HPO_4^{2-}(aq) + OH^-(aq) \quad K_b = \frac{K_w}{K_{a_3}} = \frac{1.0 \times 10^{-14}}{4.8 \times 10^{-13}} = 0.021$$

HPO_4^{2-} is the conjugate base of $H_2PO_4^-$ ($K_{a_2} = 6.2 \times 10^{-8}$).

$$HPO_4^{2-} + H_2O \ \rightleftharpoons \ H_2PO_4^- + OH^- \quad K_b = \frac{K_w}{K_{a_2}} = \frac{1.0 \times 10^{-14}}{6.2 \times 10^{-8}} = 1.6 \times 10^{-7}$$

$H_2PO_4^-$ is the conjugate base of H_3PO_4 ($K_{a_1} = 7.5 \times 10^{-3}$).

$$H_2PO_4^- + H_2O \ \rightleftharpoons \ H_3PO_4 + OH^- \quad K_b = \frac{K_w}{K_{a_1}} = \frac{1.0 \times 10^{-14}}{7.5 \times 10^{-3}} = 1.3 \times 10^{-12}$$

From the K_b values, PO_4^{3-} is the strongest base. This is expected since PO_4^{3-} is the conjugate base of the weakest acid (HPO_4^{2-}).

149. $$\text{Molality} = m = \frac{0.100 \text{ g} \times \dfrac{1 \text{ mol}}{100.0 \text{ g}}}{0.5000 \text{ kg}} = 2.00 \times 10^{-3} \text{ mol/kg} \approx 2.00 \times 10^{-3} \text{ mol/L} \quad \text{(dilute solution)}$$

$\Delta T_f = iK_f m, \ 0.0056°C = i(1.86°C/\text{molal}) (2.00 \times 10^{-3} \text{ molal}), \ i = 1.5$

If $i = 1.0$, % dissociation = 0% and if $i = 2.0$, % dissociation = 100%. Since $i = 1.5$, the weak acid is 50.% dissociated.

$$HA \ \rightleftharpoons \ H^+ + A^- \qquad K_a = \frac{[H^+][A^-]}{[HA]}$$

Since the weak acid is 50.% dissociated, then:

$$[H^+] = [A^-] = [HA]_o \times 0.50 = 2.00 \times 10^{-3} \, M \times 0.50 = 1.0 \times 10^{-3} \, M$$

$$[HA] = [HA]_o - \text{amount HA reacted} = 2.00 \times 10^{-3} \, M - 1.0 \times 10^{-3} \, M = 1.0 \times 10^{-3} \, M$$

$$K_a = \frac{[H^+][A^-]}{[HA]} = \frac{(1.0 \times 10^{-3})(1.0 \times 10^{-3})}{1.0 \times 10^{-3}} = 1.0 \times 10^{-3}$$

CHAPTER FIFTEEN

APPLICATIONS OF AQUEOUS EQUILIBRIA

Questions

13. A common ion is an ion that appears in an equilibrium reaction but came from a source other than that reaction. Addition of a common ion (H^+ or NO_2^-) to the reaction $HNO_2 \rightleftharpoons H^+ + NO_2^-$ will drive the equilibrium to the left as predicted by Le Chatelier's principle.

15. No, as long as there are both a weak acid and a weak base present, the solution will be buffered. If the concentrations are the same, the buffer will have the same capacity towards added H^+ and added OH^-. In addition, buffers with equal concentrations of weak acid and conjugate base have pH = pK_a.

17. No, since there are three colored forms, there must be two proton transfer reactions. Thus, there must be at least two acidic protons in the acid (orange) form of thymol blue.

19. The two forms of an indicator are different colors. The HIn form has one color and the In⁻ form has another color. To see only one color, that form must be in an approximately ten fold excess or greater over the other form. When the ratio of the two forms is less than 10, both colors are present. To go from $[HIn]/[In^-] = 10$ to $[HIn]/[In^-] = 0.1$ requires a change of 2 pH units (a 100-fold decrease in $[H^+]$) as the indicator changes from the HIn color to the In⁻ color.

Exercises

Buffers

21. When strong acid or strong base is added to a bicarbonate/carbonate mixture, the strong acid/base is neutralized. The reaction goes to completion, resulting in the strong acid/base being replaced with a weak acid/base, which results in a new buffer solution. The reactions are:

$$H^+(aq) + CO_3^{2-}(aq) \rightarrow HCO_3^-(aq); \quad OH^- + HCO_3^-(aq) \rightarrow CO_3^{2-}(aq) + H_2O(l)$$

23.   a. This is a weak acid problem. Let $HC_3H_5O_2$ = HOPr and $C_3H_5O_2^-$ = OPr.

$$HOPr \rightleftharpoons H^+ + OPr^- \qquad K_a = 1.3 \times 10^{-5}$$

Initial   0.100 M ~0   0
             x mol/L HOPr dissociates to reach equilibrium
Change   -x $\rightarrow$ +x +x
Equil.   0.100 - x x x

$$K_a = 1.3 \times 10^{-5} = \frac{[H^+][OPr^-]}{[HOPr]} = \frac{x^2}{0.100 - x} \approx \frac{x^2}{0.100}$$

$x = [H^+] = 1.1 \times 10^{-3}\ M$; pH = 2.96   Assumptions good by the 5% rule.

b. This is a weak base problem (Na^+ has no acidic/basic properties).

$$OPr^- + H_2O \rightleftharpoons HOPr + OH^- \qquad K_b = \frac{K_w}{K_a} = 7.7 \times 10^{-10}$$

Initial   0.100 M 0   ~0
             x mol/L OPr reacts with H_2O to reach equilibrium
Change   -x $\rightarrow$ +x +x
Equil.   0.100 - x x x

$$K_b = 7.7 \times 10^{-10} = \frac{[HOPr][OH^-]}{[OPr^-]} = \frac{x^2}{0.100 - x} \approx \frac{x^2}{0.100}$$

$x = [OH^-] = 8.8 \times 10^{-6}\ M$; pOH = 5.06; pH = 8.94   Assumptions good.

c. pure H_2O, $[H^+] = [OH^-] = 1.0 \times 10^{-7}\ M$; pH = 7.00

d. This solution contains a weak acid and its conjugate base. This is a buffer solution. We will solve for the pH through the weak acid equilibrium reaction.

$$HOPr \rightleftharpoons H^+ + OPr^- \qquad K_a = 1.3 \times 10^{-5}$$

Initial   0.100 M ~0   0.100 M
             x mol/L HOPr dissociates to reach equilibrium
Change   -x $\rightarrow$ +x +x
Equil.   0.100 - x x 0.100 + x

$$1.3 \times 10^{-5} = \frac{(0.100 + x)(x)}{0.100 - x} \approx \frac{(0.100)(x)}{0.100} = x = [H^+]$$

$[H^+] = 1.3 \times 10^{-5}\ M$; pH = 4.89   Assumptions good.

Alternatively, we can use the Henderson-Hasselbalch equation to calculate the pH of buffer solutions.

$$pH = pK_a + \log \frac{[Base]}{[Acid]} = pK_a + \log \frac{(0.100)}{(0.100)} = pK_a = -\log (1.3 \times 10^{-5}) = 4.89$$

The Henderson-Hasselbalch equation will be valid when an assumption of the type $0.1 + x \approx 0.1$ that we just made in this problem is valid. From a practical standpoint, this will almost always be true for useful buffer solutions. If the assumption is not valid, the solution will have such a low buffering capacity that it will be of no use to control the pH. Note: The Henderson-Hasselbalch equation can <u>only</u> be used to solve for the pH of buffer solutions.

25. $0.100 \ M \ HC_3H_5O_2$: percent dissociation $= \dfrac{[H^+]}{[HC_3H_5O_2]_o} \times 100 = \dfrac{1.1 \times 10^{-3} \ M}{0.100 \ M} \times 100 = 1.1\%$

$0.100 \ M \ HC_3H_5O_2 + 0.100 \ M \ NaC_3H_5O_2$: % dissociation $= \dfrac{1.3 \times 10^{-5}}{0.100} \times 100 = 1.3 \times 10^{-2}\%$

The percent dissociation of the acid decreases from 1.1% to 1.3×10^{-2} % when $C_3H_5O_2^-$ is present. This is known as the common ion effect. The presence of the conjugate base of the weak acid inhibits the acid dissociation reaction.

27.   a.   We have a weak acid ($HOPr = HC_3H_5O_2$) and a strong acid (HCl) present. The amount of H^+ donated by the weak acid will be negligible as compared to the $0.020 \ M \ H^+$ from the strong acid. To prove it let's consider the weak acid equilibrium reaction:

	HOPr	$\rightleftharpoons$	H^+	+	OPr^-	$K_a = 1.3 \times 10^{-5}$
Initial	0.100 M		0.020 M		0	
	x mol/L HOPr dissociates to reach equilibrium					
Change	$-x$	$\rightarrow$	$+x$		$+x$	
Equil.	0.100 - x		0.020 + x		x	

$$K_a = 1.3 \times 10^{-5} = \frac{(0.020 + x)(x)}{0.100 - x} \approx \frac{(0.020)(x)}{0.100}, \quad x = 6.5 \times 10^{-5} \ M$$

$[H^+] = 0.020 + x = 0.020 \ M$; pH = 1.70   Assumptions good ($x = 6.5 \times 10^{-5}$ which is $\ll 0.020$).

b.   Added H^+ reacts completely with the best base present, OPr^-. Since all species present are in the same volume of solution, we can use molarity units to do the stoichiometry part of the problem (instead of moles). The stoichiometry problem is:

	OPr^-	+	H^+	$\rightarrow$	HOPr	
Before	0.100 M		0.020 M		0	
Change	-0.020		-0.020	$\rightarrow$	+0.020	Reacts completely
After	0.080		0		0.020 M	

After reaction, a weak acid, HOPr , and its conjugate base, OPr^-, are present. This is a buffer solution. Using the Henderson-Hasselbalch equation where $pK_a = -\log (1.3 \times 10^{-5}) = 4.89$:

$$pH = pK_a + \log \frac{[Base]}{[Acid]} = 4.89 + \log \frac{(0.080)}{(0.020)} = 5.49 \qquad \text{Assumptions good.}$$

c. This is a strong acid problem. $[H^+] = 0.020 \ M$; $pH = 1.70$

d. Added H^+ reacts completely with the best base present, OPr^-.

	OPr^-	+	H^+	$\rightarrow$	$HOPr$	
Before	0.100 M		0.020 M		0.100 M	
Change	-0.020		-0.020	$\rightarrow$	+0.020	Reacts completely
After	0.080		0		0.120	

A buffer solution results (weak acid + conjugate base). Using the Henderson-Hasselbalch equation:

$$pH = pK_a + \log \frac{[Base]}{[Acid]} = 4.89 + \log \frac{(0.080)}{(0.120)} = 4.71$$

29. a. OH^- will react completely with the best acid present, $HOPr$.

	$HOPr$	+	OH^-	$\rightarrow$	OPr^-	+	H_2O	
Before	0.100 M		0.020 M		0			
Change	-0.020		-0.020	$\rightarrow$	+0.020			Reacts completely
After	0.080		0		0.020			

A buffer solution results after the reaction. Using the Henderson-Hasselbalch equation:

$$pH = pK_a + \log \frac{[Base]}{[Acid]} = 4.89 + \log \frac{(0.020)}{(0.080)} = 4.29$$

b. We have a weak base and a strong base present at the same time. The amount of OH^- added by the weak base will be negligible compared to the 0.020 M OH^- from the strong base. To prove it, let's consider the weak base equilibrium:

	OPr^-	+	H_2O	$\rightleftharpoons$	$HOPr$	+	OH^-	$K_b = 7.7 \times 10^{-10}$
Initial	0.100 M				0		0.020 M	

x mol/L OPr^- reacts with H_2O to reach equilibrium

Change	$-x$			$\rightarrow$	$+x$		$+x$
Equil.	0.100 - x				x		0.020 + x

$$K_b = 7.7 \times 10^{-10} = \frac{x(0.020 + x)}{0.100 - x} \approx \frac{x(0.020)}{0.100}, \quad x = 3.9 \times 10^{-9} \ M$$

$[OH^-] = 0.020 + x = 0.020 \ M$; $pOH = 1.70$; $pH = 12.30$ Assumptions good.

c. This is a strong base in water. $[OH^-] = 0.020 \ M$; $pOH = 1.70$; $pH = 12.30$

d. OH⁻ will react completely with HOPr, the best acid present.

$$HOPr \quad + \quad OH^- \quad \rightarrow \quad OPr^- \quad + \quad H_2O$$

Before	0.100 *M*	0.020 *M*		0.100 *M*
Change	-0.020	-0.020	$\rightarrow$	+0.020
After	0.080	0		0.120

Reacts completely

Using the Henderson-Hasselbalch equation to solve for the pH of the resulting buffer solution:

$$pH = pK_a + \log \frac{[Base]}{[Acid]} = 4.89 + \log \frac{(0.120)}{(0.080)} = 5.07$$

31.   Consider all of the results to Exercises 15.23, 15.27, and 15.29:

Solution	Initial pH	after added acid	after added base
a	2.96	1.70	4.29
b	8.94	5.49	12.30
c	7.00	1.70	12.30
d	4.89	4.71	5.07

The solution in Exercise 15.23d is a buffer; it contains both a weak acid ($HC_3H_5O_2$) and a weak base ($C_3H_5O_2^-$). Solution d shows the greatest resistance to changes in pH when either strong acid or strong base is added, which is the primary property of buffers.

33.   Major species: HNO_2, NO_2^- and Na^+. Na^+ has no acidic or basic properties. The appropriate equilibrium reaction to use is the K_a reaction of HNO_2 which contains both HNO_2 and NO_2^-. Solving the equilibrium problem (called a buffer problem):

$$HNO_2 \quad \rightleftharpoons \quad NO_2^- \quad + \quad H^+$$

Initial	1.00 *M*	1.00 *M*	~0

x mol/L HNO_2 dissociates to reach equilibrium

Change	-*x*	$\rightarrow$ +*x*	+*x*
Equil.	1.00 - *x*	1.00 + *x*	*x*

$$K_a = 4.0 \times 10^{-4} = \frac{[NO_2^-][H^+]}{[HNO_2]} = \frac{(1.00 + x)(x)}{(1.00 - x)} \approx \frac{1.00(x)}{1.00} \quad (\text{assuming } x \ll 1.00)$$

$x = 4.0 \times 10^{-4} M = [H^+]$;   Assumptions good (*x* is 4.0×10^{-2}% of 1.00).

$pH = -\log (4.0 \times 10^{-4}) = 3.40$

Note: We would get the same answer using the Henderson-Hasselbalch equation. Use whichever method you prefer.

35.   Major species after NaOH added: HNO_2, NO_2^-, Na^+ and OH^-. The OH^- from the strong base will react with the best acid present (HNO_2). Any reaction involving a strong base is assumed to go to completion. Since all species present are in the same volume of solution, we can use molarity units to do the stoichiometry part of the problem (instead of moles). The stoichiometry problem is:

	OH^-	+	HNO_2	$\rightarrow$	NO_2^-	+	H_2O
Before	0.10 mol/1.00 L		1.00 M		1.00 M		
Change	-0.10 M		-0.10 M	$\rightarrow$	+0.10 M		Reacts completely
After	0		0.90		1.10		

After all the OH^- reacts, we are left with a solution containing a weak acid (HNO_2) and its conjugate base (NO_2^-). This is what we call a buffer problem. We will solve this buffer problem using the K_a equilibrium reaction.

	HNO_2	$\rightleftharpoons$	NO_2^-	+	H^+
Initial	0.90 M		1.10 M		~0
	x mol/L HNO_2 dissociates to reach equilibrium				
Change	-x	$\rightarrow$	+x		+x
Equil.	0.90 - x		1.10 + x		x

$K_a = 4.0 \times 10^{-4} = \dfrac{(1.10 + x)(x)}{(0.90 - x)} \approx \dfrac{1.10(x)}{0.90}$, $\quad x = [H^+] = 3.3 \times 10^{-4}$ M; pH = 3.48; Assumptions good.

Note: The added NaOH to this buffer solution changes the pH only from 3.40 to 3.48. If the NaOH were added to 1.0 L of pure water, the pH would change from 7.00 to 13.00.

Major species after HCl added: HNO_2, NO_2^-, H^+, Na^+, Cl^-; The added H^+ from the strong acid will react completely with the best base present (NO_2^-).

	H^+	+	NO_2^-	$\rightarrow$	HNO_2
Before	$\dfrac{0.20 \text{ mol}}{1.00 \text{ L}}$		1.00 M		1.00 M
Change	-0.20 M		-0.20 M	$\rightarrow$	+0.20 M	Reacts completely
After	0		0.80		1.20	

After all the H^+ has reacted, we have a buffer solution (a solution containing a weak acid and its conjugate base). Solving the buffer problem:

	HNO_2	$\rightleftharpoons$	NO_2^-	+	H^+
Initial	1.20 M		0.80 M		0
Equil.	1.20 - x		0.80 + x		+x

$K_a = 4.0 \times 10^{-4} = \dfrac{(0.80 + x)(x)}{1.20 - x} \approx \dfrac{0.80(x)}{1.20}$, $\quad x = [H^+] = 6.0 \times 10^{-4}$ M; pH = 3.22; Assumptions good.

Note: The added HCl to this buffer solution changes the pH only from 3.40 to 3.22. If the HCl were added to 1.0 L of pure water, the pH would change from 7.00 to 0.70.

37.   a. $HC_2H_3O_2 \rightleftharpoons H^+ + C_2H_3O_2^-$ $K_a = 1.8 \times 10^{-5}$

Initial	0.10 M	~0	0.25 M

x mol/L $HC_2H_3O_2$ dissociates to reach equilibrium

Change	-x	$\rightarrow$ +x	+x
Equil.	0.10 - x	x	0.25 + x

$$1.8 \times 10^{-5} = \frac{x(0.25 + x)}{(0.10 - x)} \approx \frac{x(0.25)}{0.10} \quad \text{(assuming } 0.25 + x \approx 0.25 \text{ and } 0.10 - x \approx 0.10)$$

$x = [H^+] = 7.2 \times 10^{-6} \, M; \; pH = 5.14$ Assumptions good by the 5% rule.

Alternatively, we can use the Henderson-Hasselbalch equation:

$$pH = pK_a + \log \frac{[Base]}{[Acid]} \quad \text{where } pK_a = -\log(1.8 \times 10^{-5}) = 4.74$$

$$pH = 4.74 + \log \frac{(0.25)}{(0.10)} = 4.74 + 0.40 = 5.14$$

The Henderson-Hasselbalch equation will be valid when assumptions of the type $0.10 - x \approx 0.10$ that we just made are valid. From a practical standpoint, this will almost always be true for useful buffer solutions. Note: The Henderson-Hasselbalch equation can <u>only</u> be used to solve for the pH of buffer solutions.

b. $pH = 4.74 + \log \dfrac{(0.10)}{(0.25)} = 4.74 + (-0.40) = 4.34$

c. $pH = 4.74 + \log \dfrac{(0.25)}{(0.25)} = 4.74 + 0.00 = 4.74$ ($pH = pK_a$ since [acid] = [base])

d. $pH = pK_a + \log \dfrac{[base]}{[acid]}$; [base] = $[C_2H_5NH_2]$ = 0.50 M; [acid] = $[C_2H_5NH_3^+]$ = 0.25 M

$$K_a = \frac{K_w}{K_b} = \frac{1.0 \times 10^{-14}}{5.6 \times 10^{-4}} = 1.8 \times 10^{-11}$$

$$pH = -\log(1.8 \times 10^{-11}) + \log \left(\frac{0.50 \, M}{0.25 \, M} \right) = 10.74 + 0.30 = 11.04$$

e. $pH = 10.74 + \log \left(\dfrac{0.50 \, M}{0.50 \, M} \right) = 10.74 + 0.00 = 10.74$

39. $[H^+]$ added = $\dfrac{0.010 \text{ mol}}{0.25 \text{ L}}$ = 0.040 M; The added H^+ reacts completely with NH_3 to form NH_4^+.

a. NH_3 + H^+ $\rightarrow$ NH_4^+

Before	0.050 M	0.040 M		0.15 M
Change	-0.040	-0.040	$\rightarrow$	+0.040	Reacts completely
After	0.010	0		0.19

A buffer solution still exists after H^+ reacts completely. Using the Henderson-Hasselbalch equation:

$$pH = pK_a + \log \frac{[NH_3]}{[NH_4^+]} = -\log (5.6 \times 10^{-10}) + \log \left(\frac{0.010}{0.19} \right) = 9.25 + (-1.28) = 7.97$$

b. NH_3 + H^+ $\rightarrow$ NH_4^+

Before	0.50 M	0.040 M		1.50 M
Change	-0.040	-0.040	$\rightarrow$	+0.040	Reacts completely
After	0.46	0		1.54

A buffer solution still exists. $pH = pK_a + \log \dfrac{[NH_3]}{[NH_4^+]} = 9.25 + \log \left(\dfrac{0.46}{1.54} \right) = 8.73$

The two buffers differ in their capacity and not in pH (both buffers had an initial pH = 8.77). Solution b has the greater capacity since it has the largest concentration of weak acid and conjugate base. Buffers with greater capacities will be able to absorb more H^+ or OH^- added.

41. $pH = pK_a + \log \dfrac{[C_2H_3O_2^-]}{[HC_2H_3O_2]}$; $pK_a = -\log(1.8 \times 10^{-5}) = 4.74$

Since the buffer components, $C_2H_3O_2^-$ and $HC_2H_3O_2$, are both in the same volume of water, the concentration ratio of $[C_2H_3O_2^-]/[HC_2H_3O_2]$ will equal the mol ratio of mol $C_2H_3O_2^-$/mol $HC_2H_3O_2$.

$$5.00 = 4.74 + \log \frac{mol\ C_2H_3O_2^-}{mol\ HC_2H_3O_2};\ \ mol\ HC_2H_3O_2 = 0.5000\ L \times \frac{0.200\ mol}{L} = 0.100\ mol$$

$$0.26 = \log \frac{mol\ C_2H_3O_2^-}{0.100\ mol},\ \ \frac{mol\ C_2H_3O_2^-}{0.100} = 10^{0.26} = 1.8,\ \ mol\ C_2H_3O_2^- = 0.18\ mol$$

$$mass\ NaC_2H_3O_2 = 0.18\ mol\ NaC_2H_3O_2 \times \frac{82.03\ g}{mol} = 15\ g\ NaC_2H_3O_2$$

43. $C_5H_5NH^+ \rightleftharpoons H^+ + C_5H_5N$ $K_a = \dfrac{K_w}{K_b} = \dfrac{1.0 \times 10^{-14}}{1.7 \times 10^{-9}} = 5.9 \times 10^{-6}$; $pK_a = -\log (5.9 \times 10^{-6}) = 5.23$

We will use the Henderson-Hasselbalch equation to calculate the concentration ratio necessary for each buffer.

$$pH = pK_a + \log \frac{[base]}{[acid]},\ \ pH = 5.23 + \log \frac{[C_5H_5N]}{[C_5H_5NH^+]}$$

a. $4.50 = 5.23 + \log \dfrac{[C_5H_5N]}{[C_5H_5NH^+]}$ b. $5.00 = 5.23 + \log \dfrac{[C_5H_5N]}{[C_5H_5NH^+]}$

$\log \dfrac{[C_5H_5N]}{[C_5H_5NH^+]} = -0.73$ $\log \dfrac{[C_5H_5N]}{[C_5H_5NH^+]} = -0.23$

$\dfrac{[C_5H_5N]}{[C_5H_5NH^+]} = 10^{-0.73} = 0.19$ $\dfrac{[C_5H_5N]}{[C_5H_5NH^+]} = 10^{-0.23} = 0.59$

c. $5.23 = 5.23 + \log \dfrac{[C_5H_5N]}{[C_5H_5NH^+]}$ d. $5.50 = 5.23 + \log \dfrac{[C_5H_5N]}{[C_5H_5NH^+]}$

$\dfrac{[C_5H_5N]}{[C_5H_5NH^+]} = 10^{0.0} = 1.0$ $\dfrac{[C_5H_5N]}{[C_5H_5NH^+]} = 10^{0.27} = 1.9$

45. A best buffer has large and equal quantities of weak acid and conjugate base. Since [acid] = [base] for a best buffer, then $pH = pK_a + \log \dfrac{[base]}{[acid]} = pK_a + 0 = pK_a$ ($pH = pK_a$).

The best acid choice for a pH = 7.00 buffer would be the weak acid with a pK_a close to 7.0 or $K_a \approx 1 \times 10^{-7}$. HOCl is the best choice in Table 14.2 ($K_a = 3.5 \times 10^{-8}$; $pK_a = 7.46$). To make this buffer, we need to calculate the [base]/[acid] ratio.

$$7.00 = 7.46 + \log \dfrac{[base]}{[acid]}, \quad \dfrac{[OCl^-]}{[HOCl]} = 10^{-0.46} = 0.35$$

Any OCl⁻/HOCl buffer in a concentration ratio of 0.35:1 will have a pH = 7.00. One possibility is [NaOCl] = 0.35 M and [HOCl] = 1.0 M.

47. The reaction $OH^- + CH_3NH_3^+ \rightarrow CH_3NH_2 + H_2O$ goes to completion for solutions a, c and d (no reaction occurs between the species in solution b since both species are bases). After the OH^- reacts completely, there must be both $CH_3NH_3^+$ and CH_3NH_2 in solution for it to be a buffer. The important components of each solution (after the OH^- reacts completely) is/are:

a. 0.05 M CH_3NH_2 (no $CH_3NH_3^+$ remains, no buffer)

b. 0.05 M OH^- and 0.1 M CH_3NH_2 (two bases present, no buffer)

c. 0.05 M OH^- and 0.05 M CH_3NH_2 (too much OH^- added, no $CH_3NH_3^+$ remains, no buffer)

d. 0.05 M CH_3NH_2 and 0.05 M $CH_3NH_3^+$ (a buffered solution results)

Only the combination in mixture d results in a buffer. Note that the concentrations are halved from the initial values. This is because equal volumes of two solutions were added together, which halves the concentrations.

49.   When OH⁻ is added, it converts $HC_2H_3O_2$ into $C_2H_3O_2^-$: $HC_2H_3O_2 + OH^- \rightarrow C_2H_3O_2^- + H_2O$

From this reaction, the moles of $C_2H_3O_2^-$ produced <u>equal</u> the moles of OH⁻ added. Also, the total concentration of acetic acid plus acetate ion must equal 2.0 M (assuming no volume change on addition of NaOH). Summarizing for each solution:

$[C_2H_3O_2^-] + [HC_2H_3O] = 2.0 M$ and $[C_2H_3O_2^-]$ produced = [OH⁻] added

a. $pH = pK_a + \log \dfrac{[C_2H_3O_2^-]}{[HC_2H_3O_2]}$; For $pH = pK_a$, $\log \dfrac{[C_2H_3O_2^-]}{[HC_2H_3O_2]} = 0$

Therefore, $\dfrac{[C_2H_3O_2^-]}{[HC_2H_3O_2]} = 1.0$ and $[C_2H_3O_2^-] = [HC_2H_3O_2]$

Since $[C_2H_3O_2^-] + [HC_2H_3O_2] = 2.0 M$, then $[C_2H_3O_2^-] = [HC_2H_3O_2] = 1.0 M = [OH^-]$ added

To produce a 1.0 M $C_2H_3O_2^-$ solution, we need to add 1.0 mol of NaOH to 1.0 L of the 2.0 M $HC_2H_3O_2$ solution. The resultant solution will have $pH = pK_a = 4.74$.

b. $4.00 = 4.74 + \log \dfrac{[C_2H_3O_2^-]}{[HC_2H_3O_2]}$, $\dfrac{[C_2H_3O_2^-]}{[HC_2H_3O_2]} = 10^{-0.74} = 0.18$

$[C_2H_3O_2^-] = 0.18 [HC_2H_3O_2]$ or $[HC_2H_3O_2] = 5.6 [C_2H_3O_2^-]$; Since $[C_2H_3O_2^-] + [HC_2H_3O_2] = 2.0\ M$, then:

$[C_2H_3O_2^-] + 5.6 [C_2H_3O_2^-] = 2.0 M$, $[C_2H_3O_2^-] = \dfrac{2.0}{6.6} = 0.30 M = [OH^-]$ added

We need to add 0.30 mol of NaOH to 1.0 L of 2.0 M $HC_2H_3O_2$ solution to produce 0.30 M $C_2H_3O_2^-$. The resultant solution will have $pH = 4.00$.

c. $5.00 = 4.74 + \log \dfrac{[C_2H_3O_2^-]}{[HC_2H_3O_2]}$, $\dfrac{[C_2H_3O_2^-]}{[HC_2H_3O_2]} = 10^{0.26} = 1.8$

$1.8 [HC_2H_3O_2] = [C_2H_3O_2^-]$ or $[HC_2H_3O_2] = 0.56 [C_2H_3O_2^-]$; Since $[HC_2H_3O_2] + [C_2H_3O_2^-] = 2.0\ M$, then:

$1.56 [C_2H_3O_2^-] = 2.0 M$, $[C_2H_3O_2^-] = 1.3 M = [OH^-]$ added

We need to add 1.3 mol of NaOH to 1.0 L of 2.0 M $HC_2H_3O_2$ to produce a solution with $pH = 5.00$.

Acid-Base Titrations

51.

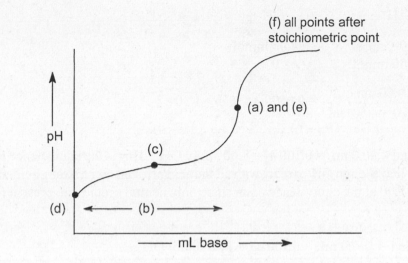

$HA + OH^- \rightarrow A^- + H_2O$; Added OH^- from the strong base converts the weak acid, HA, into its conjugate base, A^-. Initially, before any OH^- is added (point d), HA is the dominant species present. After OH^- is added, both HA and A^- are present and a buffer solution results (region b). At the equivalence point (points a and e), exactly enough OH^- has been added to convert all of the weak acid, HA, into its conjugate base, A^-. Past the equivalence point (region f), excess OH^- is present. For the answer to part b, we included almost the entire buffer region. The maximum buffer region (or the region which is the best buffer solution) is around the halfway point to equivalence (point c). At this point, enough OH^- has been added to convert exactly one-half of the weak acid present initially into its conjugate base so $[HA] = [A^-]$ and $pH = pK_a$. A best buffer has about equal concentrations of weak acid and conjugate base present.

53.   This is a strong acid ($HClO_4$) titrated by a strong base (KOH). Added OH^- from the strong base will react completely with the H^+ present from the strong acid to produce H_2O.

a.   Only strong acid present. $[H^+] = 0.200\ M$; $pH = 0.699$

b.   mmol OH^- added $= 10.0\ mL \times \dfrac{0.100\ mmol\ OH^-}{mL} = 1.00\ mmol\ OH^-$

mmol H^+ present $= 40.0\ mL \times \dfrac{0.200\ mmol\ H^+}{mL} = 8.0\ mmol\ H^+$

Note: The units mmoles are usually easier numbers to work with. The units for molarity are moles/L but are also equal to mmoles/mL.

	H^+	$+$	OH^-	$\rightarrow$	H_2O	
Before	8.00 mmol		1.00 mmol			
Change	-1.00 mmol		-1.00 mmol			Reacts completely
After	7.00 mmol		0			

The excess H^+ determines pH. $[H^+]_{excess} = \dfrac{7.00\ mmol\ H^+}{40.0\ mL + 10.0\ mL} = 0.140\ M$; $pH = 0.854$

c. mmol OH⁻ added = 40.0 mL × 0.100 M = 4.00 mmol OH⁻

$$H^+ \quad + \quad OH^- \quad \rightarrow \quad H_2O$$

Before   8.00 mmol   4.00 mmol
After   4.00 mmol   0

$$[H^+]_{excess} = \frac{4.00 \text{ mmol}}{(40.0 + 40.0) \text{ mL}} = 0.0500 \text{ } M; \text{ pH} = 1.301$$

d. mmol OH⁻ added = 80.0 mL × 0.100 M = 8.00 mmol OH⁻; This is the equivalence point since we have added just enough OH⁻ to react with all the acid present. For a strong acid-strong base titration, pH = 7.00 at the equivalence point since only neutral species are present (K^+, ClO_4^-, H_2O).

e. mmol OH⁻ added = 100.0 mL × 0.100 M = 10.0 mmol OH⁻

$$H^+ \quad + \quad OH^- \quad \rightarrow \quad H_2O$$

Before   8.00 mmol   10.0 mmol
After   0   2.0 mmol

Past the equivalence point, the pH is determined by the excess OH⁻ present.

$$[OH^-]_{excess} = \frac{2.0 \text{ mmol}}{(40.0 + 100.0) \text{ mL}} = 0.014 \text{ } M; \text{ pOH} = 1.85; \text{ pH} = 12.15$$

55.   This is a weak acid ($HC_2H_3O_2$) titrated by a strong base (KOH).

a. Only a weak acid is present. Solving the weak acid problem:

$$HC_2H_3O_2 \quad \rightleftharpoons \quad H^+ \quad + \quad C_2H_3O_2^-$$

Initial   0.200 M ~0   0
        x mol/L $HC_2H_3O_2$ dissociates to reach equilibrium
Change   -x $\rightarrow$ +x +x
Equil.   0.200 - x x x

$$K_a = 1.8 \times 10^{-5} = \frac{x^2}{0.200 - x} = \frac{x^2}{0.200}, \quad x = [H^+] = 1.9 \times 10^{-3} \text{ } M \quad \text{pH} = 2.72; \text{ Assumptions good.}$$

b. The added OH⁻ will react completely with the best acid present, $HC_2H_3O_2$.

$$\text{mmol } HC_2H_3O_2 \text{ present} = 100.0 \text{ mL} \times \frac{0.200 \text{ mmol } HC_2H_3O_2}{\text{mL}} = 20.0 \text{ mmol } HC_2H_3O_2$$

$$\text{mmol OH}^- \text{ added} = 50.0 \text{ mL} \times \frac{0.100 \text{ mmol OH}^-}{\text{mL}} = 5.00 \text{ mmol OH}^-$$

$$HC_2H_3O_2 \quad + \quad OH^- \quad \rightarrow \quad C_2H_3O_2^- \quad + \quad H_2O$$

Before	20.0 mmol	5.00 mmol	0
Change	-5.00 mmol	-5.00 mmol →	+5.00 mmol
After	15.0 mmol	0	5.00 mmol

Reacts completely

After reaction of all the strong base, we will have a buffered solution containing a weak acid ($HC_2H_3O_2$) and its conjugate base ($C_2H_3O_2^-$). We will use the Henderson-Hasselbalch equation to solve for the pH.

$$pH = pK_a + \log \frac{[C_2H_3O_2^-]}{[HC_2H_3O_2]} = -\log(1.8 \times 10^{-5}) + \log\left(\frac{5.00 \text{ mmol}/V_T}{15.0 \text{ mmol}/V_T}\right) \text{ where } V_T = \text{total}$$
volume

$$pH = 4.74 + \log\left(\frac{5.00}{15.0}\right) = 4.74 + (-0.477) = 4.26$$

Note that the total volume cancels in the Henderson-Hasselbalch equation. For the [base]/[acid] term, the mole ratio equals the concentration ratio since the components of the buffer are always in the same volume of solution.

c. mmol OH^- added = 100.0 mL $\times$ 0.100 mmol OH^-/mL = 10.0 mmol OH^-; The same amount (20.0 mmol) of $HC_2H_3O_2$ is present as before (it never changes). As before, let the OH^- react to completion, then see what remains in solution after the reaction.

$$HC_2H_3O_2 \quad + \quad OH^- \quad \rightarrow \quad C_2H_3O_2^- \quad + \quad H_2O$$

Before	20.0 mmol	10.0 mmol	0
After	10.0 mmol	0	10.0 mmol

A buffered solution results after the reaction. Since $[C_2H_3O_2^-] = [HC_2H_3O_2] = 10.0$ mmol/total volume, then pH = pK_a. This is always true at the halfway point to equivalence for a weak acid/strong base titration, pH = pK_a.

$$pH = -\log(1.8 \times 10^{-5}) = 4.74$$

d. mmol OH^- added = 150.0 mL $\times$ 0.100 M = 15.0 mmol OH^-. Added OH^- reacts completely with the weak acid.

$$HC_2H_3O_2 \quad + \quad OH^- \quad \rightarrow \quad C_2H_3O_2^- \quad + \quad H_2O$$

Before	20.0 mmol	15.0 mmol	0
After	5.0 mmol	0	15.0 mmol

We have a buffered solution after all the OH^- reacts to completion. Using the Henderson-Hasselbalch equation:

$$pH = 4.74 + \log \frac{[C_2H_3O_2^-]}{[HC_2H_3O_2]} = 4.74 + \log\left(\frac{15.0 \text{ mmol}}{5.0 \text{ mmol}}\right) \quad \text{(Total volume cancels, so we}$$
can use mol ratios.)

$$pH = 4.74 + 0.48 = 5.22$$

e.   mmol OH^- added = 200.00 mL $\times$ 0.100 M = 20.0 mmol OH^-; As before, let the added OH^- react to completion with the weak acid, then see what is in solution after this reaction.

$$HC_2H_3O_2 \quad + \quad OH^- \quad \rightarrow \quad C_2H_3O_2^- \ + \ H_2O$$

Before	20.0 mmol	20.0 mmol	0
After	0	0	20.0 mmol

This is the equivalence point. Enough OH^- has been added to exactly neutralize all the weak acid present initially. All that remains that affects the pH at the equivalence point is the conjugate base of the weak acid, $C_2H_3O_2^-$. This is a weak base equilibrium problem.

$$C_2H_3O_2^- + H_2O \ \rightleftharpoons \ HC_2H_3O_2 \ + \ OH^- \quad K_b = \frac{K_w}{K_a} = \frac{1.0 \times 10^{-14}}{1.8 \times 10^{-5}} = 5.6 \times 10^{-10}$$

Initial	20.0 mmol/300.0 mL	0	0
	x mol/L $C_2H_3O_2^-$ reacts with H_2O to reach equilibrium		
Change	$-x$ $\rightarrow$	$+x$	$+x$
Equil.	$0.0667 - x$	x	x

$$K_b = 5.6 \times 10^{-10} = \frac{x^2}{0.0667 - x} \approx \frac{x^2}{0.0667}, \quad x = [OH^-] = 6.1 \times 10^{-6} \ M$$

pOH = 5.21;   pH = 8.79;   Assumptions good.

f.   mmol OH^- added = 250.0 mL $\times$ 0.100 M = 25.0 mmol OH^-

$$HC_2H_3O_2 \quad + \quad OH^- \quad \rightarrow \quad C_2H_3O_2^- \ + \ H_2O$$

Before	20.0 mmol	25.0 mmol	0
After	0	5.0 mmol	20.0 mmol

After the titration reaction, we will have a solution containing excess OH^- and a weak base, $C_2H_3O_2^-$. When a strong base and a weak base are both present, assume the amount of OH^- added from the weak base will be minimal, i.e., the pH past the equivalence point will be determined by the amount of excess strong base.

$$[OH^-]_{excess} = \frac{5.0 \ mmol}{100.0 \ mL \ + \ 250.0 \ mL} = 0.014 \ M; \quad pOH = 1.85; \quad pH = 12.15$$

57.   We will do sample calculations for the various parts of the titration. All results are summarized in Table 15.1 at the end of Exercise 15.59.

At the beginning of the titration, only the weak acid $HC_3H_5O_3$ is present.

$$HLac \quad \rightleftharpoons \quad H^+ \quad + \quad Lac^- \quad K_a = 10^{-3.86} = 1.4 \times 10^{-4} \quad HLac = HC_3H_5O_3$$
$$Lac^- = C_3H_5O_3^-$$

Initial	0.100 M	~0	0
	x mol/L HLac dissociates to reach equilibrium		
Change	$-x$ $\rightarrow$	$+x$	$+x$
Equil.	$0.100 - x$	x	x

$$1.4 \times 10^{-4} = \frac{x^2}{0.100 - x} = \frac{x^2}{0.100}, \quad x = [H^+] = 3.7 \times 10^{-3} \ M; \quad pH = 2.43 \quad \text{Assumptions good.}$$

Up to the stoichiometric point, we calculate the pH using the Henderson-Hasselbalch equation. This is the buffer region. For example, at 4.0 mL of NaOH added:

$$\text{initial mmol HLac present} = 25.0 \text{ mL} \times \frac{0.100 \text{ mmol}}{\text{mL}} = 2.50 \text{ mmol HLac}$$

$$\text{mmol OH}^- \text{ added} = 4.0 \text{ mL} \times \frac{0.100 \text{ mmol}}{\text{mL}} = 0.40 \text{ mmol OH}^-$$

Note: The units mmol are usually easier numbers to work with. The units for molarity are moles/L but are also equal to mmoles/mL.

The 0.40 mmol added OH⁻ convert 0.40 mmoles HLac to 0.40 mmoles Lac⁻ according to the equation:

$$\text{HLac} + \text{OH}^- \rightarrow \text{Lac}^- + \text{H}_2\text{O} \qquad \text{Reacts completely}$$

mmol HLac remaining = 2.50 - 0.40 = 2.10 mmol; mmol Lac⁻ produced = 0.40 mmol

We have a buffered solution. Using the Henderson-Hasselbalch equation where $pK_a = 3.86$:

$$pH = pK_a + \log \frac{[\text{Lac}^-]}{[\text{HLac}]} = 3.86 + \log \frac{(0.40)}{(2.10)} \qquad \text{(Total volume cancels, so we can use the ratio of moles or mmoles.)}$$

$$pH = 3.86 - 0.72 = 3.14$$

Other points in the buffer region are calculated in a similar fashion. Perform a stoichiometry problem first, followed by a buffer problem. The buffer region includes all points up to 24.9 mL OH⁻ added.

At the stoichiometric point (25.0 mL OH⁻ added), we have added enough OH⁻ to convert all of the HLac (2.50 mmol) into its conjugate base, Lac⁻. All that is present is a weak base. To determine the pH, we perform a weak base calculation.

$$[\text{Lac}^-]_o = \frac{2.50 \text{ mmol}}{25.0 \text{ mL} + 25.0 \text{ mL}} = 0.0500 \ M$$

	Lac⁻ + H₂O	$\rightleftharpoons$	HLac	+	OH⁻	$K_b = \dfrac{1.0 \times 10^{-14}}{1.4 \times 10^{-4}} = 7.1 \times 10^{-11}$
Initial	0.0500 M		0		0	
	x mol/L Lac⁻ reacts with H₂O to reach equilibrium					
Change	$-x$	$\rightarrow$	$+x$		$+x$	
Equil.	0.0500 - x		x		x	

$$K_b = \frac{x^2}{0.0500 - x} \approx \frac{x^2}{0.0500} = 7.1 \times 10^{-11}$$

$x = [OH^-] = 1.9 \times 10^{-6} \, M$; pOH = 5.72; pH = 8.28   Assumptions good.

Past the stoichiometric point, we have added more than 2.50 mmol of NaOH. The pH will be determined by the excess OH^- ion present. An example of this calculation follows.

At 25.1 mL: OH^- added = 25.1 mL $\times \dfrac{0.100 \text{ mmol}}{\text{mL}}$ = 2.51 mmol OH^-; 2.50 mmol OH^- neutralizes all the weak acid present. The remainder is excess OH^-. excess OH^- = 2.51 - 2.50 = 0.01 mmol

$[OH^-]_{excess} = \dfrac{0.01 \text{ mmol}}{(25.0 + 25.1) \text{ mL}} = 2 \times 10^{-4} \, M$; pOH = 3.7; pH = 10.3

All results are listed in Table 15.1 at the end of the solution to Exercise 15.59.

59.   At beginning of the titration, only the weak base NH_3 is present. As always, solve for the pH using the K_b reaction for NH_3.

$$NH_3 + H_2O \rightleftharpoons NH_4^+ + OH^- \qquad K_b = 1.8 \times 10^{-5}$$

Initial	0.100 M	0	~0
Equil.	0.100 - x	x	x

$K_b = \dfrac{x^2}{0.100 - x} = \dfrac{x^2}{0.100} = 1.8 \times 10^{-5}$

$x = [OH^-] = 1.3 \times 10^{-3} \, M$; pOH = 2.89; pH = 11.11   Assumptions good.

In the buffer region (4.0 - 24.9 mL), we can use the Henderson-Hasselbalch equation:

$$K_a = \frac{1.0 \times 10^{-14}}{1.8 \times 10^{-5}} = 5.6 \times 10^{-10}; \quad pK_a = 9.25; \quad pH = 9.25 + \log \frac{[NH_3]}{[NH_4^+]}$$

We must determine the amounts of NH_3 and NH_4^+ present after the added H^+ reacts completely with the NH_3. For example, after 8.0 mL HCl are added:

initial mmol NH_3 present = 25.0 mL $\times \dfrac{0.100 \text{ mmol}}{\text{mL}}$ = 2.50 mmol NH_3

mmol H^+ added = 8.0 mL $\times \dfrac{0.100 \text{ mmol}}{\text{mL}}$ = 0.80 mmol H^+

Added H^+ reacts with NH_3 to completion: $NH_3 + H^+ \rightarrow NH_4^+$

mmol NH_3 remaining = 2.50 - 0.80 = 1.70 mmol; mmol NH_4^+ produced = 0.80 mmol

pH = 9.25 + $\log \dfrac{1.70}{0.80}$ = 9.58   (Mole ratios can be used since the total volume cancels.)

Other points in the buffer region are calculated in similar fashion. All results are summarized in Table 15.1 at the end of this Exercise.

At the stoichiometric point (25.0 mL H^+ added), just enough HCl has been added to convert all of the weak base (NH_3) into its conjugate acid (NH_4^+). Perform a weak acid calculation. $[NH_4^+]_o = 2.50$ mmol/50.0 mL $= 0.0500\ M$

$$NH_4^+ \rightleftharpoons H^+ + NH_3 \qquad K_a = 5.6 \times 10^{-10}$$

Initial	$0.0500\ M$	0	0
Equil.	$0.0500 - x$	x	x

$$5.6 \times 10^{-10} = \frac{x^2}{0.0500 - x} = \frac{x^2}{0.0500}, \quad x = [H^+] = 5.3 \times 10^{-6}\ M;\ \ pH = 5.28 \quad \text{Assumptions good.}$$

Beyond the stoichiometric point, the pH is determined by the excess H^+. For example, at 28.0 mL of H^+ added:

$$\text{mmol } H^+ \text{ added} = 28.0\ mL \times \frac{0.100\ \text{mmol}}{mL} = 2.80\ \text{mmol } H^+$$

$$\text{Excess mmol } H^+ = 2.80\ \text{mmol} - 2.50\ \text{mmol} = 0.30\ \text{mmol excess } H^+$$

$$[H^+]_{excess} = \frac{0.30\ \text{mmol}}{(25.0 + 28.0)\ mL} = 5.7 \times 10^{-3}\ M;\ \ pH = 2.24$$

All results are summarized in Table 15.1 that follows.

Table 15.1: Summary of pH Results for Exercises 15.57 and 15.59 (Graph follows)

Titrant mL	Exercise 15.57	Exercise 15.59
0.0	2.43	11.11
4.0	3.14	9.97
8.0	3.53	9.58
12.5	3.86	9.25
20.0	4.46	8.65
24.0	5.24	7.87
24.5	5.6	7.6
24.9	6.3	6.9
25.0	8.28	5.28
25.1	10.3	3.7
26.0	11.30	2.71
28.0	11.75	2.24
30.0	11.96	2.04

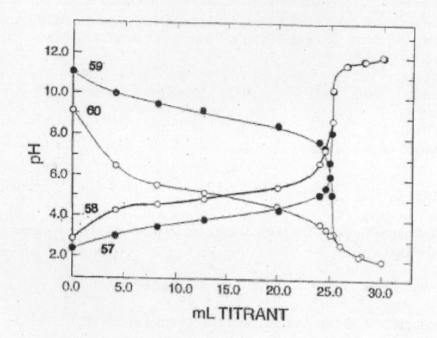

mL TITRANT

61.   a. This is a weak acid/strong base titration. At the halfway point to equivalence, [weak acid] = [conjugate base], so pH = pK_a (always true for a weak acid/strong base titration).

pH = -log (6.4 × 10^{-5}) = 4.19

mmol $HC_7H_5O_2$ present = 100. mL × 0.10 M = 10. mmol $HC_7H_5O_2$. For the equivalence point, 10. mmol of OH⁻ must be added. The volume of OH⁻ added to reach the equivalence point is:

$$10. \text{ mmol OH}^- \times \frac{1 \text{ mL}}{0.10 \text{ mmol OH}^-} = 1.0 \times 10^2 \text{ mL OH}^-$$

At the equivalence point, 10. mmol of $HC_7H_5O_2$ is neutralized by 10. mmol of OH⁻ to produce 10. mmol of $C_7H_5O_2^-$. This is a weak base. The total volume of the solution is 100.0 mL + 1.0 × 10^2 mL = 2.0 × 10^2 mL. Solving the weak base equilibrium problem:

$$C_7H_5O_2^- + H_2O \rightleftharpoons HC_7H_5O_2 + OH^- \quad K_b = \frac{K_w}{K_a} = \frac{1.0 \times 10^{-14}}{6.4 \times 10^{-5}} = 1.6 \times 10^{-10}$$

Initial   10. mmol/2.0 × 10^2 mL   0   0
Equil.   0.050 - x x x

$$K_b = 1.6 \times 10^{-10} = \frac{x^2}{0.050 - x} \approx \frac{x^2}{0.050}, \quad x = [\text{OH}^-] = 2.8 \times 10^{-6} \, M$$

pOH = 5.55; pH = 8.45   Assumptions good.

    b. At the halfway point to equivalence for a weak base/strong acid titration, pH = pK_a since [weak base] = [conjugate acid].

$$K_a = \frac{K_w}{K_b} = \frac{1.0 \times 10^{-14}}{5.6 \times 10^{-4}} = 1.8 \times 10^{-11}; \quad pH = pK_a = -\log(1.8 \times 10^{-11}) = 10.74$$

For the equivalence point (mmol acid added = mmol base present):

mmol $C_2H_5NH_2$ present = 100.0 mL × 0.10 M = 10. mmol $C_2H_5NH_2$

mL HNO_3 added = 10. mmol H^+ × $\dfrac{1 \text{ mL}}{0.20 \text{ mmol H}^+}$ = 50. mL H^+

The strong acid added completely converts the weak base into its conjugate acid. Therefore, at the equivalence point, $[C_2H_5NH_3^+]_o$ = 10. mmol/(100.0 + 50.) mL = 0.067 M. Solving the weak acid equilibrium problem:

$$C_2H_5NH_3^+ \quad \rightleftharpoons \quad H^+ \quad + \quad C_2H_5NH_2$$

Initial   0.067 M 0   0
Equil.   0.067 - x x x

$$K_a = 1.8 \times 10^{-11} = \frac{x^2}{0.067 - x} \approx \frac{x^2}{0.067}, \quad x = [H^+] = 1.1 \times 10^{-6} \, M$$

pH = 5.96; Assumption good.

c. In a strong acid/strong base titration, the halfway point has no special significance other than exactly one-half of the original amount of acid present has been neutralized.

mmol H^+ present = 100.0 mL × 0.50 M = 50. mmol H^+

mL OH^- added = 25. mmol OH^- × $\dfrac{1 \text{ mL}}{0.25 \text{ mmol}}$ = 1.0 × 10² mL OH^-

$$H^+ \quad + \quad OH^- \quad \rightarrow \quad H_2O$$

Before   50. mmol   25 mmol
After   25 mmol   0

$$[H^+]_{excess} = \frac{25 \text{ mmol}}{(100.0 + 1.0 \times 10^2) \text{ mL}} = 0.13 \, M; \quad pH = 0.89$$

At the equivalence point of a strong acid/strong base titration, only neutral species are present (Na^+, Cl^-, H_2O), so the pH = 7.00.

63.   At equivalence point: 16.00 mL × 0.125 mmol/mL = 2.00 mmol OH^- added. There must be 2.00 mmol HX present initially.

2.00 mL NaOH added = 2.00 mL × 0.125 mmol/mL = 0.250 mmol OH⁻; 0.250 mmol of OH⁻ added will convert 0.250 mmol HX into 0.250 mmol X⁻. Remaining HX = 2.00 − 0.250 = 1.75 mmol HX; This is a buffer solution where $[H^+] = 10^{-6.912} = 1.22 \times 10^{-7} \, M$. Let V_T = total volume of solution (which cancels). Assuming the initial and equilibrium HX and X⁻ concentrations are the same (a very good assumption for buffer solutions):

$$K_a = \frac{[H^+][X^-]}{[HX]} = \frac{1.22 \times 10^{-7}\,(0.250/V_T)}{1.75/V_T} = \frac{1.22 \times 10^{-7}(0.250)}{1.75} = 1.74 \times 10^{-8}$$

Note: We could also solve for K_a using the Henderson-Hasselbach equation

Indicators

65. $HIN \rightleftharpoons In^- + H^+$ $K_a = \dfrac{[In^-][H^+]}{[HIn]} = 1.0 \times 10^{-9}$

   a. In a very acid solution, the HIn form dominates, so the solution will be yellow.

   b. The color change occurs when the concentration of the more dominant form is approximately ten times as great as the less dominant form of the indicator.

   $\dfrac{[HIn]}{[In^-]} = \dfrac{10}{1}$; $K_a = 1.0 \times 10^{-9} = \left(\dfrac{1}{10}\right)[H^+]$, $[H^+] = 1 \times 10^{-8} \, M$; pH = 8.0 at color change

   c. This is way past the equivalence point (100.0 mL OH⁻ added), so the solution is very basic and the In⁻ form of the indicator dominates. The solution will be blue.

67.   When choosing an indicator, we want the color change of the indicator to occur approximately at the pH of the equivalence point. Since the pH generally changes very rapidly at the equivalence point, we don't have to be exact. This is especially true for strong acid/strong base titrations. Some choices where color change occurs at about the pH of the equivalence point are:

Exercise	pH at eq. pt.	Indicator
15.53	7.00	bromthymol blue or phenol red
15.55	8.79	o-cresolphthalein or phenolphthalein

69.

Exercise	pH at eq. pt.	Indicator
15.57	8.28	phenolphthalein
15.59	5.28	bromcresol green

71.   pH > 5 for bromcresol green to be blue. pH < 8 for thymol blue to be yellow. The pH is between 5 and 8.

Solubility Equilibria

73.   a. $AgC_2H_3O_2(s) \rightleftharpoons Ag^+(aq) + C_2H_3O_2^-(aq)$ $K_{sp} = [Ag^+][C_2H_3O_2^-]$

b. $Al(OH)_3(s) \rightleftharpoons Al^{3+}(aq) + 3\ OH^-(aq)$ $K_{sp} = [Al^{3+}][OH^-]^3$

c. $Ca_3(PO_4)_2(s) \rightleftharpoons 3\ Ca^{2+}(aq) + 2\ PO_4^{3-}(aq)$ $K_{sp} = [Ca^{2+}]^3[PO_4^{3-}]^2$

75.   In our setup, s = solubility of the ionic solid in mol/L. This is defined as the maximum amount of a salt which can dissolve. Since solids do not appear in the K_{sp} expression, we do not need to worry about their initial and equilibrium amounts.

a. $CaC_2O_4(s)$ $\rightleftharpoons$ $Ca^{2+}(aq)$ + $C_2O_4^{2-}(aq)$

Initial		0	0
	s mol/L of $CaC_2O_4(s)$ dissolves to reach equilibrium		
Change	$-s$ $\rightarrow$	$+s$	$+s$
Equil.		s	s

From the problem, $s = 4.8 \times 10^{-5}$ mol/L.

$K_{sp} = [Ca^{2+}][C_2O_4^{2-}] = (s)(s) = s^2$, $K_{sp} = (4.8 \times 10^{-5})^2 = 2.3 \times 10^{-9}$

b. $BiI_3(s)$ $\rightleftharpoons$ $Bi^{3+}(aq)$ + $3\ I^-(aq)$

Initial		0	0
	s mol/L of $BiI_3(s)$ dissolves to reach equilibrium		
Change	$-s$ $\rightarrow$	$+s$	$+3s$
Equil.		s	$3s$

$K_{sp} = [Bi^{3+}][I^-]^3 = (s)(3s)^3 = 27\ s^4$, $K_{sp} = 27(1.32 \times 10^{-5})^4 = 8.20 \times 10^{-19}$

77. $PbBr_2(s)$ $\rightleftharpoons$ $Pb^{2+}(aq)$ + $2\ Br^-(aq)$

Initial		0	0
	s mol/L of $PbBr_2(s)$ dissolves to reach equilibrium		
Change	$-s$ $\rightarrow$	$+s$	$+2s$
Equil.		s	$2s$

From the problem, $s = [Pb^{2+}] = 2.14 \times 10^{-2}\ M$. So:

$K_{sp} = [Pb^{2+}][Br^-]^2 = s(2s)^2 = 4s^3$, $K_{sp} = 4(2.14 \times 10^{-2})^3 = 3.92 \times 10^{-5}$

79.   In our setups, s = solubility in mol/L. Since solids do not appear in the K_{sp} expression, we do not need to worry about their initial or equilibrium amounts.

a. $Ag_3PO_4(s)$ $\rightleftharpoons$ $3\ Ag^+(aq)$ $+$ $PO_4^{3-}(aq)$

Initial   0   0

                        s mol/L of $Ag_3PO_4(s)$ dissolves to reach equilibrium

Change $-s$ $\rightarrow$ $+3s$ $+s$
Equil. $3s$ s

$K_{sp} = 1.8 \times 10^{-18} = [Ag^+]^3\ [PO_4^{3-}] = (3s)^3(s) = 27\ s^4$

$27\ s^4 = 1.8 \times 10^{-18}$, $s = (6.7 \times 10^{-20})^{1/4} = 1.6 \times 10^{-5}$ mol/L = molar solubility

b. $CaCO_3(s)$ $\rightleftharpoons$ $Ca^{2+}(aq)$ $+$ $CO_3^{2-}(aq)$

Initial s = solubility (mol/L)   0   0
Equil. s s

$K_{sp} = 8.7 \times 10^{-9} = [Ca^{2+}]\ [CO_3^{2-}] = s^2$, $s = 9.3 \times 10^{-5}$ mol/L

c. $Hg_2Cl_2(s)$ $\rightleftharpoons$ $Hg_2^{2+}(aq)$ $+$ $2\ Cl^-(aq)$

Initial s = solubility (mol/L)   0   0
Equil. s $2s$

$K_{sp} = 1.1 \times 10^{-18} = [Hg_2^{2+}]\ [Cl^-]^2 = (s)(2s)^2 = 4s^3$, $s = 6.5 \times 10^{-7}$ mol/L

81.   Let s = solubility of $Co(OH)_3$ in mol/L. Note: Since solids do not appear in the K_{sp} expression, we do not need to worry about their initial or equilibrium amounts.

                    $Co(OH)_3(s)$ $\rightleftharpoons$ $Co^{3+}(aq)$ $+$ $3\ OH^-(aq)$

Initial   0 $1.0 \times 10^{-7}\ M$ from water

                        s mol/L of $Co(OH)_3(s)$ dissolves to reach equilibrium = molar solubility

Change $-s$ $\rightarrow$ $+s$ $+3s$
Equil. s $1.0 \times 10^{-7} + 3s$

$K_{sp} = 2.5 \times 10^{-43} = [Co^{3+}]\ [OH^-]^3 = (s)(1.0 \times 10^{-7} + 3s)^3 \approx s(1.0 \times 10^{-7})^3$

$s = \dfrac{2.5 \times 10^{-43}}{1.0 \times 10^{-21}} = 2.5 \times 10^{-22}$ mol/L   Assumption good $(1.0 \times 10^{-7} + 3s \approx 1.0 \times 10^{-7})$.

83.   a. Since both solids dissolve to produce 3 ions in solution, we can compare values of K_{sp} to determine relative molar solubility. Since the K_{sp} for CaF_2 is smaller, $CaF_2(s)$ is less soluble (in mol/L).

       b. We must calculate molar solubilities since each salt yields a different number of ions when it dissolves.

$$Ca_3(PO_4)_2(s) \rightleftharpoons 3\ Ca^{2+}(aq) + 2\ PO_4^{3-}(aq) \qquad K_{sp} = 1.3 \times 10^{-32}$$

Initial	s = solubility (mol/L)	0	0
Equil.		$3s$	$2s$

$$K_{sp} = [Ca^{2+}]^3\ [PO_4^{3-}]^2 = (3s)^3(2s)^2 = 108\ s^5, \quad s = (1.3 \times 10^{-32}/108)^{1/5} = 1.6 \times 10^{-7}\ \text{mol/L}$$

$$FePO_4(s) \rightleftharpoons Fe^{3+}(aq) + PO_4^{3-}(aq) \qquad K_{sp} = 1.0 \times 10^{-22}$$

Initial	s = solubility (mol/L)	0	0
Equil.		s	s

$$K_{sp} = [Fe^{3+}]\ [PO_4^{3-}] = s^2, \quad s = \sqrt{1.0 \times 10^{-22}} = 1.0 \times 10^{-11}\ \text{mol/L}$$

Since the molar solubility of $FePO_4$ is smaller, $FePO_4$ is less soluble (in mol/L).

85.   a. $Fe(OH)_3(s) \rightleftharpoons Fe^{3+}(aq) + 3\ OH^-(aq)$

Initial	0	$1 \times 10^{-7}\ M$ from water

s mol/L of $Fe(OH)_3(s)$ dissolves to reach equilibrium = molar solubility

Change	$-s$	$\rightarrow$ $+s$	$+3s$
Equil.		s	$1 \times 10^{-7} + 3s$

$$K_{sp} = 4 \times 10^{-38} = [Fe^{3+}]\ [OH^-]^3 = (s)(1 \times 10^{-7} + 3s)^3 \approx s(1 \times 10^{-7})^3$$

$$s = 4 \times 10^{-17}\ \text{mol/L} \qquad \text{Assumption good } (3s << 1 \times 10^{-7}).$$

   b. $Fe(OH)_3(s) \rightleftharpoons Fe^{3+}(aq) + 3\ OH^-(aq)$ pH = 5.0 so $[OH^-] = 1 \times 10^{-9}\ M$

Initial	0	$1 \times 10^{-9}\ M$ (buffered)

s mol/L dissolves to reach equilibrium

Change	$-s$	$\rightarrow$ $+s$	-----	(assume no pH change in buffer)
Equil.		s	1×10^{-9}	

$$K_{sp} = 4 \times 10^{-38} = [Fe^{3+}]\ [OH^-]^3 = (s)(1 \times 10^{-9})^3, \quad s = 4 \times 10^{-11}\ \text{mol/L} = \text{molar solubility}$$

   c. $Fe(OH)_3(s) \rightleftharpoons Fe^{3+}(aq) + 3\ OH^-(aq)$ pH = 11.0 so $[OH^-] = 1 \times 10^{-3}\ M$

Initial	0	0.001 M (buffered)

s mol/L dissolves to reach equilibrium

Change	$-s$	$\rightarrow$ $+s$	-----	(assume no pH change)
Equil.		s	0.001	

$$K_{sp} = 4 \times 10^{-38} = [Fe^{3+}]\ [OH^-]^3 = (s)(0.001)^3, \quad s = 4 \times 10^{-29}\ \text{mol/L} = \text{molar solubility}$$

Note: As $[OH^-]$ increases, solubility decreases. This is the common ion effect.

87.
$$Ca_3(PO_4)_2(s) \rightleftharpoons 3\ Ca^{2+}(aq) + 2\ PO_4^{3-}(aq)$$

Initial		0	0.20 M

s mol/L of $Ca_3(PO_4)_2(s)$ dissolves to reach equilibrium

Change	$-s$ $\rightarrow$	$+3s$	$+2s$
Equil.		$3s$	$0.20 + 2s$

$$K_{sp} = 1.3 \times 10^{-32} = [Ca^{2+}]^3\ [PO_4^{3-}]^2 = (3s)^3(0.20 + 2s)^2$$

Assuming $0.20 + 2s \approx 0.20$: $1.3 \times 10^{-32} = (3s)^3(0.20)^2 = 27\ s^3(0.040)$

s = molar solubility = 2.3×10^{-11} mol/L; Assumption good.

89.   If the anion in the salt can act as a base in water, the solubility of the salt will increase as the solution becomes more acidic. Added H^+ will react with the base, forming the conjugate acid. As the basic anion is removed, more of the salt will dissolve to replenish the basic anion. The salts with basic anions are Ag_3PO_4, $CaCO_3$, $CdCO_3$ and $Sr_3(PO_4)_2$. Hg_2Cl_2 and PbI_2 do not have any pH dependence since Cl^- and I^- are terrible bases (the conjugate bases of a strong acids).

$$Ag_3PO_4(s) + H^+(aq) \rightarrow 3\ Ag^+(aq) + HPO_4^{2-}(aq) \xrightarrow{\text{excess } H^+} 3\ Ag^+(aq) + H_3PO_4(aq)$$

$$CaCO_3(s) + H^+ \rightarrow Ca^{2+} + HCO_3^- \xrightarrow{\text{excess } H^+} Ca^{2+} + H_2CO_3\ [H_2O(l) + CO_2(g)]$$

$$CdCO_3(s) + H^+ \rightarrow Cd^{2+} + HCO_3^- \xrightarrow{\text{excess } H^+} Cd^{2+} + H_2CO_3\ [H_2O(l) + CO_2(g)]$$

$$Sr_3(PO_4)_2(s) + 2\ H^+ \rightarrow 3\ Sr^{2+} + 2\ HPO_4^{2-} \xrightarrow{\text{excess } H^+} 3\ Sr^{2+} + 2\ H_3PO_4$$

91.   Potentially, $BaSO_4(s)$ could form if Q is greater than K_{sp}.

$$BaSO_4(s) \rightleftharpoons Ba^{2+}(aq) + SO_4^{2-}(aq) \quad K_{sp} = 1.5 \times 10^{-9}$$

To calculate Q, we need the initial concentrations of Ba^{2+} and SO_4^{2-}.

$$[Ba^{2+}]_o = \frac{\text{mmoles Ba}^{2+}}{\text{total mL solution}} = \frac{75.0\ \text{mL} \times \dfrac{0.020\ \text{mmol Ba}^{2+}}{\text{mL}}}{75.0\ \text{mL} \times 125\ \text{mL}} = 0.0075\ M$$

$$[SO_4^{2-}]_o = \frac{\text{mmoles SO}_4^{2-}}{\text{total mL solution}} = \frac{125\ \text{mL} \times \dfrac{0.040\ \text{mmol SO}_4^{2-}}{\text{mL}}}{200.\ \text{mL}} = 0.025\ M$$

$$Q = [Ba^{2+}]_o[SO_4^{2-}]_o = (0.0075\ M)(0.025\ M) = 1.9 \times 10^{-4}$$

$Q > K_{sp}$ (1.5×10^{-9}) so $BaSO_4(s)$ will form.

93.   The concentrations of ions are large, so Q will be greater than K_{sp} and $BaC_2O_4(s)$ will form. To solve this problem, we will assume that the precipitation reaction goes to completion; then we will solve an equilibrium problem to get the actual ion concentrations.

$$100. \text{ mL} \times \frac{0.200 \text{ mmol K}_2\text{C}_2\text{O}_4}{\text{mL}} = 20.0 \text{ mmol K}_2\text{C}_2\text{O}_4$$

$$150. \text{ mL} \times \frac{0.250 \text{ mmol BaBr}_2}{\text{mL}} = 37.5 \text{ mmol BaBr}_2$$

$$Ba^{2+}(aq) + C_2O_4^{2-}(aq) \rightarrow BaC_2O_4(s) \qquad K = 1/K_{sp} \gg 1$$

Before	37.5 mmol	20.0 mmol	0
Change	-20.0	-20.0 $\rightarrow$	+20.0	Reacts completely (K is large)
After	17.5	0	20.0

New initial concentrations (after complete precipitation) are: $[Ba^{2+}] = \dfrac{17.5 \text{ mmol}}{250. \text{ mL}} = 7.00 \times 10^{-2} M$

$$[K^+] = \frac{2(20.0 \text{ mmol})}{250. \text{ mL}} = 0.160 \ M; \ [Br^-] = \frac{2(37.5 \text{ mmol})}{250. \text{ mL}} = 0.300 \ M$$

For K^+ and Br^-, these are also the final concentrations. For Ba^{2+} and $C_2O_4^{2-}$, we need to perform an equilibrium calculation.

$$BaC_2O_4(s) \rightleftharpoons Ba^{2+}(aq) + C_2O_4^{2-}(aq) \qquad K_{sp} = 2.3 \times 10^{-8}$$

Initial		0.0700 M	0
	s mol/L of $BaC_2O_4(s)$ dissolves to reach equilibrium		
Equil.		0.0700 + s	s

$$K_{sp} = 2.3 \times 10^{-8} = [Ba^{2+}][C_2O_4^{2-}] = (0.0700 + s)(s) \approx 0.0700 \ s$$

$s = [C_2O_4^{2-}] = 3.3 \times 10^{-7}$ mol/L; $[Ba^{2+}] = 0.0700 \ M$ Assumption good ($s \ll 0.0700$).

95. $Ag_3PO_4(s) \rightleftharpoons 3 Ag^+(aq) + PO_4^{3-}(aq)$; When Q is greater than K_{sp}, precipitation will occur. We will calculate the $[Ag^+]_o$ necessary for $Q = K_{sp}$. Any $[Ag^+]_o$ greater than this calculated number will cause precipitation of $Ag_3PO_4(s)$. In this problem, $[PO_4^{3-}]_o = [Na_3PO_4]_o = 1.0 \times 10^{-5} M$.

$$K_{sp} = 1.8 \times 10^{-18}; \ Q = 1.8 \times 10^{-18} = [Ag^+]_o^3 [PO_4^{3-}]_o = [Ag^+]_o^3 (1.0 \times 10^{-5} M)$$

$$[Ag^+]_o = \left(\frac{1.8 \times 10^{-18}}{1.0 \times 10^{-5}} \right)^{1/3}, \ [Ag^+]_o = 5.6 \times 10^{-5} M$$

When $[Ag^+]_o = [AgNO_3]_o$ is greater than $5.6 \times 10^{-5} M$, precipitation of $Ag_3PO_4(s)$ will occur.

Complex Ion Equilibria

97. a.
$$Co^{2+} + NH_3 \rightleftharpoons CoNH_3^{2+} \qquad\qquad K_1$$
$$CoNH_3^{2+} + NH_3 \rightleftharpoons Co(NH_3)_2^{2+} \qquad K_2$$
$$Co(NH_3)_2^{2+} + NH_3 \rightleftharpoons Co(NH_3)_3^{2+} \qquad K_3$$
$$Co(NH_3)_3^{2+} + NH_3 \rightleftharpoons Co(NH_3)_4^{2+} \qquad K_4$$
$$Co(NH_3)_4^{2+} + NH_3 \rightleftharpoons Co(NH_3)_5^{2+} \qquad K_5$$
$$Co(NH_3)_5^{2+} + NH_3 \rightleftharpoons Co(NH_3)_6^{2+} \qquad K_6$$

$$Co^{2+} + 6\,NH_3 \rightleftharpoons Co(NH_3)_6^{2+} \qquad\qquad K_f = K_1K_2K_3K_4K_5K_6$$

Note: The various K's are included for your information. Each NH_3 adds with a corresponding K value associated with that reaction. The overall formation constant, K_f, for the overall reaction is equal to the product of all the stepwise K values.

b.
$$Ag^+ + NH_3 \rightleftharpoons AgNH_3^+ \qquad\qquad K_1$$
$$AgNH_3^+ + NH_3 \rightleftharpoons Ag(NH_3)_2^+ \qquad K_2$$

$$Ag^+ + 2\,NH_3 \rightleftharpoons Ag(NH_3)_2^+ \qquad\qquad K_f = K_1K_2$$

99. $Hg^{2+}(aq) + 2\,I^-(aq) \rightarrow HgI_2(s)$, orange ppt.; $HgI_2(s) + 2\,I^-(aq) \rightarrow HgI_4^{2-}(aq)$, soluble complex ion

101. The formation constant for HgI_4^{2-} is an extremely large number. Because of this, we will let the Hg^{2+} and I^- ions present initially react to completion, and then solve an equilibrium problem to determine the Hg^{2+} concentration.

$$Hg^{2+} \quad + \quad 4\,I^- \quad \rightleftharpoons \quad HgI_4^{2-} \qquad K = 1.0 \times 10^{30}$$

	Hg^{2+}	I^-		HgI_4^{2-}	
Before	0.010 M	0.78 M		0	
Change	-0.010	-0.040	$\rightarrow$	+0.010	Reacts completely (K large)
After	0	0.74		0.010	New initial conditions

x mol/L HgI_4^{2-} dissociates to reach equilibrium

Change	+x	+4x	$\leftarrow$	-x	
Equil.	x	0.74 + 4x		0.010 - x	

$$K = 1.0 \times 10^{30} = \frac{[HgI_4^{2-}]}{[Hg^{2+}][I^-]^4} = \frac{(0.010 - x)}{(x)(0.74 + 4x)^4}; \quad \text{Making normal assumptions:}$$

$$1.0 \times 10^{30} = \frac{(0.010)}{(x)(0.74)^4}, \quad x = [Hg^{2+}] = 3.3 \times 10^{-32}\,M \quad \text{Assumptions good.}$$

Note: 3.3×10^{-32} mol/L corresponds to one Hg^{2+} ion per 5×10^7 L. It is very reasonable to approach the equilibrium in two steps. The reaction does essentially go to completion.

103.   a. $AgI(s) \rightleftharpoons Ag^+(aq) + I^-(aq)$ $K_{sp} = [Ag^+][I^-] = 1.5 \times 10^{-16}$

       Initial s = solubility (mol/L)   0   0
       Equil. s s

       $K_{sp} = 1.5 \times 10^{-16} = s^2$, $s = 1.2 \times 10^{-8}$ mol/L

       b. $AgI(s) \rightleftharpoons Ag^+ + I^-$ $K_{sp} = 1.5 \times 10^{-16}$
                $Ag^+ + 2\, NH_3 \rightleftharpoons Ag(NH_3)_2^+$ $K_f = 1.7 \times 10^7$
       _____

       $AgI(s) + 2\, NH_3(aq) \rightleftharpoons Ag(NH_3)_2^+(aq) + I^-(aq)$ $K = K_{sp} \times K_f = 2.6 \times 10^{-9}$

                $AgI(s) + 2\, NH_3 \rightleftharpoons Ag(NH_3)_2^+ + I^-$

       Initial   3.0 M 0   0
                   s mol/L of AgBr(s) dissolves to reach equilibrium = molar solubility
       Equil.   3.0 - 2s s s

       $$K = \frac{[Ag(NH_3)_2^+][I^-]}{[NH_3]^2} = \frac{s^2}{(3.0 - 2s)^2} = 2.6 \times 10^{-9} \approx \frac{s^2}{(3.0)^2},\ s = 1.5 \times 10^{-4}\ \text{mol/L}$$

       Assumption good.

       c.   The presence of NH_3 increases the solubility of AgI. Added NH_3 removes Ag^+ from
            solution by forming the complex ion, $Ag(NH_3)_2^+$. As Ag^+ is removed, more AgI(s) will dissolve
            to replenish the Ag^+ concentration.

105.   Test tube 1: added Cl^- reacts with Ag^+ to form a silver chloride precipitate. The net ionic equation
       is $Ag^+(aq) + Cl^-(aq) \rightarrow AgCl(s)$. Test tube 2: added NH_3 reacts with Ag^+ ions to form a soluble
       complex ion, $Ag(NH_3)_2^+$. As this complex ion forms, Ag^+ is removed from the solution, which causes
       the AgCl(s) to dissolve. When enough NH_3 is added, all of the silver chloride precipitate will
       dissolve. The equation is $AgCl(s) + 2\, NH_3(aq) \rightarrow Ag(NH_3)_2^+(aq) + Cl^-(aq)$. Test tube 3: added
       H^+ reacts with the weak base, NH_3, to form NH_4^+. As NH_3 is removed from the $Ag(NH_3)_2^+$
       complex ion, Ag^+ ions are released to solution and can then react with Cl^- to reform AgCl(s).   The
       equations are $Ag(NH_3)_2^+(aq) + 2\, H^+(aq) \rightarrow Ag^+(aq) + 2\, NH_4^+(aq)$ and $Ag^+(aq) + Cl^-(aq) \rightarrow AgCl(s)$.

Additional Exercises

107. $NH_3 + H_2O \rightleftharpoons NH_4^+ + OH^-$ $K_b = \dfrac{[NH_4^+][OH^-]}{[NH_3]}$; Taking the -log of the K_b expression:

       $$-\log K_b = -\log [OH^-] - \log \frac{[NH_4^+]}{[NH_3]},\quad -\log [OH^-] = -\log K_b + \log \frac{[NH_4^+]}{[NH_3]}$$

       $$pOH = pK_b + \log \frac{[NH_4^+]}{[NH_3]}\ \text{ or }\ pOH = pK_b + \log \frac{[Acid]}{[Base]}$$

109.   A best buffer is when pH $\approx$ pK$_a$ since these solutions have about equal concentrations of weak acid and conjugate base. Therefore, choose combinations that yield a buffer where pH $\approx$ pK$_a$, i.e., look at the acids available and choose the one whose pK$_a$ is closest to the pH.

a.   Potassium fluoride + HCl will yield a buffer consisting of HF (pK$_a$ = 3.14) and F$^-$.

b.   Benzoic acid + NaOH will yield a buffer consisting of benzoic acid (pK$_a$ = 4.19) and benzoate anion.

c.   Sodium acetate + acetic acid (pK$_a$ = 4.74) is the best choice for pH = 5.0 buffer since acetic acid has a pK$_a$ value closest to 5.0.

d.   HOCl and NaOH: This is the best choice to produce a conjugate acid/base pair with pH = 7.0. This mixture would yield a buffer consisting of HOCl (pK$_a$ = 7.46) and OCl$^-$. Actually, the best choice for a pH = 7.0 buffer is an equimolar mixture of ammonium chloride and sodium acetate. NH$_4^+$ is a weak acid (K$_a$ = 5.6 $\times$ 10^{-10}) and C$_2$H$_3$O$_2^-$ is a weak base (K$_b$ = 5.6 $\times$ 10^{-10}). A mixture of the two will give a buffer at pH = 7.0 since the weak acid and weak base are the same strengths (K$_a$ for NH$_4^+$ = K$_b$ for C$_2$H$_3$O$_2^-$). NH$_4$C$_2$H$_3$O$_2$ is commercially available, and its solutions are used for pH = 7.0 buffers.

e.   Ammonium chloride + NaOH will yield a buffer consisting of NH$_4^+$ (pK$_a$ = 9.26) and NH$_3$.

111. a.   HC$_2$H$_3$O$_2$ + OH$^-$ $\rightleftharpoons$ C$_2$H$_3$O$_2^-$ + H$_2$O

$$K_{eq} = \frac{[C_2H_3O_2^-]}{[HC_2H_3O_2][OH^-]} \times \frac{[H^+]}{[H^+]} = \frac{K_{a, HC_2H_3O_2}}{K_w} = \frac{1.8 \times 10^{-5}}{1.0 \times 10^{-14}} = 1.8 \times 10^9$$

b.   C$_2$H$_3$O$_2^-$ + H$^+$ $\rightleftharpoons$ HC$_2$H$_3$O$_2$ $K_{eq} = \dfrac{[HC_2H_3O_2]}{[H^+][C_2H_3O_2^-]} = \dfrac{1}{K_{a, HC_2H_3O_2}} = 5.6 \times 10^4$

c.   HCl + NaOH $\rightarrow$ NaCl + H$_2$O

Net ionic equation is: H$^+$ + OH$^-$ $\rightleftharpoons$ H$_2$O; $K_{eq} = \dfrac{1}{K_w} = 1.0 \times 10^{14}$

113. In the final solution: [H$^+$] = 10$^{-2.15}$ = 7.1 $\times$ 10^{-3} M

Beginning mmol HCl = 500.0 mL $\times$ 0.200 mmol/mL = 100. mmol HCl

Amount of HCl that reacts with NaOH = 1.50 $\times$ 10^{-2} mmol/mL $\times$ V

$$\frac{7.1 \times 10^{-3} \text{ mmol}}{\text{mL}} = \frac{\text{final mmol H}^+}{\text{total volume}} = \frac{100. - 0.0150 \text{ V}}{500.0 + \text{V}}$$

3.6 + 7.1 $\times$ 10^{-3} V = 100. - 1.50 $\times$ 10^{-2} V, 2.21 $\times$ 10^{-2} V = 100. - 3.6

V = 4.36 $\times$ 10^3 mL = 4.36 L = 4.4 L NaOH

115. $HA + OH^- \rightarrow A^- + H_2O$ where HA = acetylsalicylic acid (assuming a monoprotic acid)

mmol HA present = 27.36 mL $OH^- \times \dfrac{0.5106 \text{ mmol } OH^-}{\text{mL } OH^-} \times \dfrac{1 \text{ mmol } HA}{\text{mmol } OH^-} = 13.97$ mmol HA

Molar mass of HA = $\dfrac{\text{grams}}{\text{mol}} = \dfrac{2.51 \text{ g HA}}{13.97 \times 10^{-3} \text{ mol HA}} = 180.$ g/mol

To determine the K_a value, use the pH data. After complete neutralization of acetylsalicylic acid by OH^-, we have 13.97 mmol of A^- produced from the neutralization reaction. A^- will react completely with the added H^+ and reform acetylsalicylic acid, HA.

mmol H^+ added = 13.68 mL $\times \dfrac{0.5106 \text{ mmol } H^+}{\text{mL}} = 6.985$ mmol H^+

	A^-	+	H^+	$\rightarrow$	HA	
Before	13.97 mmol		6.985 mmol		0	
Change	-6.985		-6.985	$\rightarrow$	+6.985	Reacts completely
After	6.985 mmol		0		6.985 mmol	

We have back titrated this solution to the halfway point to equivalence where pH = pK_a (assuming HA is a weak acid). We know this because after H^+ reacts completely, equal mmol of HA and A^- are present, which only occurs at the halfway point to equivalence. Assuming acetylsalicylic acid is a weak acid, then pH = pK_a = 3.48. $K_a = 10^{-3.48} = 3.3 \times 10^{-4}$

117. 50.0 mL $\times 0.100$ M = 5.00 mmol NaOH initially

at pH = 10.50, pOH = 3.50, $[OH^-] = 10^{-3.50} = 3.2 \times 10^{-4}$ M

mmol OH^- remaining = 3.2×10^{-4} mmol/mL $\times 73.75$ mL = 2.4×10^{-2} mmol

mmol OH^- that reacted = 5.00 - 0.024 = 4.98 mmol

Since the weak acid is monoprotic, 23.75 mL of the weak acid solution contains 4.98 mmol HA.

$[HA]_o = \dfrac{4.98 \text{ mmol}}{23.75 \text{ mL}} = 0.210$ M

119.

	$Ca_5(PO_4)_3OH(s)$	$\rightleftharpoons$	$5 Ca^{2+}$	+	$3 PO_4^{3-}$	+	OH^-
Initial	s = solubility (mol/L)		0		0		1.0×10^{-7} from water
Equil.			$5s$		$3s$		$s + 1.0 \times 10^{-7} \approx s$

$K_{sp} = 6.8 \times 10^{-37} = [Ca^{2+}]^5 [PO_4^{3-}]^3 [OH^-] = (5s)^5(3s)^3(s)$

$6.8 \times 10^{-37} = (3125)(27)s^9$, $s = 2.7 \times 10^{-5}$ mol/L   Assumption is good.

The solubility of hydroxyapatite will increase as the solution gets more acidic since both phosphate and hydroxide can react with H^+.

$$Ca_5(PO_4)_3F(s) \quad \rightleftharpoons \quad 5\ Ca^{2+} \ + \ 3\ PO_4^{3-} \ + \ F^-$$

Initial	s = solubility (mol/L)	0	0	0
Equil.		$5s$	$3s$	s

$K_{sp} = 1 \times 10^{-60} = (5s)^5(3s)^3(s) = (3125)(27)s^9, \ s = 6 \times 10^{-8}$ mol/L

The hydroxyapatite in tooth enamel is converted to the less soluble fluorapatite by fluoride-treated water. The less soluble fluorapatite is more difficult to remove, making teeth less susceptible to decay.

121.   a. $Pb(OH)_2(s) \quad \rightleftharpoons \quad Pb^{2+}(aq) \ + \ 2\ OH^-(aq)$

Initial	s = solubility (mol/L)	0	1.0×10^{-7} M from water
Equil.		s	$1.0 \times 10^{-7} + 2s$

$K_{sp} = 1.2 \times 10^{-15} = [Pb^{2+}][OH^-]^2 = s(1.0 \times 10^{-7} + 2s)^2 \approx s(2s^2) = 4s^3$

$s = [Pb^{2+}] = 6.7 \times 10^{-6}$ M;   Assumption to ignore OH^- from water is good by the 5% rule.

        b. $Pb(OH)_2(s) \ \rightleftharpoons \ Pb^{2+}(aq) \ + \ 2\ OH^-(aq)$

Initial	0	0.10 M	pH = 13.00, $[OH^-]$ = 0.10 M
	s mol/L $Pb(OH)_2(s)$ dissolves to reach equilibrium		
Equil.	s	0.10	(buffered solution)

$1.2 \times 10^{-15} = (s)(0.10)^2, \ s = [Pb^{2+}] = 1.2 \times 10^{-13}$ M

    c.   We need to calculate the Pb^{2+} concentration in equilibrium with $EDTA^{4-}$. Since K is large for the formation of $PbEDTA^{2-}$, let the reaction go to completion; then solve an equilibrium problem to get the Pb^{2+} concentration.

$$Pb^{2+} \ + \ EDTA^{4-} \ \rightleftharpoons \ PbEDTA^{2-} \qquad K = 1.1 \times 10^{18}$$

Before	0.010 M	0.050 M		0	
	0.010 mol/L Pb^{2+} reacts completely (large K)				
Change	-0.010	-0.010	$\rightarrow$	+0.010	Reacts completely
After	0	0.040		0.010	New initial conditions
	x mol/L $PbEDTA^{2-}$ dissociates to reach equilibrium				
Equil.	x	0.040 + x		0.010 - x	

$1.1 \times 10^{18} = \dfrac{(0.010 - x)}{(x)(0.040 + x)} \approx \dfrac{(0.010)}{x(0.040)}, \ x = [Pb^{2+}] = 2.3 \times 10^{-19}$ M Assumptions good.

Now calculate the solubility quotient for $Pb(OH)_2$ to see if precipitation occurs. The concentration of OH^- is 0.10 M since we have a solution buffered at pH = 13.00.

$$Q = [Pb^{2+}]_o [OH^-]_o^2 = (2.3 \times 10^{-19})(0.10)^2 = 2.3 \times 10^{-21} < K_{sp} \ (1.2 \times 10^{-15})$$

$Pb(OH)_2(s)$ will not form since Q is less than K_{sp}.

Challenge Problems

123.   mmol $HC_3H_5O_2$ present initially $= 45.0 \text{ mL} \times \dfrac{0.750 \text{ mmol}}{\text{mL}} = 33.8 \text{ mmol } HC_3H_5O_2$

mmol $C_3H_5O_2^-$ present initially $= 55.0 \text{ mL} \times \dfrac{0.700 \text{ mmol}}{\text{mL}} = 38.5 \text{ mmol } C_3H_5O_2^-$

The initial pH of the buffer is:

$$pH = pK_a + \log \frac{[C_3H_5O_2^-]}{[HC_3H_5O_2]} = -\log (1.3 \times 10^{-5}) + \log \frac{\dfrac{38.5 \text{ mmol}}{100.0 \text{ mL}}}{\dfrac{33.8 \text{ mmol}}{100.0 \text{ mL}}} = 4.89 + \log \frac{38.5}{33.8} = 4.95$$

Note: Since the buffer components are in the same volume of solution, we can use the mol (or mmol) ratio in the Henderson-Hasselbalch equation to solve for pH instead of using the concentration ratio of $[C_3H_5O_2^-]/[HC_3H_5O_2]$. The total volume always cancels for buffer solutions.

When NaOH is added, the pH will increase, and the added OH^- will convert $HC_3H_5O_2$ into $C_3H_5O_2^-$. The pH after addition of OH^- increases by 2.5%, so the resulting pH is:

4.95 + 0.025 (4.95) = 5.07

At this pH, a buffer solution still exists and the mmol ratio between $C_3H_5O_2^-$ and $HC_3H_5O_2$ is:

$$pH = pK_a + \log \frac{\text{mmol } C_3H_5O_2^-}{\text{mmol } HC_3H_5O_2}, \ \ 5.07 = 4.89 + \log \frac{\text{mmol } C_3H_5O_2^-}{\text{mmol } HC_3H_5O_2}$$

$$\frac{\text{mmol } C_3H_5O_2^-}{\text{mmol } HC_3H_5O_2} = 10^{0.18} = 1.5$$

Let x = mmol OH^- added to increase pH to 5.07. Since OH^- will essentially react to completion with $HC_3H_5O_2$, the setup for the problem using mmol is:

	$HC_3H_5O_2$	+	OH^-	→	$C_3H_5O_2^-$	
Before	33.8 mmol		x mmol		38.5 mmol	
Change	$-x$		$-x$	→	$+x$	Reacts completely
After	33.8 - x		0		38.5 + x	

Solving for x:

$$\frac{\text{mmol } C_3H_5O_2^-}{\text{mmol } HC_3H_5O_2} = 1.5 = \frac{38.5 + x}{33.8 - x}, \ \ 1.5 (33.8 - x) = 38.5 + x, \ \ x = 4.9 \text{ mmol } OH^- \text{ added}$$

The volume of NaOH necessary to raise the pH by 2.5% is:

$$4.9 \text{ mmol NaOH} \times \frac{1 \text{ mL}}{0.10 \text{ mmol NaOH}} = 49 \text{ mL}$$

49 mL of 0.10 M NaOH must be added to increase the pH by 2.5%.

125.   For HOCl, $K_a = 3.5 \times 10^{-8}$ and $pK_a = -\log(3.5 \times 10^{-8}) = 7.46$; This will be a buffer solution since the pH is close to the pK_a value.

$$pH = pK_a + \log \frac{[OCl^-]}{[HOCl]}, \quad 8.00 = 7.46 + \log \frac{[OCl^-]}{[HOCl]}, \quad \frac{[OCl^-]}{[HOCl]} = 10^{0.54} = 3.5$$

1.00 L $\times$ 0.0500 M = 0.0500 mol HOCl initially. Added OH^- converts HOCl into OCl^-. The total moles of OCl^- and HOCl must equal 0.0500 mol. Solving where n = moles:

$$n_{OCl^-} + n_{HOCl} = 0.0500 \text{ and } n_{OCl^-} = 3.5 \, n_{HOCl}$$

$$4.5 \, n_{HOCl} = 0.0500, \quad n_{HOCl} = 0.011 \text{ mol}; \quad n_{OCl^-} = 0.039 \text{ mol}$$

We need to add 0.039 mol NaOH to produce 0.039 mol OCl^-.

$$0.039 \text{ mol } OH^- = V \times 0.0100 \, M, \quad V = 3.9 \text{ L NaOH}$$

127.   The first titration plot (from 0-100.0 mL) corresponds to the titration of H_2A by OH^-. The reaction is $H_2A + OH^- \rightarrow HA^- + H_2O$. After all of the H_2A has been reacted, the second titration (from 100.0 - 200.0 mL) corresponds to the titration of HA^- by OH^-. The reaction is $HA^- + OH^- \rightarrow A^{2-} + H_2O$.

a.   At 100.0 mL of NaOH, just enough OH^- has been added to react completely with all of the H_2A present (mol OH^- added = mol H_2A present initially). From the balanced equation, the mol of HA^- produced will equal the mol of H_2A present initially. Since mol HA^- present at 100.0 mL OH^- added equals the mol of H_2A present initially, exactly 100.0 mL more of NaOH must be added to react with all of the HA^-. The volume of NaOH added to reach the second equivalence point equals 100.0 mL + 100.0 mL = 200.0 mL.

b. $H_2A + OH^- \rightarrow HA^- + H_2O$ is the reaction occurring from 0-100.0 mL NaOH added.

i.   Since no reaction has taken place, H_2A and H_2O are the major species.

ii.   Adding OH^- converts H_2A into HA^-. The major species up to 100.0 mL NaOH added are H_2A, HA^-, H_2O, and Na^+.

iii.   At 100.0 mL NaOH added, mol of OH^- = mol H_2A, so all of the H_2A present initially has been converted into HA^-. The major species are HA^-, H_2O, and Na^+.

iv. Between 100.0 and 200.0 mL NaOH added, the OH⁻ converts HA⁻ into A²⁻. The major species are HA⁻, A²⁻, H₂O, and Na⁺.

v. At the second equivalence point (200.0 mL), just enough OH⁻ has been added to convert all of the HA⁻ into A²⁻. The major species are A²⁻, H₂O, and Na⁺.

vi. Past 200.0 mL NaOH added, excess OH⁻ is present. The major species are OH⁻, A²⁻, H₂O, and Na⁺.

c. 50.0 mL of NaOH added correspond to the first halfway point to equivalence. Exactly one-half of the H₂A present initially has been converted into its conjugate base HA⁻, so [H₂A] = [HA⁻] in this buffer solution.

$$H_2A \rightleftharpoons HA^- + H^+ \qquad K_{a_1} = \frac{[HA^-][H^+]}{[H_2A]}$$

When [HA⁻] = [H₂A], then $K_{a_1} = [H^+]$ or $pK_{a_1} = pH$.

Here, pH = 4.0 so $pK_{a_1} = 4.0$ and $K_{a_1} = 10^{-4.0} = 1 \times 10^{-4}$.

150.0 mL of NaOH added correspond to the second halfway point to equivalence where [HA⁻] = [A²⁻] in this buffer solution.

$$HA^- \rightleftharpoons A^{2-} + H^+ \qquad K_{a_2} = \frac{[A^{2-}][H^+]}{[HA^-]}$$

When [A²⁻] = [HA⁻], then $K_{a_2} = [H^+]$ or $pK_{a_2} = pH$.

Here, pH = 8.0 so $pK_{a_2} = 8.0$ and $K_{a_2} = 10^{-8.0} = 1 \times 10^{-8}$.

129. a. $$SrF_2(s) \rightleftharpoons Sr^{2+}(aq) + 2\,F^-(aq)$$

Initial	0	0
s mol/L SrF₂ dissolves to reach equilibrium		
Equil.	s	$2s$

$$[Sr^{2+}][F^-]^2 = K_{sp} = 7.9 \times 10^{-10} = 4s^3, \quad s = 5.8 \times 10^{-4} \text{ mol/L}$$

b. Greater, because some of the F⁻ would react with water:

$$F^- + H_2O \rightleftharpoons HF + OH^- \qquad K_b = \frac{K_w}{K_a(HF)} = 1.4 \times 10^{-11}$$

This lowers the concentration of F⁻, forcing more SrF₂ to dissolve.

c. $SrF_2(s) \rightleftharpoons Sr^{2+} + 2\ F^-$ $K_{sp} = 7.9 \times 10^{-10} = [Sr^{2+}]\ [F^-]^2$

Let s = solubility = $[Sr^{2+}]$, then $2s$ = total F^- concentration.

Since F^- is a weak base, some of the F^- is converted into HF. Therefore:

  total F^- concentration = $2s = [F^-] + [HF]$.

$HF \rightleftharpoons H^+ + F^-$ $K_a = 7.2 \times 10^{-4} = \dfrac{[H^+]\ [F^-]}{[HF]} = \dfrac{1.0 \times 10^{-2}\ [F^-]}{[HF]}$ (since pH = 2.00 buffer)

$7.2 \times 10^{-2} = \dfrac{[F^-]}{[HF]}$, $[HF] = 14\ [F^-]$; Solving:

   $[Sr^{2+}] = s$; $2s = [F^-] + [HF] = [F^-] + 14\ [F^-]$, $2s = 15\ [F^-]$, $[F^-] = 2s/15$

$7.9 \times 10^{-10} = [Sr^{2+}]\ [F^-]^2 = (s) \left(\dfrac{2s}{15} \right)^2$, $s = 3.5 \times 10^{-3}$ mol/L

CHAPTER SIXTEEN

SPONTANEITY, ENTROPY, AND FREE ENERGY

Questions

7.   a. A spontaneous process is one that occurs without any outside intervention.

        b. Entropy is a measure of disorder or randomness.

        c. The system is the portion of the universe in which we are interested.

        d. The surroundings are everything else in the universe besides the system.

9.   Living organisms need an external source of energy to carry out these processes. Green plants use the energy from sunlight to produce glucose from carbon dioxide and water by photosynthesis. In the human body, the energy released from the metabolism of glucose helps drive the synthesis of proteins. For all processes combined, ΔS_{univ} must be greater than zero (second law).

11.   Dispersion increases the entropy of the universe since the more widely something is dispersed, the greater the disorder. We must do work to overcome this disorder. In terms of the second law, it would be more advantageous to prevent contamination of the environment than to clean it up later. As a substance disperses, we have a much larger area that must be decontaminated.

13. $w_{max} = \Delta G$; When ΔG is negative, the magnitude of ΔG is equal to the maximum possible useful work obtainable from the process (at constant T and P). When ΔG is positive, the magnitude of ΔG is equal to the minimum amount of work that must be expended to make the process spontaneous. Due to waste energy (heat) in any real process, the amount of useful work obtainable from a spontaneous process is always less than w_{max} and, for a nonspontaneous reaction, an amount of work greater than w_{max} must be applied to make the process spontaneous.

Exercises

Spontaneity, Entropy, and the Second Law of Thermodynamics: Free Energy

15.   a, b and c; From our own experiences, salt water, colored water and rust form without any outside intervention. A bedroom, however, spontaneously gets cluttered. It takes an outside energy source to clean a bedroom.

17.   We draw all of the possible arrangements of the two particles in the three levels.

2 kJ	__	__	x	__	x	xx
1 kJ	__	x	__	xx	x	__
0 kJ	xx	x	x	__	__	__

Total E =   0 kJ   1 kJ   2 kJ   2 kJ   3 kJ   4 kJ

The most likely total energy is 2 kJ.

19.   a. H_2 at 100°C and 0.5 atm; Higher temperature and lower pressure means greater volume and hence, greater positional entropy.

       b. N_2 at STP has the greater volume.   c. $H_2O(l)$ is more disordered than $H_2O(s)$.

21.   a.   To boil a liquid requires heat. Hence, this is an endothermic process. All endothermic processes decrease the entropy of the surroundings (ΔS_{surr} is negative).

       b.   This is an exothermic process. Heat is released when gas molecules slow down enough to form the solid. In exothermic processes, the entropy of the surroundings increases (ΔS_{surr} is positive).

23. $\Delta G = \Delta H - T\Delta S$; When ΔG is negative, the process will be spontaneous.

       a. $\Delta G = \Delta H - T\Delta S = 25 \times 10^3$ J - (300. K)(5.0 J/K) = 24,000 J, Not spontaneous

       b. $\Delta G = 25,000$ J - (300. K)(100. J/K) = -5000 J, Spontaneous

       c.   Without calculating ΔG, we know this reaction will be spontaneous at all temperatures. ΔH is negative and ΔS is positive (-$T\Delta S$ < 0). ΔG will always be less than zero with these sign combinations for ΔH and ΔS.

       d. $\Delta G = -1.0 \times 10^4$ J - (200. K)(-40. J/K) = -2000 J, Spontaneous

25.   At the boiling point, $\Delta G = 0$ so $\Delta H = T\Delta S$.

$$\Delta S = \frac{\Delta H}{T} = \frac{27.5 \text{ kJ/mol}}{(273 + 35) \text{ K}} = 8.93 \times 10^{-2} \text{ kJ/K} \cdot \text{mol} = 89.3 \text{ J/K} \cdot \text{mol}$$

27.   a. $NH_3(s) \rightarrow NH_3(l)$; $\Delta G = \Delta H - T\Delta S = 5650$ J/mol - 200. K (28.9 J/K·mol)

          $\Delta G = 5650$ J/mol - 5780 J/mol = -130 J/mol

          Yes, NH_3 will melt since $\Delta G < 0$ at this temperature.

       b.   At the melting point, $\Delta G = 0$ so $T = \dfrac{\Delta H}{\Delta S} = \dfrac{5650 \text{ J/mol}}{28.9 \text{ J/K} \cdot \text{mol}} = 196$ K.

Chemical Reactions: Entropy Changes and Free Energy

29.   a.   Decrease in disorder; $\Delta S°(-)$ b.   Increase in disorder; $\Delta S°(+)$

c.   Decrease in disorder $(\Delta n < 0)$; $\Delta S°(-)$ d.   Increase in disorder $(\Delta n > 0)$; $\Delta S°(+)$

For c and d, concentrate on the gaseous products and reactants. When there are more gaseous product molecules than gaseous reactant molecules $(\Delta n > 0)$, then $\Delta S°$ will be positive (disorder increases). When Δn is negative, then $\Delta S°$ is negative (disorder decreases).

31.   a. $C_{graphite}(s)$; Diamond is a more ordered structure than graphite.

b. $C_2H_5OH(g)$; The gaseous state is more disordered than the liquid state.

c. $CO_2(g)$; The gaseous state is more disordered than the solid state.

33.   a. $2 H_2S(g) + SO_2(g) \rightarrow 3 S_{rhombic}(s) + 2 H_2O(g)$: Since there are more molecules of reactant gases compared to product molecules of gas $(\Delta n = 2 - 3 < 0)$, then $\Delta S°$ will be negative.
$\Delta S° = \Sigma n_p S°_{products} - \Sigma n_r S°_{reactants}$

$\Delta S° = [3 \text{ mol } S_{rhombic}(s) (32 \text{ J/K·mol}) + 2 \text{ mol } H_2O(g) (189 \text{ J/K·mol})]$

$- [2 \text{ mol } H_2S(g) (206 \text{ J/K·mol}) + 1 \text{ mol } SO_2(g) (248 \text{ J/K·mol})]$

$\Delta S° = 474 \text{ J/K} - 660. \text{ J/K} = -186 \text{ J/K}$

b. $2 SO_3(g) \rightarrow 2 SO_2(g) + O_2(g)$; Since Δn of gases is positive $(\Delta n = 3-2)$, $\Delta S°$ will be positive.

$\Delta S = 2 \text{ mol}(248 \text{ J/K·mol}) + 1 \text{ mol}(205 \text{ J/K·mol}) - [2 \text{ mol}(257 \text{ J/K·mol})] = 187 \text{ J/K}$

c. $Fe_2O_3(s) + 3 H_2(g) \rightarrow 2 Fe(s) + 3 H_2O(g)$; Since Δn of gases = 0 $(\Delta n = 3 - 3)$, we can't easily predict if $\Delta S°$ will be positive of negative.

$\Delta S = 2 \text{ mol}(27 \text{ J/K·mol}) + 3 \text{ mol}(189 \text{ J/K·mol}) - [1 \text{ mol}(90. \text{ J/K·mol}) + 3 \text{ mol}(131 \text{ J/K·mol})]$

$\Delta S = 138 \text{ J/K}$

35. $C_2H_2(g) + 4 F_2(g) \rightarrow 2 CF_4(g) + H_2(g)$; $\Delta S° = 2 S°_{CF_4} + S°_{H_2} - [S°_{C_2H_2} + 4 S°_{F_2}]$

$-358 \text{ J/K} = (2 \text{ mol}) S°_{CF_4} + 131 \text{ J/K} - [201 \text{ J/K} + 4(203 \text{ J/K})]$, $S°_{CF_4} = 262 \text{ J/K·mol}$

37.   a. $S_{rhombic} \rightarrow S_{monoclinic}$; This phase transition is spontaneous $(\Delta G < 0)$ at temperatures above 95°C. $\Delta G = \Delta H - T\Delta S$; For ΔG to be negative only above a certain temperature, ΔH is positive and ΔS is positive (see Table 16.5 of text).

b.   Since ΔS is positive, $S_{rhombic}$ is the more ordered crystalline structure.

39. a. When a bond is formed, energy is released, so ΔH is negative. Since there are more reactant molecules of gas than product molecules of gas ($\Delta n < 0$), ΔS will be negative.

b. $\Delta G = \Delta H - T\Delta S$; For this reaction to be spontaneous ($\Delta G < 0$), the favorable enthalpy term must dominate. The reaction will be spontaneous at low temperatures where the ΔH term dominates.

41. a.

	$CH_4(g)$	$+$	$2 O_2(g)$	$\rightarrow$	$CO_2(g)$	$+$	$2 H_2O(g)$	
ΔH_f°	-75 kJ/mol		0		-393.5		-242	
ΔG_f°	-51 kJ/mol		0		-394		-229	Data from Appendix 4
S°	186 J/K•mol		205		214		189	

$\Delta H^\circ = \Sigma n_p \Delta H_{f,\,products}^\circ - \Sigma n_r \Delta H_{f,\,reactants}^\circ$; $\Delta S^\circ = \Sigma n_p S_{products}^\circ - \Sigma n_r S_{reactants}^\circ$

$\Delta H^\circ = 2 \text{ mol}(-242 \text{ kJ/mol}) + 1 \text{ mol}(-393.5 \text{ kJ/mol}) - [1 \text{ mol}(-75 \text{ kJ/mol})] = -803 \text{ kJ}$

$\Delta S^\circ = 2 \text{ mol}(189 \text{ J/K•mol}) + 1 \text{ mol}(214 \text{ J/K•mol})$

$\qquad\qquad\qquad - [1 \text{ mol}(186 \text{ J/K•mol}) + 2 \text{ mol}(205 \text{ J/K•mol})] = -4 \text{ J/K}$

There are two ways to get ΔG°. We can use $\Delta G^\circ = \Delta H^\circ - T\Delta S^\circ$ (be careful of units):

$\Delta G^\circ = \Delta H^\circ - T\Delta S^\circ = -803 \times 10^3 \text{ J} - (298 \text{ K})(-4 \text{ J/K}) = -8.018 \times 10^5 \text{ J} = -802 \text{ kJ}$

or we can use ΔG_f° values where $\Delta G^\circ = \Sigma n_p \Delta G_{f,\,products}^\circ - \Sigma n_r \Delta G_{f,\,reactants}^\circ$:

$\Delta G^\circ = 2 \text{ mol}(-229 \text{ kJ/mol}) + 1 \text{ mol}(-394 \text{ kJ/mol}) - [1 \text{ mol}(-51 \text{ kJ/mol})]$

$\Delta G^\circ = -801 \text{ kJ}$ (Answers are the same within round-off error.)

b.

	$6 CO_2(g)$	$+$	$6 H_2O(l)$	$\rightarrow$	$C_6H_{12}O_6(s)$	$+$	$6 O_2(g)$
ΔH_f°	-393.5 kJ/mol		-286		-1275		0
S°	214 J/K•mol		70.		212		205

$\Delta H^\circ = -1275 - [6(-286) + 6(-393.5)] = 2802 \text{ kJ}$

$\Delta S^\circ = 6(205) + 212 - [6(214) + 6(70.)] = -262 \text{ J/K}$

$\Delta G^\circ = 2802 \text{ kJ} - (298 \text{ K})(-0.262 \text{ kJ/K}) = 2880. \text{ kJ}$

c. $P_4O_{10}(s) + 6 H_2O(l) \rightarrow 4 H_3PO_4(s)$

ΔH_f° (kJ/mol)	-2984	-286	-1279
S° (J/K•mol)	229	70.	110.

$\Delta H^\circ = 4 \text{ mol}(-1279 \text{ kJ/mol}) - [1 \text{ mol}(-2984 \text{ kJ/mol}) + 6 \text{ mol}(-286 \text{ kJ/mol})] = -416 \text{ kJ}$

$\Delta S^\circ = 4(110.) - [229 + 6(70.)] = -209 \text{ J/K}$

$\Delta G^\circ = \Delta H^\circ - T\Delta S^\circ = -416 \text{ kJ} - (298 \text{ K})(-0.209 \text{ kJ/K}) = -354 \text{ kJ}$

d. $HCl(g) + NH_3(g) \rightarrow NH_4Cl(s)$

ΔH_f° (kJ/mol)	-92	-46	-314
S° (J/K•mol)	187	193	96

$\Delta H^\circ = -314 - [-92 - 46] = -176 \text{ kJ}; \quad \Delta S^\circ = 96 - [187 + 193] = -284 \text{ J/K}$

$\Delta G^\circ = \Delta H^\circ - T\Delta S^\circ = -176 \text{ kJ} - (298 \text{ K})(-0.284 \text{ kJ/K}) = -91 \text{ kJ}$

43. $CH_4(g) + CO_2(g) \rightarrow CH_3CO_2H(l)$

$\Delta H^\circ = -484 - [-75 + (-393.5)] = -16 \text{ kJ}; \quad \Delta S^\circ = 160 - [186 + 214] = -240. \text{ J/K}$

$\Delta G^\circ = \Delta H^\circ - T\Delta S^\circ = -16 \text{ kJ} - (298 \text{ K})(-0.240 \text{ kJ/K}) = 56 \text{ kJ}$

This reaction is spontaneous only at temperatures below $T = \Delta H^\circ/\Delta S^\circ = 67 \text{ K}$ (where the favorable ΔH° term will dominate, giving a negative ΔG° value). This is not practical. Substances will be in condensed phases, and rates will be very slow at this extremely low temperature.

$CH_3OH(g) + CO(g) \rightarrow CH_3CO_2H(l)$

$\Delta H^\circ = -484 - [-110.5 + (-201)] = -173 \text{ kJ}; \quad \Delta S^\circ = 160 - [198 + 240.] = -278 \text{ J/K}$

$\Delta G^\circ = -173 \text{ kJ} - (298 \text{ K})(-0.278 \text{ kJ/K}) = -90. \text{ kJ}$

This reaction also has a favorable enthalpy and an unfavorable entropy term. This reaction is spontaneous at temperatures below $T = \Delta H^\circ/\Delta S^\circ = 622 \text{ K}$. The reaction of CH_3OH and CO will be preferred. It is spontaneous at high enough temperatures that the rates of reaction should be reasonable.

45. $SO_3(g) \rightarrow SO_2(g) + 1/2\ O_2(g)$ $\Delta G° = -1/2\ (-142\ kJ)$
           $S(s) + 3/2\ O_2(g) \rightarrow SO_3(g)$ $\Delta G° = -371\ kJ$
           _____

           $S(s) + O_2(g) \rightarrow SO_2(g)$ $\Delta G° = 71\ kJ - 371\ kJ = -300.\ kJ$

47. $\Delta G° = \Sigma n_p \Delta G°_{f\ products} - \Sigma n_r \Delta G°_{f\ reactants}$, $-374\ kJ = -1105\ kJ - \Delta G°_{f,\ SF_4}$, $\Delta G°_{f,\ SF_4} = -731\ kJ/mol$

49. $\Delta G° = \Sigma n_p \Delta G°_{f\ products} - \Sigma n_r \Delta G°_{f\ reactants}$

       $\Delta G° = 1\ mol(-2698\ kJ/mol) + 6\ mol(-237\ kJ/mol) - [4\ mol(13\ kJ/mol) + 8\ mol(0)] = -4172\ kJ$

Free Energy: Pressure Dependence and Equilibrium

51. $\Delta G = \Delta G° + RT\ \ln Q$; For this reaction: $\Delta G = \Delta G° + RT\ \ln \dfrac{P_{NO_2} \times P_{O_2}}{P_{NO} \times P_{O_3}}$

       $\Delta G° = 1\ mol(52\ kJ/mol) + 1\ mol(0) - [1\ mol(87\ kJ/mol) + 1\ mol(163\ kJ/mol)] = -198\ kJ$

       $\Delta G = -198\ kJ + \dfrac{8.3145\ J/K \cdot mol}{1000\ J/kJ}\ (298\ K)\ \ln \dfrac{(1.00 \times 10^{-7})\ (1.00 \times 10^{-3})}{(1.00 \times 10^{-6})\ (2.00 \times 10^{-6})}$

       $\Delta G = -198\ kJ + 9.69\ kJ = -188\ kJ$

53. $\Delta G = \Delta G° + RT\ \ln Q = \Delta G° + RT\ \ln \dfrac{P_{N_2O_4}}{P_{NO_2}^2}$

       $\Delta G° = 1\ mol(98\ kJ/mol) - 2\ mol(52\ kJ/mol) = -6\ kJ$

   a.   These are standard conditions, so $\Delta G = \Delta G°$ since $Q = 1$ and $\ln Q = 0$. Since $\Delta G°$ is negative, then the forward reaction is spontaneous. The reaction shifts right to reach equilibrium.

   b. $\Delta G = -6 \times 10^3\ J + 8.3145\ J/K \cdot mol\ (298\ K)\ \ln \dfrac{0.50}{(0.21)^2}$

       $\Delta G = -6 \times 10^3\ J + 6.0 \times 10^3\ J = 0$

       Since $\Delta G = 0$, this reaction is at equilibrium (no shift).

   c. $\Delta G = -6 \times 10^3\ J + 8.3145\ J/K \cdot mol\ (298\ K)\ \ln \dfrac{1.6}{(0.29)^2}$

       $\Delta G = -6 \times 10^3\ J + 7.3 \times 10^3\ J = 1.3 \times 10^3\ J = 1 \times 10^3\ J$

       Since ΔG is positive, the reverse reaction is spontaneous, so the reaction shifts to the left to reach equilibrium.

55.   At 25.0°C: $\Delta G° = \Delta H° - T\Delta S° = -58.03 \times 10^3$ J/mol $- (298.2$ K$)(-176.6$ J/K•mol$)$
$$= -5.37 \times 10^3 \text{ J/mol}$$

$\Delta G° = -RT \ln K, \ln K = \dfrac{-\Delta G°}{RT} = \dfrac{-(-5.37 \times 10^3 \text{ J/mol})}{(8.3145 \text{ J/K} \bullet \text{mol})(298.2 \text{ K})} = 2.166,\ K = e^{2.166} = 8.72$

At 100.0°C: $\Delta G° = -58.03 \times 10^3$ J/mol $- (373.2$ K$)(-176.6$ J/K•mol$) = 7.88 \times 10^3$ J/mol

$\ln K = \dfrac{-(7.88 \times 10^3 \text{ J/mol})}{(8.3145 \text{ J/K} \bullet \text{mol})(373.2 \text{ K})} = -2.540,\ K = e^{-2.540} = 0.0789$

Note: When determining exponents, we will round off after the calculation is complete. This helps eliminate excessive round-off error.

57.   a. $\Delta G° = -RT \ln K = -\dfrac{8.3145 \text{ J}}{\text{K mol}} (298 \text{ K}) \ln (1.00 \times 10^{-14}) = 7.99 \times 10^4 \text{ J} = 79.9 \text{ kJ/mol}$

b. $\Delta G°_{313} = -RT \ln K = -\dfrac{8.3145 \text{ J}}{\text{K mol}} (313 \text{ K}) \ln (2.92 \times 10^{-14}) = 8.11 \times 10^4 \text{ J} = 81.1 \text{ kJ/mol}$

59. $K = \dfrac{P_{NF_3}^2}{P_{N_2} \times P_{F_2}^3} = \dfrac{(0.48)^2}{0.021(0.063)^3} = 4.4 \times 10^4$

$\Delta G°_{800} = -RT \ln K = -8.3145$ J/K•mol $(800.$ K$) \ln (4.4 \times 10^4) = -7.1 \times 10^4$ J/mol $= -71$ kJ/mol

61.   The equation $\ln K = \dfrac{-\Delta H°}{R}\left(\dfrac{1}{T}\right) + \dfrac{\Delta S°}{R}$ is in the form of a straight line equation $(y = mx + b)$. A

graph of ln K vs. 1/T will yield a straight line with slope $= m = -\Delta H°/R$ and a y-intercept $= b = \Delta S°/R$.

From the plot:

$\text{slope} = \dfrac{\Delta y}{\Delta x} = \dfrac{0 - 40.}{3.0 \times 10^{-3} \text{ K}^{-1} - 0} = -1.3 \times 10^4 \text{ K}$

$-1.3 \times 10^4 \text{ K} = -\Delta H°/R,\ \Delta H° = 1.3 \times 10^4 \text{ K} \times 8.3145 \text{ J/K} \bullet \text{mol} = 1.1 \times 10^5 \text{ J/mol}$

y-intercept $= 40. = \Delta S°/R,\ \Delta S° = 40. \times 8.3145$ J/K • mol $= 330$ J/K • mol

As seen here, when $\Delta H°$ is positive, the slope of the ln K vs. 1/T plot is negative. When $\Delta H°$ is negative, as in an exothermic process, the slope of the ln K vs. 1/T plot will be positive (slope $= -\Delta H°/R$).

Additional Exercises

63.   It appears the sum of the two processes has no net change. This is not so. By the second law of thermodynamics, ΔS_{univ} must have increased even though it looks as if we have gone through a cyclic process.

65.   The introduction of mistakes is an effect of entropy. The purpose of redundant information is to provide a control to check the "correctness" of the transmitted information.

67.   S (monoclinic) $\rightarrow$ S (rhombic); $\Delta H° = 0 - 0.30 = -0.30$ kJ; $\Delta S° = 31.73 - 32.55 = -0.82$ J/K

At the conversion temperature: $\Delta G° = 0$ so $\Delta H° = T\Delta S°$; $T = \dfrac{\Delta H°}{\Delta S°} = \dfrac{-3.0 \times 10^2 \text{ J}}{-0.82 \text{ J/K}} = 370$ K

69.
$$\begin{array}{ll} \text{HgbO}_2 \qquad\; \rightarrow \text{Hgb} \;+\; \text{O}_2 & \Delta G° = -(-70 \text{ kJ}) \\ \text{Hgb} \;+\; \text{CO} \rightarrow \text{HgbCO} & \Delta G° = -80 \text{ kJ} \end{array}$$

$$\text{HgbO}_2 + \text{CO} \rightarrow \text{HgbCO} + \text{O}_2 \quad \Delta G° = -10 \text{ kJ}$$

$$\Delta G° = -RT\ln K, \quad K = \exp\left(\frac{-\Delta G°}{RT}\right) = \exp\left(\frac{-(-10 \times 10^3 \text{ J})}{(8.3145 \text{ J/K} \cdot \text{mol})(298 \text{ K})}\right) = 60$$

71. $\text{HF(aq)} \rightleftharpoons \text{H}^+\text{(aq)} + \text{F}^-\text{(aq)}$; $\Delta G = \Delta G° + RT \ln \dfrac{[\text{H}^+][\text{F}^-]}{[\text{HF}]}$

$\Delta G° = -RT \ln K = -(8.3145 \text{ J/K} \bullet \text{mol}) (298 \text{ K}) \ln (7.2 \times 10^{-4}) = 1.8 \times 10^4$ J/mol

a.   The concentrations are all at standard conditions, so $\Delta G = \Delta G° = 1.8 \times 10^4$ J/mol (since $Q = 1.0$ and $\ln Q = 0$). Since $\Delta G°$ is positive, then the reaction shifts left to reach equilibrium.

b. $\Delta G = 1.8 \times 10^4$ J/mol $+ (8.3145 \text{ J/K} \bullet \text{mol}) (298 \text{ K}) \ln \dfrac{(2.7 \times 10^{-2})^2}{0.98}$

$\Delta G = 1.8 \times 10^4$ J/mol $- 1.8 \times 10^4$ J/mol $= 0$

Since $\Delta G = 0$, then the reaction is at equilibrium (no shift).

c. $\Delta G = 1.8 \times 10^4 + 8.3145 \, (298) \ln \dfrac{(1.0 \times 10^{-5})^2}{1.0 \times 10^{-5}} = -1.1 \times 10^4$ J/mol; shifts right

d. $\Delta G = 1.8 \times 10^4 + 8.3145 \, (298) \ln \dfrac{7.2 \times 10^{-4}(0.27)}{0.27} = 1.8 \times 10^4 - 1.8 \times 10^4 = 0$; at equilibrium

e. $\Delta G = 1.8 \times 10^4 + 8.3145 \, (298) \ln \dfrac{1.0 \times 10^{-3}(0.67)}{0.52} = 2 \times 10^3$ J/mol; shifts left

73.   a. $\Delta G° = -RT \ln K$, $K = \exp(-\Delta G°/RT) = \exp\left(\dfrac{-(-30{,}500 \text{ J})}{8.3145 \text{ J/K} \bullet \text{mol} \times 298 \text{ K}}\right) = 2.22 \times 10^5$

b. $C_6H_{12}O_6(s) + 6\ O_2(g) \rightarrow 6\ CO_2(g) + 6\ H_2O(l)$

$\Delta G° = 6\ mol(-394\ kJ/mol) + 6\ mol(-237\ kJ/mol) - 1\ mol(-911\ kJ/mol) = -2875\ kJ$

$$\frac{2875\ kJ}{mol\ glucose} \times \frac{1\ mol\ ATP}{30.5\ kJ} = \frac{94.3\ mol\ ATP}{mol\ glucose} ;\ 94.3\ molecules\ ATP/molecule\ glucose$$

This is an overstatement. The assumption that all of the free energy goes into this reaction is false. Actually, only 38 moles of ATP are produced by the metabolism of one mole of glucose.

75. ΔS is more favorable for reaction two than for reaction one, resulting in $K_2 > K_1$. In reaction one, seven particles in solution are forming one particle. In reaction two, four particles form one particle, which results in a smaller decrease in disorder than for reaction one.

Challenge Problems

77. $3\ O_2(g) \rightleftharpoons 2\ O_3(g)$; $\Delta H° = 2(143\ kJ) = 286\ kJ$; $\Delta G° = 2(163\ kJ) = 326\ kJ$

$$\ln K = \frac{-\Delta G°}{RT} = \frac{-326 \times 10^3\ J}{(8.3145\ J/K \bullet mol)\ (298\ K)} = -131.573,\ K = e^{-131.573} = 7.22 \times 10^{-58}$$

We need the value of K at 230. K. From Section 16.8 of the text: $\ln K = \dfrac{-\Delta H°}{RT} + \dfrac{\Delta S°}{R}$

For two sets of K and T:

$$\ln K_1 = \frac{-\Delta H°}{R}\left(\frac{1}{T_1}\right) + \frac{\Delta S°}{R};\ \ln K_2 = \frac{-\Delta H°}{R}\left(\frac{1}{T_2}\right) + \frac{\Delta S°}{R}$$

Subtracting the first expression from the second:

$$\ln K_2 - \ln K_1 = \frac{\Delta H°}{R}\left(\frac{1}{T_1} - \frac{1}{T_2}\right)\ or\ \ln\frac{K_2}{K_1} = \frac{\Delta H°}{R}\left(\frac{1}{T_1} - \frac{1}{T_2}\right)$$

Let $K_2 = 7.22 \times 10^{-58}$, $T_2 = 298$; $K_1 = K_{230}$, $T_1 = 230.$ K; $\Delta H° = 286 \times 10^3\ J$

$$\ln\frac{7.22 \times 10^{-58}}{K_{230}} = \frac{286 \times 10^3}{8.3145}\left(\frac{1}{230.} - \frac{1}{298}\right) = 34.13$$

$$\frac{7.22 \times 10^{-58}}{K_{230}} = e^{34.13} = 6.6 \times 10^{14},\ K_{230} = 1.1 \times 10^{-72}$$

$$K_{230} = 1.1 \times 10^{-72} = \frac{P_{O_3}^2}{P_{O_2}^3} = \frac{P_{O_3}^2}{(1.0 \times 10^{-3}\ atm)^3},\ P_{O_3} = 3.3 \times 10^{-41}\ atm$$

The volume occupied by one molecule of ozone is:

$$V = \frac{nRT}{P} = \frac{(1/6.022 \times 10^{23}\ \text{mol})\ (0.08206\ \text{L atm/mol}\cdot\text{K})\ (230.\ \text{K})}{(3.3 \times 10^{-41}\ \text{atm})},\quad V = 9.5 \times 10^{17}\ \text{L}$$

Equilibrium is probably not maintained under these conditions. When only two ozone molecules are in a volume of 9.5×10^{17} L, the reaction is not at equilibrium. Under these conditions, $Q > K$ and the reaction shifts left. But with only 2 ozone molecules in this huge volume, it is extremely unlikely they will collide with each other. In these conditions, the concentration of ozone is not large enough to maintain equilibrium.

79.   a.   From the plot, the activation energy of the reverse reaction is $E_a + (-\Delta G^\circ) = E_a - \Delta G^\circ$ (ΔG° is a negative number as drawn in the diagram).

$$k_f = A \exp\left(\frac{-E_a}{RT}\right) \text{ and } k_r = A \exp\left(\frac{-(E_a - \Delta G^\circ)}{RT}\right),\quad \frac{k_f}{k_r} = \frac{A \exp\left(\dfrac{-E_a}{RT}\right)}{A \exp\left(\dfrac{-(E_a - \Delta G^\circ)}{RT}\right)}$$

If the A factors are equal: $\dfrac{k_f}{k_r} = \exp\left(\dfrac{-E_a}{RT} + \dfrac{(E_a - \Delta G^\circ)}{RT}\right) = \exp\left(\dfrac{-\Delta G^\circ}{RT}\right)$

From $\Delta G^\circ = -RT \ln K$, $K = \exp\left(\dfrac{-\Delta G^\circ}{RT}\right)$; Since K and $\dfrac{k_f}{k_r}$ are both equal to the same expression, then $K = \dfrac{k_f}{k_r}$.

   b.   A catalyst will lower the activation energy for both the forward and reverse reaction (but not change ΔG°). Therefore, a catalyst must increase the rate of both the forward and reverse reactions.

81.   a. $\Delta G^\circ = G_B^\circ - G_A^\circ = 11{,}718 - 8996 = 2722$ J

$$K = \exp\left(\frac{-\Delta G^\circ}{RT}\right) = \exp\left(\frac{-2722\ \text{J}}{(8.3145\ \text{J/K}\cdot\text{mol})\ (298\ \text{K})}\right) = 0.333$$

   b.   Since $Q = 1.00 > K$, reaction shifts left. Let x = atm of B(g) which reacts to reach equilibrium.

$$A(g) \quad \rightleftharpoons \quad B(g)$$

Initial   1.00 atm   1.00 atm
Equil.   1.00 + x 1.00 - x

$K = \dfrac{P_B}{P_A} = \dfrac{1.00 - x}{1.00 + x} = 0.333$, $\ 1.00 - x = 0.333 + 0.333\,x$, $\ x = 0.50$ atm

$P_B = 1.00 - 0.50 = 0.50$ atm; $\ P_A = 1.00 + 0.50 = 1.50$ atm

c. $\Delta G = \Delta G° + RT \ln Q = \Delta G° + RT \ln (P_B/P_A)$

$\Delta G = 2722$ J $+ (8.3145)(298) \ln (0.50/1.50) = 2722$ J $- 2722$ J $= 0$ (carrying extra sig. figs.)

83. $K_p = P_{CO_2}$; To insure Ag_2CO_3 from decomposing, P_{CO_2} should be greater than K_p.

From Exercise 16.61, $\ln K = \dfrac{-\Delta H°}{RT} + \dfrac{\Delta S°}{R}$. For two conditions of K and T, the equation is:

$$\ln \frac{K_2}{K_1} = \frac{\Delta H°}{R} \left(\frac{1}{T_1} - \frac{1}{T_2} \right)$$

Let $T_1 = 25°C = 298$ K, $K_1 = 6.23 \times 10^{-3}$ torr; $T_2 = 110.°C = 383$ K, $K_2 = ?$

$$\ln \frac{K_2}{6.23 \times 10^{-3} \text{ torr}} = \frac{79.14 \times 10^3 \text{ J/mol}}{8.3145 \text{ J/K•mol}} \left(\frac{1}{298 \text{ K}} - \frac{1}{383 \text{ K}} \right)$$

$$\ln \frac{K_2}{6.23 \times 10^{-3}} = 7.1, \quad \frac{K_2}{6.23 \times 10^{-3}} = e^{7.1} = 1.2 \times 10^3, \quad K_2 = 7.5 \text{ torr}$$

To prevent decomposition of Ag_2CO_3, the partial pressure of CO_2 should be greater than 7.5 torr.

85.   Use the thermodynamic data to calculate the boiling point of the solvent.

At boiling point, $\Delta G = 0 = \Delta H - T\Delta S$, $\Delta H = T\Delta S$, $T = \dfrac{\Delta H}{\Delta S} = \dfrac{33.90 \times 10^3 \text{ J/mol}}{95.95 \text{ J/K•mol}} = 353.3$ K

$\Delta T = K_b m$, $(355.4$ K $- 353.3$ K$) = 2.5$ K kg/mol (m), $m = \dfrac{2.1}{2.5} = 0.84$ mol/kg

mass solvent $= 150.$ mL $\times \dfrac{0.879 \text{ g}}{\text{mL}} \times \dfrac{1 \text{ kg}}{1000 \text{ g}} = 0.132$ kg

mass solute $= 0.132$ kg solvent $\times \dfrac{0.84 \text{ mol solute}}{\text{kg solvent}} \times \dfrac{142 \text{ g}}{\text{mol}} = 15.7$ g $= 16$ g solute

CHAPTER SEVENTEEN

ELECTROCHEMISTRY

Review of Oxidation - Reduction Reactions

13.   Oxidation: increase in oxidation number; loss of electrons

Reduction: decrease in oxidation number; gain of electrons

15.   The species oxidized shows an increase in oxidation numbers and is called the reducing agent. The species reduced shows a decrease in oxidation numbers and is called the oxidizing agent. The pertinent oxidation numbers are listed by the substance oxidized and the substance reduced.

	Redox?	Ox. Agent	Red. Agent	Substance Oxidized	Substance Reduced
a.	Yes	H_2O	CH_4	CH_4 (C, -4 → +2)	H_2O (H, +1 → 0)
b.	Yes	$AgNO_3$	Cu	Cu (0 → +2)	$AgNO_3$ (Ag, +1 → 0)
c.	Yes	HCl	Zn	Zn (0 → +2)	HCl (H, +1 → 0)
d.	No; There is no change in any of the oxidation numbers.				

Questions

17.   In a galvanic cell, a spontaneous redox reaction occurs which produces an electric current. In an electrolytic cell, electricity is used to force a redox reaction that is not spontaneous to occur.

19.   a. Cathode: The electrode at which reduction occurs.

b. Anode: The electrode at which oxidation occurs.

c. Oxidation half-reaction: The half-reaction in which electrons are products. In a galvanic cell, the oxidation half-reaction always occurs at the anode.

d. Reduction half-reaction: The half-reaction in which electrons are reactants. In a galvanic cell, the reduction half-reaction always occurs at the cathode.

21.   As a battery discharges, E_{cell} decreases, eventually reaching zero. A charged battery is not at equilibrium. At equilibrium, $E_{cell} = 0$ and $\Delta G = 0$. We get no work out of an equilibrium system. A battery is useful to us because it can do work as it approaches equilibrium.

23.   Both fuel cells and batteries are galvanic cells. However, fuel cells, unlike batteries, have the reactants continuously supplied and can produce a current indefinitely.

Exercises

Galvanic Cells, Cell Potentials, Standard Reduction Potentials, and Free Energy

25.   A typical galvanic cell diagram is:

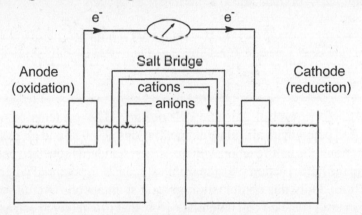

The diagram for all cells will look like this. The contents of each half-cell compartment will be identified for each reaction, with all solute concentrations at 1.0 M and all gases at 1.0 atm. For Exercises 17.25 and 17.26, the flow of ions through the salt bridge was not asked for in the questions. If asked, however, cations always flow into the cathode compartment, and anions always flow into the anode compartment. This is required to keep each compartment electrically neutral.

a.   Table 17.1 of the text lists balanced reduction half-reactions for many substances. For this overall reaction, we need the Cl_2 to Cl^- reduction half-reaction and the Cr^{3+} to $Cr_2O_7^{2-}$ oxidation half-reaction. Manipulating these two half-reactions gives the overall balanced equation.

$$(Cl_2 + 2\ e^- \rightarrow 2\ Cl^-) \times 3$$
$$7\ H_2O + 2\ Cr^{3+} \rightarrow Cr_2O_7^{2-} + 14\ H^+ + 6\ e^-$$

$$7\ H_2O(l) + 2\ Cr^{3+}(aq) + 3\ Cl_2(g) \rightarrow Cr_2O_7^{2-}(aq) + 6\ Cl^-(aq) + 14\ H^+(aq)$$

The contents of each compartment are:

Cathode: Pt electrode; Cl_2 bubbled into solution, Cl^- in solution

Anode: Pt electrode; Cr^{3+}, H^+, and $Cr_2O_7^{2-}$ in solution

We need a nonreactive metal to use as the electrode in each case, since all the reactants and products are in solution. Pt is a common choice. Another possibility is graphite.

b. $Cu^{2+} + 2 e^- \rightarrow Cu$

             $Mg \rightarrow Mg^{2+} + 2e^-$

_____

$Cu^{2+}(aq) + Mg(s) \rightarrow Cu(s) + Mg^{2+}(aq)$

Cathode: Cu electrode; Cu^{2+} in solution;   Anode: Mg electrode; Mg^{2+} in solution

27.   To determine E° for the overall cell reaction, we must add the standard reduction potential to the standard oxidation potential ($E_{cell}^{\circ} = E_{red}^{\circ} + E_{ox}^{\circ}$). Reference Table 17.1 for values of standard reduction potentials. Remember that $E_{ox}^{\circ} = -E_{red}^{\circ}$ and that standard potentials are not multiplied by the integer used to obtain the overall balanced equation.

25a. $E_{cell}^{\circ} = E_{Cl_2 \rightarrow Cl^-}^{\circ} + E_{Cr^{3+} \rightarrow Cr_2O_7^{2-}}^{\circ} = 1.36 \text{ V} + (-1.33 \text{ V}) = 0.03 \text{ V}$

25b. $E_{cell}^{\circ} = E_{Cu^{2+} \rightarrow Cu}^{\circ} + E_{Mg \rightarrow Mg^{2+}}^{\circ} = 0.34 \text{ V} + 2.37 \text{ V} = 2.71 \text{ V}$

29.   Reference Exercise 17.25 for a typical galvanic cell design. The contents of each half-cell compartment are identified below with all solute concentrations at 1.0 M and all gases at 1.0 atm. For each pair of half-reactions, the half-reaction with the largest standard reduction potential will be the cathode reaction, and the half-reaction with the smallest reduction potential will be reversed to become the anode reaction. Only this combination gives a spontaneous overall reaction, i.e., a reaction with a positive overall standard cell potential. Note that in a galvanic cell as illustrated in Exercise 17.25, the cations in the salt bridge migrate to the cathode, and the anions migrate to the anode.

a. $Cl_2 + 2 e^- \rightarrow 2 Cl^-$ E° = 1.36 V

             $2 Br^- \rightarrow Br_2 + 2 e^-$ -E° = -1.09 V

_____

$Cl_2(g) + 2 Br^-(aq) \rightarrow Br_2(aq) + 2 Cl^-(aq)$ E_{cell}° = 0.27 V

The contents of each compartment are:

    Cathode: Pt electrode; $Cl_2(g)$ bubbled in, Cl^- in solution

    Anode: Pt electrode; Br_2 and Br^- in solution

b. $(2 e^- + 2 H^+ + IO_4^- \rightarrow IO_3^- + H_2O) \times 5$ E° = 1.60 V

             $(4 H_2O + Mn^{2+} \rightarrow MnO_4^- + 8 H^+ + 5 e^-) \times 2$ -E° = -1.51 V

_____

$10 H^+ + 5 IO_4^- + 8 H_2O + 2 Mn^{2+} \rightarrow 5 IO_3^- + 5 H_2O + 2 MnO_4^- + 16 H^+$ E_{cell}° = 0.09 V

This simplifies to:

$$3\ H_2O(l) + 5\ IO_4^-(aq) + 2\ Mn^{2+}(aq) \rightarrow 5\ IO_3^-(aq) + 2\ MnO_4^-(aq) + 6\ H^+(aq) \quad E^\circ_{cell} = 0.09\ V$$

Cathode: Pt electrode; IO_4^-, IO_3^-, and H_2SO_4 (as a source of H^+) in solution

Anode: Pt electrode; Mn^{2+}, MnO_4^- and H_2SO_4 in solution

31.   In standard line notation, the anode is listed first and the cathode is listed last. A double line separates the two compartments. By convention, the electrodes are on the ends with all solutes and gases towards the middle. A single line is used to indicate a phase change. We also include all concentrations.

25a. $Pt \mid Cr^{3+}\ (1.0\ M),\ Cr_2O_7^{2-}\ (1.0\ M),\ H^+\ (1.0\ M)\ \|\ Cl_2\ (1.0\ atm)\mid Cl^-\ (1.0\ M)\mid Pt$

25b. $Mg \mid Mg^{2+}\ (1.0\ M)\ \|\ Cu^{2+}\ (1.0\ M)\mid Cu$

29a. $Pt \mid Br^-\ (1.0\ M),\ Br_2\ (1.0\ M)\ \|\ Cl_2\ (1.0\ atm)\mid Cl^-\ (1.0\ M)\mid Pt$

29b. $Pt \mid Mn^{2+}\ (1.0\ M),\ MnO_4^-\ (1.0\ M),\ H^+\ (1.0\ M)\ \|\ IO_4^-\ (1.0\ M),\ H^+\ (1.0\ M),\ IO_3^-\ (1.0\ M)\mid Pt$

33.   Locate the pertinent half-reactions in Table 17.1, and then figure which combination will give a positive standard cell potential. In all cases, the anode compartment contains the species with the smallest standard reduction potential. For part a, the copper compartment is the anode, and in part b, the cadmium compartment is the anode.

a. $Au^{3+} + 3\ e^- \rightarrow Au$ $E^\circ = 1.50\ V$
           $(Cu^+ \rightarrow Cu^{2+} + e^-) \times 3$ $-E^\circ = -0.16\ V$
           _____

           $Au^{3+}(aq) + 3\ Cu^+(aq) \rightarrow Au(s) + 3\ Cu^{2+}(aq)$ $E^\circ_{cell} = 1.34\ V$

b. $(VO_2^+ + 2\ H^+ + e^- \rightarrow VO^{2+} + H_2O) \times 2$ $E^\circ = 1.00\ V$
           $Cd \rightarrow Cd^{2+} + 2e^-$ $-E^\circ = 0.40\ V$
           _____

           $2\ VO_2^+(aq) + 4\ H^+(aq) + Cd(s) \rightarrow 2\ VO^{2+}(aq) + 2\ H_2O(l) + Cd^{2+}(aq) \quad E^\circ_{cell} = 1.40\ V$

35.   a. $2\ Ag^+ + 2\ e^- \rightarrow 2\ Ag$ $E^\circ = 0.80\ V$
              $Cu \rightarrow Cu^{2+} + 2\ e^-$ $-E^\circ = -0.34\ V$
              _____

              $2\ Ag^+ + Cu \rightarrow Cu^{2+} + 2\ Ag$ $E^\circ_{cell} = 0.46\ V$ Spontaneous at standard conditions ($E^\circ_{cell} > 0$).

       b. $Zn^{2+} + 2\ e^- \rightarrow Zn$ $E^\circ = -0.76\ V$
              $Ni \rightarrow Ni^{2+} + 2\ e^-$ $-E^\circ = 0.23\ V$
              _____

              $Zn^{2+} + Ni \rightarrow Zn + Ni^{2+}$ $E^\circ_{cell} = -0.53\ V$ Not spontaneous at standard conditions ($E^\circ_{cell} < 0$).

8

37.
$$Cl_2 + 2\ e^- \rightarrow 2\ Cl^- \qquad\qquad E° = 1.36\ V$$
$$(ClO_2^- \rightarrow ClO_2 + e^-) \times 2 \qquad -E° = -0.954\ V$$

$$2\ ClO_2^-(aq) + Cl_2(g) \rightarrow 2\ ClO_2(aq) + 2\ Cl^-(aq) \qquad E°_{cell} = 0.41\ V = 0.41\ J/C$$

$$\Delta G° = -nFE°_{cell} = -(2\ mol\ e^-)(96,485\ C/mol\ e^-)(0.41\ J/C) = -7.9 \times 10^4\ J = -79\ kJ$$

39. Since the cells are at standard conditions, $w_{max} = \Delta G = \Delta G° = -nFE°_{cell}$. See Exercise 17.33 for the balanced overall equations and for $E°_{cell}$.

33a. $w_{max} = -(3\ mol\ e^-)(96,485\ C/mol\ e^-)(1.34\ J/C) = -3.88 \times 10^5\ J = -388\ kJ$

33b. $w_{max} = -(2\ mol\ e^-)(96,485\ C/mol\ e^-)(1.40\ J/C) = -2.70 \times 10^5\ J = -270.\ kJ$

41. $CH_3OH(l) + 3/2\ O_2(g) \rightarrow CO_2(g) + 2\ H_2O(l) \qquad \Delta G° = 2(-237) + (-394) - [-166] = -702\ kJ$

The balanced half-reactions are:

$$H_2O + CH_3OH \rightarrow CO_2 + 6\ H^+ + 6\ e^- \quad and \quad O_2 + 4\ H^+ + 4\ e^- \rightarrow 2\ H_2O$$

For 3/2 mol O_2, 6 moles of electrons will be transferred (n = 6).

$$\Delta G° = -nFE°, \quad E° = \frac{-\Delta G°}{nF} = \frac{-(-702,000\ J)}{(6\ mol\ e^-)(96,485\ C/mol\ e^-)} = 1.21\ J/C = 1.21\ V$$

43. Good oxidizing agents are easily reduced. Oxidizing agents are on the left side of the reduction half-reactions listed in Table 17.1. We look for the largest, most positive standard reduction potentials to correspond to the best oxidizing agents. The ordering from worst to best oxidizing agents is:

$$K^+ \quad < \quad H_2O \quad < \quad Cd^{2+} \quad < \quad I_2 \quad < \quad AuCl_4^- \quad < \quad IO_3^-$$
$$E°(V) \quad -2.92 \qquad -0.83 \qquad -0.40 \qquad 0.54 \qquad 0.99 \qquad 1.20$$

45.   a. $2\ H^+ + 2\ e^- \rightarrow H_2 \quad E° = 0.00\ V; \quad Cu \rightarrow Cu^{2+} + 2\ e^- \quad -E° = -0.34\ V$

$E°_{cell} = -0.34\ V$; No, H^+ cannot oxidize Cu to Cu^{2+} at standard conditions ($E°_{cell} < 0$).

b. $Fe^{3+} + e^- \rightarrow Fe^{2+} \quad E° = 0.77\ V; \quad 2\ I^- \rightarrow I_2 + 2\ e^- \quad -E° = -0.54\ V$

$E°_{cell} = 0.77 - 0.54 = 0.23\ V$; Yes, Fe^{3+} can oxidize I^- to I_2.

c. $H_2 \rightarrow 2\ H^+ + 2\ e^- \quad -E° = 0.00\ V; \quad Ag^+ + e^- \rightarrow Ag \quad E° = 0.80\ V$

$E°_{cell} = 0.80\ V$; Yes, H_2 can reduce Ag^+ to Ag at standard conditions ($E°_{cell} > 0$).

d. $Fe^{2+} \rightarrow Fe^{3+} + e^- \quad -E° = -0.77\ V; \quad Cr^{3+} + e^- \rightarrow Cr^{2+} \quad E° = -0.50\ V$

$E°_{cell} = -0.50 - 0.77 = -1.27\ V$; No, Fe^{2+} cannot reduce Cr^{3+} to Cr^{2+} at standard conditions.

47.   a. $2\,Br^- \rightarrow Br_2 + 2\,e^-$ $-E^\circ = -1.09\,V$; $2\,Cl^- \rightarrow Cl_2 + 2\,e^-$ $-E^\circ = -1.36\,V$; $E^\circ > 1.09\,V$ to oxidize
           Br^-; $E^\circ < 1.36\,V$ to not oxidize Cl^-; $Cr_2O_7^{2-}$, O_2, MnO_2, and IO_3^- are all possible since when all
           of these oxidizing agents are coupled with Br^-, $E^\circ_{cell} > 0$, and when coupled with Cl^-, $E^\circ_{cell} < 0$
           (assuming standard conditions).

      b. $Mn \rightarrow Mn^{2+} + 2\,e^-$ $-E^\circ = 1.18$; $Ni \rightarrow Ni^{2+} + 2\,e^-$ $-E^\circ = 0.23\,V$;   Any oxidizing agent with
           $-0.23\,V > E^\circ > -1.18\,V$ will work. $PbSO_4$, Cd^{2+}, Fe^{2+}, Cr^{3+}, Zn^{2+} and H_2O will be able to oxidize
           Mn but not Ni (assuming standard conditions).

49. $ClO^- + H_2O + 2\,e^- \rightarrow 2\,OH^- + Cl^-$ $E^\circ = 0.90\,V$
           $2\,NH_3 + 2\,OH^- \rightarrow N_2H_4 + 2\,H_2O + 2\,e^-$ $-E^\circ = 0.10\,V$
           _____
           $ClO^-(aq) + 2\,NH_3(aq) \rightarrow Cl^-(aq) + N_2H_4(aq) + H_2O(l)$ $E^\circ_{cell} = 1.00\,V$

           Since E°_{cell} is positive for this reaction, then at standard conditions ClO^- can spontaneously oxidize
           NH_3 to the somewhat toxic N_2H_4.

The Nernst Equation

51. $H_2O_2 + 2\,H^+ + 2\,e^- \rightarrow 2\,H_2O$ $E^\circ = 1.78\,V$
                     $(Ag \rightarrow Ag^+ + e^-) \times 2$ $-E^\circ = -0.80\,V$
           _____
           $H_2O_2(aq) + 2\,H^+(aq) + 2\,Ag(s) \rightarrow 2\,H_2O(l) + 2\,Ag^+(aq)$ $E^\circ_{cell} = 0.98\,V$

      a.   A galvanic cell is based on spontaneous chemical reactions. At standard conditions, this reaction
           produces a voltage of 0.98 V. Any change in concentration that increases the tendency of the
           forward reaction to occur will increase the cell potential. Conversely, any change in
           concentration that decreases the tendency of the forward reaction to occur (increases the
           tendency of the reverse reaction to occur) will decrease the cell potential. Using Le Chatelier's
           principle, increasing the reactant concentrations of H_2O_2 and H^+ from 1.0 M to 2.0 M will drive
           the forward reaction further to right (will further increase the tendency of the forward reaction
           to occur). Therefore, E_{cell} will be greater than E°_{cell}.

      b.   Here, we decreased the reactant concentration of H^+ and increased the product concentration
           of Ag^+ from the standard conditions. This decreases the tendency of the forward reaction to
           occur which will decrease E_{cell} as compared to E°_{cell} ($E_{cell} < E^\circ_{cell}$).

53.   For concentration cells, the driving force for the reaction is the difference in ion concentrations
      between the anode and cathode. In order to equalize the ion concentrations, the anode always has
      the smaller ion concentration. The general setup for this concentration cell is:

           Cathode: $Ag^+(x\,M) + e^- \rightarrow Ag$ $E^\circ = 0.80\,V$
           Anode: $Ag \rightarrow Ag^+(y\,M) + e^-$ $-E^\circ = -0.80\,V$
           _____
                $Ag^+(cathode,\,x\,M) \rightarrow Ag^+(anode,\,y\,M)$ $E^\circ_{cell} = 0.00\,V$

$$E_{cell} = E^\circ_{cell} - \frac{0.0591}{n} \log Q = \frac{-0.0591}{1} \log \frac{[Ag^+]_{anode}}{[Ag^+]_{cathode}}$$

For each concentration cell, we will calculate the cell potential using the above equation. Remember that the anode always has the smaller ion concentration.

a. Since both compartments are at standard conditions ($[Ag^+] = 1.0 \, M$) then $E_{cell} = E^\circ_{cell} = 0 \, V$. No voltage is produced since no reaction occurs. Concentration cells only produce a voltage when the ion concentrations are <u>not</u> equal.

b. Cathode = $2.0 \, M \, Ag^+$; Anode = $1.0 \, M \, Ag^+$; Electron flow is always from the anode to the cathode, so electrons flow to the right in the diagram.

$$E_{cell} = \frac{-0.0591}{n} \log \frac{[Ag^+]_{anode}}{[Ag^+]_{cathode}} = \frac{-0.0591}{1} \log \frac{1.0}{2.0} = 0.018 \, V$$

c. Cathode = $1.0 \, M \, Ag^+$; Anode = $0.10 \, M \, Ag^+$; Electrons flow to the left in the diagram.

$$E_{cell} = \frac{-0.0591}{n} \log \frac{[Ag^+]_{anode}}{[Ag^+]_{cathode}} = \frac{-0.0591}{1} \log \frac{0.10}{1.0} = 0.059 \, V$$

d. Cathode = $1.0 \, M \, Ag^+$; Anode = $4.0 \times 10^{-5} \, M \, Ag^+$; Electrons flow to the left in the diagram.

$$E_{cell} = \frac{-0.0591}{1} \log \frac{4.0 \times 10^{-5}}{1.0} = 0.26 \, V$$

e. Since the ion concentrations are the same, then $\log ([Ag^+]_{anode}/[Ag^+]_{cathode}) = \log (1.0) = 0$ and $E_{cell} = 0$. No electron flow occurs.

55.
$$5 \, e^- + 8 \, H^+ + MnO_4^- \rightarrow Mn^{2+} + 4 \, H_2O \qquad\qquad E^\circ = 1.51 \, V$$
$$(Fe^{2+} \rightarrow Fe^{3+} + e^-) \times 5 \qquad\qquad -E^\circ = -0.77 \, V$$
_____
$$8 \, H^+(aq) + MnO_4^-(aq) + 5 \, Fe^{2+}(aq) \rightarrow 5 \, Fe^{3+}(aq) + Mn^{2+}(aq) + 4 \, H_2O(l) \qquad E^\circ_{cell} = 0.74 \, V$$

$$E_{cell} = E^\circ_{cell} - \frac{0.0591}{n} \log Q = 0.74 \, V - \frac{0.0591}{5} \log \frac{[Fe^{3+}]^5 [Mn^{2+}]}{[Fe^{2+}]^5 [MnO_4^-] [H^+]^8} \; ; \; pH = 4.0 \text{ so } H^+ = 1 \times 10^{-4} \, M$$

$$E_{cell} = 0.74 - \frac{0.0591}{5} \log \frac{(1 \times 10^{-6})^5 (1 \times 10^{-6})}{(1 \times 10^{-3})^5 (1 \times 10^{-2}) (1 \times 10^{-4})^8}$$

$$E_{cell} = 0.74 - \frac{0.0591}{5} \log (1 \times 10^{13}) = 0.74 \, V - 0.15 \, V = 0.59 \, V = 0.6 \, V \text{ (1 sig. fig. due to concentrations)}$$

Yes, $E_{cell} > 0$ so the reaction will occur as written.

57.
$$Cu^{2+} + 2 \, e^- \rightarrow Cu \qquad\qquad E^\circ = 0.34 \, V$$
$$Zn \rightarrow Zn^{2+} + 2 \, e^- \qquad\qquad -E^\circ = 0.76 \, V$$
_____
$$Cu^{2+}(aq) + Zn(s) \rightarrow Zn^{2+}(aq) + Cu(s) \qquad E^\circ_{cell} = 1.10 \, V$$

Since Zn^{2+} is a product in the reaction, the Zn^{2+} concentration increases from $1.00\,M$ to $1.20\,M$. This means that the reactant concentration of Cu^{2+} must decrease from $1.00\,M$ to $0.80\,M$ (from the 1:1 mol ratio in the balanced reaction).

$$E_{cell} = E_{cell}^{\circ} - \frac{0.0591}{n}\, \log Q = 1.10\,V - \frac{0.0591}{2}\, \log \frac{[Zn^{2+}]}{[Cu^{2+}]}$$

$$E_{cell} = 1.10\,V - \frac{0.0591}{2}\, \log \frac{1.20}{0.80} = 1.10\,V - 0.0052\,V = 1.09\,V$$

59. $Cu^{2+}(aq) + H_2(g) \rightarrow 2\,H^+(aq) + Cu(s)$ $E_{cell}^{\circ} = 0.34\,V - 0.00V = 0.34\,V$; $n = 2$ mol electrons

Since $P_{H_2} = 1.0$ atm and $[H^+] = 1.0\,M$: $E_{cell} = E_{cell}^{\circ} - \frac{0.0591}{2}\, \log \frac{1}{[Cu^{2+}]}$

a. $E_{cell} = 0.34\,V - \frac{0.0591}{2}\, \log \frac{1}{2.5 \times 10^{-4}} = 0.34\,V - 0.11V = 0.23\,V$

b. $0.195\,V = 0.34\,V - \frac{0.0591}{2}\, \log \frac{1}{[Cu^{2+}]}$, $\log \frac{1}{[Cu^{2+}]} = 4.91$, $[Cu^{2+}] = 10^{-4.91} = 1.2 \times 10^{-5}\,M$

Note: When determining exponents, we will carry extra significant figures.

61. $Cu^{2+}(aq) + H_2(g) \rightarrow 2\,H^+(aq) + Cu(s)$ $E_{cell}^{\circ} = 0.34\,V - 0.00\,V = 0.34\,V$; $n = 2$

Since $P_{H_2} = 1.0$ atm and $[H^+] = 1.0\,M$: $E_{cell} = E_{cell}^{\circ} - \frac{0.0591}{2}\, \log \frac{1}{[Cu^{2+}]}$

Use the K_{sp} expression to calculate the Cu^{2+} concentration in the cell.

$Cu(OH)_2(s) \rightleftharpoons Cu^{2+}(aq) + 2\,OH^-(aq)$ $K_{sp} = 1.6 \times 10^{-19} = [Cu^{2+}]\,[OH^-]^2$

From problem, $[OH^-] = 0.10\,M$, so: $[Cu^{2+}] = \dfrac{1.6 \times 10^{-19}}{(0.10)^2} = 1.6 \times 10^{-17}\,M$

$$E_{cell} = E_{cell}^{\circ} - \frac{0.0591}{2}\, \log \frac{1}{[Cu^{2+}]} = 0.34\,V - \frac{0.0591}{2}\, \log \frac{1}{1.6 \times 10^{-17}} = 0.34 - 0.50 = -0.16\,V$$

Since $E_{cell} < 0$, the forward reaction is not spontaneous, but the reverse reaction is spontaneous. The Cu electrode becomes the anode and $E_{cell} = 0.16\,V$ for the reverse reaction. The cell reaction is: $2\,H^+(aq) + Cu(s) \rightarrow Cu^{2+}(aq) + H_2(g)$.

63. See Exercises 17.25, 17.27 and 17.29 for balanced reactions and standard cell potentials. Balanced reactions are necessary to determine n, the moles of electrons transferred.

25a. $7\,H_2O + 2\,Cr^{3+} + 3\,Cl_2 \rightarrow Cr_2O_7^{2-} + 6\,Cl^- + 14\,H^+$ $E_{cell}^{\circ} = 0.03\,V = 0.03$ J/C

$\Delta G^{\circ} = -nFE_{cell}^{\circ} = -(6$ mol $e^-)(96{,}485$ C/mol $e^-)(0.03$ J/C$) = -1.7 \times 10^4$ J $= -20$ kJ

$$E_{cell} = E°_{cell} - \frac{0.0591}{n} \log Q: \text{ At equilibrium, } E_{cell} = 0 \text{ and } Q = K, \text{ so } E°_{cell} = \frac{0.0591}{n} \log K$$

$$\log K = \frac{nE°}{0.0591} = \frac{6(0.03)}{0.0591} = 3.05, \quad K = 10^{3.05} = 1 \times 10^3$$

Note: When determining exponents, we will round off to the correct number of significant figures after the calculation is complete in order to help eliminate excessive round-off error.

25b. $\Delta G° = -(2 \text{ mol } e^-)(96{,}485 \text{ C/mol } e^-)(2.71 \text{ J/C}) = -5.23 \times 10^5 \text{ J} = -523 \text{ kJ}$

$$\log K = \frac{2(2.71)}{0.0591} = 91.709, \quad K = 5.12 \times 10^{91}$$

29a. $\Delta G° = -(2 \text{ mol } e^-)(96{,}485 \text{ C/mol } e^-)(0.27 \text{ J/C}) = -5.21 \times 10^4 \text{ J} = -52 \text{ kJ}$

$$\log K = \frac{2(0.27)}{0.0591} = 9.14, \quad K = 1.4 \times 10^9$$

29b. $\Delta G° = -(10 \text{ mol } e^-)(96{,}485 \text{ C/mol } e^-)(0.09 \text{ J/C}) = -8.7 \times 10^4 \text{ J} = -90 \text{ kJ}$

$$\log K = \frac{10(0.09)}{0.0591} = 15.23, \quad K = 2 \times 10^{15}$$

65.   a.   Possible reaction: $I_2(s) + 2 Cl^-(aq) \rightarrow 2 I^-(aq) + Cl_2(g)$ $E°_{cell} = 0.54 \text{ V} - 1.36 \text{ V} = -0.82 \text{ V}$
This reaction is not spontaneous at standard conditions since $E°_{cell} < 0$. No reaction occurs.

b.   Possible reaction: $Cl_2(g) + 2 I^-(aq) \rightarrow I_2(s) + 2 Cl^-(aq)$ $E°_{cell} = 0.82 \text{ V};$ This reaction is spontaneous at standard conditions since $E°_{cell} > 0$. The reaction will occur.

$Cl_2(g) + 2 I^-(aq) \rightarrow I_2(s) + 2 Cl^-(aq)$ $E°_{cell} = 0.82 \text{ V} = 0.82 \text{ J/C}$

$$\Delta G° = -nFE°_{cell} = -(2 \text{ mol } e^-)(96{,}485 \text{ C/mol } e^-)(0.82 \text{ J/C}) = -1.6 \times 10^5 \text{ J} = -160 \text{ kJ}$$

$$E° = \frac{0.0591}{n} \log K, \quad \log K = \frac{nE°}{0.0591} = \frac{2(0.82)}{0.0591} = 27.75, \quad K = 10^{27.75} = 5.6 \times 10^{27}$$

c.   Possible reaction: $2 Ag(s) + Cu^{2+}(aq) \rightarrow Cu(s) + 2 Ag^+(aq)$ $E°_{cell} = -0.46 \text{ V};$ No reaction occurs.

d. Fe^{2+} can be oxidized or reduced. The other species present are H^+, SO_4^{2-}, H_2O, and O_2 from air. Only O_2 in the presence of H^+ has a large enough standard reduction potential to oxidize Fe^{2+} to Fe^{3+} (resulting in $E°_{cell} > 0$). All other combinations, including the possible reduction of Fe^{2+}, give negative cell potentials. The spontaneous reaction is:

$4 Fe^{2+}(aq) + 4 H^+(aq) + O_2(g) \rightarrow 4 Fe^{3+}(aq) + 2 H_2O(l)$ $E°_{cell} = 1.23 - 0.77 = 0.46 \text{ V}$

$$\Delta G° = -nFE°_{cell} = -(4 \text{ mol } e^-)(96{,}485 \text{ C/mol } e^-)(0.46 \text{ J/C})(1 \text{ kJ}/1000 \text{ J}) = -180 \text{ kJ}$$

$$\log K = \frac{4(0.46)}{0.0591} = 31.13, \quad K = 1.3 \times 10^{31}$$

67. a.
$$Au^{3+} + 3\,e^- \rightarrow Au \qquad\qquad E° = 1.50\ V$$
$$(Tl \rightarrow Tl^+ + e^-) \times 3 \qquad -E° = 0.34\ V$$

$$Au^{3+}(aq) + 3\,Tl(s) \rightarrow Au(s) + 3\,Tl^+(aq) \qquad E°_{cell} = 1.84\ V$$

b. $\Delta G° = -nFE°_{cell} = -(3\ \text{mol } e^-)(96{,}485\ \text{C/mol } e^-)(1.84\ \text{J/C}) = -5.33 \times 10^5\ J = -533\ kJ$

$$\log K = \frac{nE°}{0.0591} = \frac{3(1.84)}{0.0591} = 93.401, \quad K = 10^{93.401} = 2.52 \times 10^{93}$$

c. $E_{cell} = 1.84\ V - \dfrac{0.0591}{3} \log \dfrac{[Tl^+]^3}{[Au^{3+}]} = 1.84 - \dfrac{0.0591}{3} \log \dfrac{(1.0 \times 10^{-4})^3}{1.0 \times 10^{-2}}$

$E_{cell} = 1.84 - (-0.20) = 2.04\ V$

69.
$$CdS + 2\,e^- \rightarrow Cd + S^{2-} \qquad\qquad E° = -1.21\ V$$
$$Cd \rightarrow Cd^{2+} + 2\,e^- \qquad\qquad -E° = 0.402\ V$$

$$CdS(s) \rightarrow Cd^{2+}(aq) + S^{2-}(aq) \quad E°_{cell} = -0.81\ V \qquad K_{sp} = ?$$

$$\log K_{sp} = \frac{nE°}{0.0591} = \frac{2(-0.81)}{0.0591} = -27.41, \quad K_{sp} = 10^{-27.41} = 3.9 \times 10^{-28}$$

71.
$$Ag^+ + e^- \rightarrow Ag \qquad\qquad E° = 0.80\ V$$
$$Ag + 2\,S_2O_3^{2-} \rightarrow Ag(S_2O_3)_2^{3-} + e^- \qquad -E° = -0.017\ V$$

$$Ag^+(aq) + 2\,S_2O_3^{2-}(aq) \rightarrow Ag(S_2O_3)_2^{3-}(aq) \qquad E°_{cell} = 0.78\ V \quad K = ?$$

For this overall reaction, $E°_{cell} = \dfrac{0.0591}{n} \log K$

$$\log K = \frac{nE°}{0.0591} = \frac{(1)(0.78)}{0.0591} = 13.20, \quad K = 10^{13.20} = 1.6 \times 10^{13}$$

Electrolysis

73. a. $Al^{3+} + 3\,e^- \rightarrow Al$; 3 mol e^- are needed to produce 1 mol Al from Al^{3+}.

$$1.0 \times 10^3\ \text{g Al} \times \frac{1\ \text{mol Al}}{26.98\ \text{g Al}} \times \frac{3\ \text{mol } e^-}{\text{mol Al}} \times \frac{96{,}485\ \text{C}}{\text{mol } e^-} \times \frac{1\ s}{100.0\ \text{C}} = 1.07 \times 10^5\ s = 30.\ \text{hours}$$

b. $1.0\ \text{g Ni} \times \dfrac{1\ \text{mol Ni}}{58.69\ \text{g Ni}} \times \dfrac{2\ \text{mol } e^-}{\text{mol Ni}} \times \dfrac{96{,}485\ \text{C}}{\text{mol } e^-} \times \dfrac{1\ s}{100.0\ \text{C}} = 33\ s$

c. $5.0\ \text{mol Ag} \times \dfrac{1\ \text{mol } e^-}{\text{mol Ag}} \times \dfrac{96{,}485\ \text{C}}{\text{mol } e^-} \times \dfrac{1\ s}{100.0\ \text{C}} = 4.8 \times 10^3\ s = 1.3\ \text{hours}$

75. $15\ A = \dfrac{15\ \text{C}}{s} \times \dfrac{60\ s}{\text{min}} \times \dfrac{60\ \text{min}}{h} = 5.4 \times 10^4\ \text{C of charge passed in 1 hour}$

a. $5.4 \times 10^4 \, C \times \dfrac{1 \, mol \, e^-}{96,485 \, C} \times \dfrac{1 \, mol \, Co}{2 \, mol \, e^-} \times \dfrac{58.93 \, g \, Co}{mol \, Co} = 16 \, g \, Co$

b. $5.4 \times 10^4 \, C \times \dfrac{1 \, mol \, e^-}{96,485 \, C} \times \dfrac{1 \, mol \, Hf}{4 \, mol \, e^-} \times \dfrac{178.5 \, g \, Hf}{mol \, Hf} = 25 \, g \, Hf$

c. $2 \, I^- \rightarrow I_2 + 2 \, e^-$; $5.4 \times 10^4 \, C \times \dfrac{1 \, mol \, e^-}{96,485 \, C} \times \dfrac{1 \, mol \, I_2}{2 \, mol \, e^-} \times \dfrac{253.8 \, g \, I_2}{mol \, I_2} = 71 \, g \, I_2$

d. $CrO_3(l) \rightarrow Cr^{6+} + 3 \, O^{2-}$; 6 mol e$^-$ are needed to produce 1 mol Cr from molten CrO_3.

$5.4 \times 10^4 \, C \times \dfrac{1 \, mol \, e^-}{96,485 \, C} \times \dfrac{1 \, mol \, Cr}{6 \, mol \, e^-} \times \dfrac{52.00 \, g \, Cr}{mol \, Cr} = 4.9 \, g \, Cr$

77. $1397 \, s \times \dfrac{6.50 \, C}{s} \times \dfrac{1 \, mol \, e^-}{96,485 \, C} \times \dfrac{1 \, mol \, M}{3 \, mol \, e^-} = 3.14 \times 10^{-2} \, mol \, M$ where M = unknown metal

Molar mass $= \dfrac{1.41 \, g \, M}{3.14 \times 10^{-2} \, mol \, M} = \dfrac{44.9 \, g}{mol}$; The element is scandium. Sc forms 3+ ions.

79. F_2 is produced at the anode: $2 \, F^- \rightarrow F_2 + 2 \, e^-$

$2.00 \, h \times \dfrac{60 \, min}{h} \times \dfrac{60 \, s}{min} \times \dfrac{10.0 \, C}{s} \times \dfrac{1 \, mol \, e^-}{96,485 \, C} = 0.746 \, mol \, e^-$

$0.746 \, mol \, e^- \times \dfrac{1 \, mol \, F_2}{2 \, mol \, e^-} = 0.373 \, mol \, F_2$; $PV = nRT$, $V = \dfrac{nRT}{P}$

$V = \dfrac{(0.373 \, mol) \, (0.08206 \, L \bullet atm/K \bullet mol) \, (298 \, K)}{1.00 \, atm} = 9.12 \, L \, F_2$

K is produced at the cathode: $K^+ + e^- \rightarrow K$

$0.746 \, mol \, e^- \times \dfrac{1 \, mol \, K}{mol \, e^-} \times \dfrac{39.10 \, g \, K}{mol \, K} = 29.2 \, g \, K$

81. $\dfrac{150. \times 10^3 \, g \, C_6H_8N_2}{h} \times \dfrac{1 \, h}{60 \, min} \times \dfrac{1 \, min}{60 \, s} \times \dfrac{1 \, mol \, C_6H_8N_2}{108.14 \, g \, C_6H_8N_2} \times \dfrac{2 \, mol \, e^-}{mol \, C_6H_8N_2} \times \dfrac{96,485 \, C}{mol \, e^-}$

$= 7.44 \times 10^4 \, C/s$ or a current of $7.44 \times 10^4 \, A$

83. $2.30 \, min \times \dfrac{60 \, s}{min} = 138 \, s$; $138 \, s \times \dfrac{2.00 \, C}{s} \times \dfrac{1 \, mol \, e^-}{96,485 \, C} \times \dfrac{1 \, mol \, Ag}{mol \, e^-} = 2.86 \times 10^{-3} \, mol \, Ag$

$[Ag^+] = 2.86 \times 10^{-3} \, mol \, Ag^+/0.250 \, L = 1.14 \times 10^{-2} \, M$

85. $Au^{3+} + 3\ e^- \rightarrow Au$ $E° = 1.50\ V$ $Ni^{2+} + 2\ e^- \rightarrow Ni$ $E° = -0.23\ V$
       $Ag^+ + e^- \rightarrow Ag$ $E° = 0.80\ V$ $Cd^{2+} + 2\ e^- \rightarrow Cd$ $E° = -0.40\ V$

       $2\ H_2O + 2e^- \rightarrow H_2 + 2\ OH^-$ $E° = -0.83\ V$

Au(s) will plate out first since it has the most positive reduction potential, followed by Ag(s), which is followed by Ni(s), and finally Cd(s) will plate out last since it has the most negative reduction potential of the metals listed.

87.   Reduction occurs at the cathode, and oxidation occurs at the anode. First, determine all the species present, then look up pertinent reduction and/or oxidation potentials in Table 17.1 for all these species. The cathode reaction will be the reaction with the most positive reduction potential, and the anode reaction will be the reaction with the most positive oxidation potential.

a.   Species present: Ni^{2+} and Br^-; Ni^{2+} can be reduced to Ni, and Br^- can be oxidized to Br_2 (from Table 17.1). The reactions are:

           Cathode: $Ni^{2+} + 2e^- \rightarrow Ni$ $E° = -0.23\ V$
           Anode: $2\ Br^- \rightarrow Br_2 + 2\ e^-$ $-E° = -1.09\ V$

b.   Species present: Al^{3+} and F^-; Al^{3+} can be reduced, and F^- can be oxidized. The reactions are:

           Cathode: $Al^{3+} + 3\ e^- \rightarrow Al$ $E° = -1.66\ V$
           Anode: $2\ F^- \rightarrow F_2 + 2\ e^-$ $-E° = -2.87\ V$

c.   Species present: Mn^{2+} and I^-; Mn^{2+} can be reduced, and I^- can be oxidized. The reactions are:

           Cathode: $Mn^{2+} + 2\ e^- \rightarrow Mn$ $E° = -1.18\ V$
           Anode: $2\ I^- \rightarrow I_2 + 2\ e^-$ $-E° = -0.54\ V$

Additional Exercises

89.   The half-reaction for the SCE is:

       $Hg_2Cl_2 + 2\ e^- \rightarrow 2\ Hg + 2\ Cl^-$ $E_{SCE} = 0.242\ V$

For a spontaneous reaction to occur, E_{cell} must be positive. Using the standard reduction potentials in Table 17.1 and the given SCE potential, deduce which combination will produce a positive overall cell potential.

a. $Cu^{2+} + 2\ e^- \rightarrow Cu$ $E° = 0.34\ V$

       $E_{cell} = 0.34 - 0.242 = 0.10\ V$; SCE is the anode.

b. $Fe^{3+} + e^- \rightarrow Fe^{2+}$ $E° = 0.77\ V$

       $E_{cell} = 0.77 - 0.242 = 0.53\ V$; SCE is the anode.

c. $AgCl + e^- \rightarrow Ag + Cl^-$ $E° = 0.22$ V

$E_{cell} = 0.242 - 0.22 = 0.02$ V; SCE is the cathode.

d. $Al^{3+} + 3\ e^- \rightarrow Al$ $E° = -1.66$ V

$E_{cell} = 0.242 + 1.66 = 1.90$ V; SCE is the cathode.

e. $Ni^{2+} + 2\ e^- \rightarrow Ni$ $E° = -0.23$ V

$E_{cell} = 0.242 + 0.23 = 0.47$ V; SCE is the cathode.

91. $2\ Ag^+(aq) + Cu(s) \rightarrow Cu^{2+}(aq) + 2\ Ag(s)$ $E°_{cell} = 0.80 - 0.34$ V $= 0.46$ V; A galvanic cell produces a voltage as the forward reaction occurs. Any stress that increases the tendency of the forward reaction to occur will increase the cell potential, while a stress that decreases the tendency of the forward reaction to occur will decrease the cell potential.

a. Added Cu^{2+} (a product ion) will decrease the tendency of the forward reaction to occur, which will decrease the cell potential.

b. Added NH_3 removes Cu^{2+} in the form of $Cu(NH_3)_4^{2+}$. Removal of a product ion will increase the tendency of the forward reaction to occur, which will increase the cell potential.

c. Added Cl^- removes Ag^+ in the form of $AgCl(s)$. Removal of a reactant ion will decrease the tendency of the forward reaction to occur, which will decrease the cell potential.

d. $Q_1 = \dfrac{[Cu^{2+}]_o}{[Ag^+]_o^2}$; As the volume of solution is doubled, each concentration is halved.

$Q_2 = \dfrac{1/2\ [Cu^{2+}]_o}{(1/2\ [Ag^+]_o)^2} = \dfrac{2[Cu^{2+}]_o}{[Ag^+]_o^2} = 2\ Q_1$

The reaction quotient is doubled as the concentrations are halved. Since reactions are spontaneous when $Q < K$ and since Q increases when the solution volume doubles, the reaction is closer to equilibrium, which will decrease the cell potential.

e. Since $Ag(s)$ is not a reactant in this spontaneous reaction, and since solids do not appear in the reaction quotient expressions, replacing the silver electrode with a platinum electrode will have no effect on the cell potential.

93.   a. $\Delta G° = \Sigma n_p \Delta G°_{f,\ products} - \Sigma n_r \Delta G°_{f,\ reactants} = 2(-480.) + 3(86) - [3(-40.)] = -582$ kJ

From oxidation numbers, $n = 6$. $\Delta G° = -nFE°$, $E° = \dfrac{-\Delta G°}{nF} = \dfrac{-(-582,000\ J)}{6(96,485)\ C} = 1.01$ V

$\log K = \dfrac{nE°}{0.0591} = \dfrac{6(1.01)}{0.0591} = 102.538$, $K = 10^{102.538} = 3.45 \times 10^{102}$

b. $\qquad$ $2 e^- + Ag_2S \rightarrow 2 Ag + S^{2-}) \times 3$ $\qquad\qquad$ $E^°_{Ag_2S} = ?$

$\qquad\qquad\qquad$ $(Al \rightarrow Al^{3+} + 3 e^-) \times 2$ $\qquad\qquad$ $-E^° = 1.66 \text{ V}$

$3 Ag_2S(s) + 2 Al(s) \rightarrow 6 Ag(s) + 3 S^{2-}(aq) + 2 Al^{3+}(aq)$ $\quad$ $E^°_{cell} = 1.01 \text{ V} = E^°_{Ag_2S} + 1.66\text{V}$

$$E^°_{Ag_2S} = 1.01 \text{ V} - 1.66 \text{ V} = -0.65 \text{ V}$$

95. From Exercise 17.27a: $3 Cl_2(g) + 2 Cr^{3+}(aq) + 7 H_2O(l) \rightleftharpoons 14 H^+(aq) + Cr_2O_7^{2-}(aq) + 6 Cl^-(aq)$
$E^°_{cell} = 0.03 \text{ V}$

$$E_{cell} = E^°_{cell} - \frac{0.0591}{6} \log \frac{[Cr_2O_7^{2-}][H^+]^{14}[Cl^-]^6}{[Cr^{3+}]^2 P_{Cl_2}^3}$$

When $K_2Cr_2O_7$ and Cl^- are added to concentrated H_2SO_4, Q becomes a large number due to $[H^+]^{14}$ term. The log of a large number is positive. E_{cell} becomes negative, which means the reverse reaction becomes spontaneous. The pungent fumes were $Cl_2(g)$.

97. Consider the strongest oxidizing agent combined with the strongest reducing agent from Table 17.1:

$\qquad\qquad$ $F_2 + 2 e^- \rightarrow 2 F^-$ $\qquad\qquad$ $E^° = 2.87 \text{ V}$

$\qquad\qquad$ $(Li \rightarrow Li^+ + e^-) \times 2$ $\qquad\qquad$ $E^° = 3.05 \text{ V}$

$F_2(g) + 2 Li(s) \rightarrow 2 Li^+(aq) + 2 F^-(aq)$ $\qquad$ $E^°_{cell} = 5.92 \text{ V}$

The claim is impossible. The strongest oxidizing agent and reducing agent when combined only give an $E^°_{cell}$ value of about 6 V.

99. a. $O_2 + 2 H_2O + 4 e^- \rightarrow 4 OH^-$ $\qquad\qquad$ $E^° = 0.40 \text{ V}$

$\qquad\qquad$ $(H_2 + 2 OH^- \rightarrow 2 H_2O + 2 e^-) \times 2$ $\qquad$ $-E^° = 0.83 \text{ V}$

$\qquad$ $2 H_2(g) + O_2(g) \rightarrow 2 H_2O(l)$ $\qquad\qquad$ $E^°_{cell} = 1.23 \text{ V} = 1.23 \text{ J/C}$

Since standard conditions are assumed, then $w_{max} = \Delta G^°$ for 2 mol H_2O produced.

$\Delta G^° = -nFE^°_{cell} = -(4 \text{ mol } e^-)(96,485 \text{ C/mol } e^-)(1.23 \text{ J/C}) = -475,000 \text{ J} = -475 \text{ kJ}$

For 1.00×10^3 g H_2O produced, w_{max} is:

$$1.00 \times 10^3 \text{ g } H_2O \times \frac{1 \text{ mol } H_2O}{18.02 \text{ g } H_2O} \times \frac{-475 \text{ kJ}}{2 \text{ mol } H_2O} = -13,200 \text{ kJ} = w_{max}$$

The work done can be no larger than the free energy change. The best that could happen is that all of the free energy released would go into doing work, but this does not occur in any real process since there is always waste energy in a real process. Fuel cells are more efficient in converting chemical energy into electrical energy; they are also less massive. The major disadvantage is that they are expensive. In addition, $H_2(g)$ and $O_2(g)$ are an explosive mixture if ignited; much more so than fossil fuels.

101. $(CO + O^{2-} \rightarrow CO_2 + 2\,e^-) \times 2$

$O_2 + 4\,e^- \rightarrow 2\,O^{2-}$

$2\,CO + O_2 \rightarrow 2\,CO_2$

$\Delta G = -nFE, \quad E = \dfrac{-\Delta G}{nF} = \dfrac{-(-380 \times 10^3\ J)}{(4\ \text{mol e}^-)(96{,}485\ C/\text{mol e}^-)} = 0.98\ V$

103. $\text{mol e}^- = 50.0\ \text{min} \times \dfrac{60\ s}{\text{min}} \times \dfrac{2.50\ C}{s} \times \dfrac{1\ \text{mol e}^-}{96{,}485\ C} = 7.77 \times 10^{-2}\ \text{mol e}^-$

$\text{mol Ru} = 2.618\ \text{g Ru} \times \dfrac{1\ \text{mol Ru}}{101.1\ \text{g Ru}} = 2.590 \times 10^{-2}\ \text{mol Ru}$

$\dfrac{\text{mol e}^-}{\text{mol Ru}} = \dfrac{7.77 \times 10^{-2}\ \text{mol e}^-}{2.590 \times 10^{-2}\ \text{mol Ru}} = 3.00;$ The charge on the ruthenium ions is $+3$ $(Ru^{3+} + 3\,e^- \rightarrow Ru)$.

Challenge Problems

105. $\Delta G^\circ = -nFE^\circ = \Delta H^\circ - T\Delta S^\circ, \quad E^\circ = \dfrac{T\Delta S^\circ}{nF} - \dfrac{\Delta H^\circ}{nF}$

If we graph E° vs. T we should get a straight line ($y = mx + b$). The slope of the line is equal to $\Delta S^\circ/nF$, and the y-intercept is equal to $-\Delta H^\circ/nF$. From the equation above, E° will have a small temperature dependence when ΔS° is close to zero.

107. $(Ag^+ + e^- \rightarrow Ag) \times 2 \qquad E^\circ = 0.80\ V$

$Pb \rightarrow Pb^{2+} + 2\,e^- \qquad -E^\circ = -(-0.13)$

$2\,Ag^+ + Pb \rightarrow 2\,Ag + Pb^{2+} \quad E^\circ_{cell} = 0.93\ V$

$E = E^\circ - \dfrac{0.0591}{n} \log \dfrac{[Pb^{2+}]}{[Ag^+]^2}, \quad 0.83\ V = 0.93\ V - \dfrac{0.0591}{2} \log \dfrac{(1.8)}{[Ag^+]^2}$

$\log \dfrac{(1.8)}{[Ag^+]^2} = \dfrac{0.10(2)}{0.0591} = 3.4, \quad \dfrac{(1.8)}{[Ag^+]^2} = 10^{3.4}, \quad [Ag^+] = 0.027\ M$

$$Ag_2SO_4(s) \rightleftharpoons 2\,Ag^+(aq) + SO_4^{2-}(aq) \qquad K_{sp} = [Ag^+]^2[SO_4^{2-}]$$

Initial $s = $ solubility (mol/L)   0   0
Equil. $2s$ s

From problem: $2s = 0.027\ M, \quad s = 0.027/2$

$K_{sp} = (2s)^2(s) = (0.027)^2(0.027/2) = 9.8 \times 10^{-6}$

109. $2 H^+ + 2 e^- \rightarrow H_2$ $E° = 0.000\ V$

                       $Fe \rightarrow Fe^{2+} + 2 e^-$ $-E° = -(-0.440V)$

$$2\ H^+(aq) + Fe(s) \rightarrow H_2(g) + Fe^{2+}(aq) \qquad E°_{cell} = 0.440\ V$$

$$E_{cell} = E°_{cell} - \frac{0.0591}{n} \log Q, \text{ where } n = 2 \text{ and } Q = \frac{P_{H_2} \times [Fe^{2+}]}{[H^+]^2}$$

To determine K_a for the weak acid, first use the electrochemical data to determine the H^+ concentration in the half-cell containing the weak acid.

$$0.333\ V = 0.440\ V - \frac{0.0591}{2} \log \frac{1.00\ (1.00 \times 10^{-3})}{[H^+]^2}$$

$$\frac{0.107(2)}{0.0591} = \log \frac{1.00 \times 10^{-3}}{[H^+]^2}, \quad \frac{1.00 \times 10^{-3}}{[H^+]^2} = 10^{3.621} = 4.18 \times 10^3, \quad [H^+] = 4.89 \times 10^{-4}\ M$$

Now we can solve for the K_a value of the weak acid HA through the normal setup for a weak acid problem.

$$HA \quad \rightleftharpoons \quad H^+ \quad + \quad A^- \qquad K_a = \frac{[H^+][A^-]}{[HA]}$$

Initial   1.00 M ~0   0
Equil.   1.00 - x x x

$$K_a = \frac{x^2}{1.00 - x} \text{ where } x = [H^+] = 4.89 \times 10^{-4}\ M, \quad K_a = \frac{(4.89 \times 10^{-4})^2}{1.00 - 4.89 \times 10^{-4}} = 2.39 \times 10^{-7}$$

111.   a. $E_{cell} = E_{ref} + 0.05916\ pH$, $0.480\ V = 0.250\ V + 0.05916\ pH$

$$pH = \frac{0.480 - 0.250}{0.05916} = 3.888; \quad \text{Uncertainty} = \pm 1\ mV = \pm 0.001\ V$$

$$pH_{max} = \frac{0.481 - 0.250}{0.05916} = 3.905; \quad pH_{min} = \frac{0.479 - 0.250}{0.05916} = 3.871$$

So, if the uncertainty in potential is $\pm 0.001\ V$, the uncertainty in pH is ± 0.017 or about ± 0.02 pH units. For this measurement, $[H^+] = 10^{-3.888} = 1.29 \times 10^{-4}\ M$. For an error of $+1$ mV, $[H^+] = 10^{-3.905} = 1.24 \times 10^{-4}\ M$. For an error of -1 mV, $[H^+] = 10^{-3.871} = 1.35 \times 10^{-4}\ M$. So, the uncertainty in $[H^+]$ is $\pm 0.06 \times 10^{-4}\ M = \pm 6 \times 10^{-6}\ M$.

   b.   From part a, we will be within ± 0.02 pH units if we measure the potential to the nearest ± 0.001 V (1 mV).

113.   a. $(Ag^+ + e^- \rightarrow Ag) \times 2$ $E° = 0.80\ V$
                    $Cu \rightarrow Cu^{2+} + 2 e^-$ $-E° = -0.34\ V$

$$2\ Ag^+(aq) + Cu(s) \rightarrow 2\ Ag(s) + Cu^{2+}(aq) \quad E°_{cell} = 0.46\ V$$

$$E_{cell} = E_{cell}^\circ - \frac{0.0591}{n} \log Q \text{ where } n = 2 \text{ and } Q = \frac{[Cu^{2+}]}{[Ag^+]^2}$$

To calculate E_{cell}, we need to use the K_{sp} data to determine $[Ag^+]$.

$$AgCl(s) \rightleftharpoons Ag^+(aq) + Cl^-(aq) \quad K_{sp} = 1.6 \times 10^{-10} = [Ag^+][Cl^-]$$

Initial s = solubility (mol/L)   0   0
Equil. s s

$$K_{sp} = 1.6 \times 10^{-10} = s^2, \quad s = [Ag^+] = 1.3 \times 10^{-5} \text{ mol/L}$$

$$E_{cell} = 0.46 \text{ V} - \frac{0.0591}{2} \log \frac{2.0}{(1.3 \times 10^{-5})^2} = 0.46 \text{ V} - 0.30 = 0.16 \text{ V}$$

b. $$Cu^{2+}(aq) + 4 NH_3(aq) \rightleftharpoons Cu(NH_4)_4^{2+}(aq) \quad K = 1.0 \times 10^{13} = \frac{[Cu(NH_3)_4^{2+}]}{[Cu^{2+}][NH_3]^4}$$

Since K is very large for the formation of $Cu(NH_3)_4^{2+}$, the forward reaction is dominant. At equilibrium, essentially all of the 2.0 M Cu^{2+} will react to form 2.0 M $Cu(NH_3)_4^{2+}$. This reaction requires 8.0 M NH_3 to react with all of the Cu^{2+} in the balanced equation. Therefore, the mol of NH_3 added to 1.0 L solution will be larger than 8.0 mol since some NH_3 must be present at equilibrium. In order to calculate how much NH_3 is present at equilibrium, we need to use the electrochemical data to determine the Cu^{2+} concentration.

$$E_{cell} = E_{cell}^\circ - \frac{0.0591}{n} \log Q, \quad 0.52 \text{ V} = 0.46 \text{ V} - \frac{0.0591}{2} \log \frac{[Cu^{2+}]}{(1.3 \times 10^{-5})^2}$$

$$\log \frac{[Cu^{2+}]}{(1.3 \times 10^{-5})^2} = \frac{-0.06(2)}{0.0591} = -2.03, \quad \frac{[Cu^{2+}]}{(1.3 \times 10^{-5})^2} = 10^{-2.03} = 9.3 \times 10^{-3}$$

$[Cu^{2+}] = 1.6 \times 10^{-12} = 2 \times 10^{-12} M$ (We carried extra significant figures in the calculation.)

Note: Our assumption that the 2.0 M Cu^{2+} essentially reacts to completion is excellent as only 2×10^{-12} M Cu^{2+} remains after this reaction. Now we can solve for the equilibrium $[NH_3]$.

$$K = 1.0 \times 10^{13} = \frac{[Cu(NH_3)_4^{2+}]}{[Cu^{2+}][NH_3]^4} = \frac{(2.0)}{(2 \times 10^{-12})[NH_3]^4}, \quad [NH_3] = 0.6 \ M$$

Since 1.0 L of solution is present, then 0.6 mol NH_3 remains at equilibrium. The total mol of NH_3 added is 0.6 mol plus the 8.0 mol NH_3 necessary to form 2.0 M $Cu(NH_3)_4^{2+}$. Therefore, 8.0 + 0.6 = 8.6 mol NH_3 were added.

CHAPTER EIGHTEEN

THE NUCLEUS: A CHEMIST'S VIEW

Questions

1. Fission: Splitting of a heavy nucleus into two (or more) lighter nuclei.

   Fusion: Combining two light nuclei to form a heavier nucleus.

   The maximum binding energy per nucleon occurs at Fe. Nuclei smaller than Fe become more stable by fusing to form heavier nuclei closer in mass to Fe. Nuclei larger than Fe form more stable nuclei by splitting to form lighter nuclei closer in mass to Fe.

3. The assumptions are that the ^{14}C level in the atmosphere is constant or that the ^{14}C level at the time the plant died can be calculated. A constant ^{14}C level is a poor assumption, and accounting for variation is complicated. Another problem is that some of the material must be destroyed to determine the ^{14}C level.

5. No, coal-fired power plants also pose risks. A partial list of risks is:

Coal	Nuclear
Air pollution	Radiation exposure to workers
Coal mine accidents	Disposal of wastes
Health risks to miners	Meltdown
(black lung disease)	Terrorists
	Public fear

7. For fusion reactions, a collision of sufficient energy must occur between two positively charged particles to initiate the reaction. This requires high temperatures. In fission, an electrically neutral neutron collides with the positively charged nucleus. This has a much lower activation energy.

Exercises

Radioactive Decay and Nuclear Transformations

9.   All nuclear reactions must be charge balanced and mass balanced. To charge balance, balance the sum of the atomic numbers on each side of the reaction, and to mass balance, balance the sum of the mass numbers on each side of the reaction.

   a. $^{51}_{24}Cr + ^{0}_{-1}e \rightarrow ^{51}_{23}V$

   b. $^{131}_{53}I \rightarrow ^{0}_{-1}e + ^{131}_{54}Xe$

11.   a. $^{68}_{31}Ga + ^{0}_{-1}e \rightarrow ^{68}_{30}Zn$ b. $^{62}_{29}Cu \rightarrow ^{0}_{+1}e + ^{62}_{28}Ni$

   c. $^{212}_{87}Fr \rightarrow ^{4}_{2}He + ^{208}_{85}At$ d. $^{129}_{51}Sb \rightarrow ^{0}_{-1}e + ^{129}_{52}Te$

13. $^{247}_{97}Bk \rightarrow ^{207}_{82}Pb + ?\ ^{4}_{2}He + ?\ ^{0}_{-1}e$; The change in mass number (247 - 207 = 40) is due exclusively to the alpha particles. A change in mass number of 40 requires 10 $^{4}_{2}He$ particles to be produced. The atomic number only changes by 97 - 82 = 15. The 10 alpha particles change the atomic number by 20, so 5 $^{0}_{-1}e$ (5 beta particles) are produced in the decay series of ^{247}Bk to ^{207}Pb.

15. $^{53}_{26}Fe$ has too many protons. It will undergo either positron production, electron capture and/or alpha particle production. $^{59}_{26}Fe$ has too many neutrons and will undergo beta particle production. (See Table 18.2 of the text.)

17.   a. $^{249}_{98}Cf + ^{18}_{8}O \rightarrow ^{263}_{106}Sg + 4\ ^{1}_{0}n$ b. $^{259}_{104}Rf;\ ^{263}_{106}Sg \rightarrow ^{4}_{2}He + ^{259}_{104}Rf$

Kinetics of Radioactive Decay

19.   All radioactive decay follows first-order kinetics where $t_{1/2} = (\ln 2)/k$.

   $$t_{1/2} = \frac{\ln 2}{k} = \frac{0.693}{1.0 \times 10^{-3}\ h^{-1}} = 690\ h$$

21.   Kr-81 is most stable since it has the longest half-life while Kr-73 is hottest (least stable) since it has the shortest half-life.

12.5% of each isotope will remain after 3 half-lives:

$$100\% \xrightarrow[t_{1/2}]{} 50\% \xrightarrow[t_{1/2}]{} 25\% \xrightarrow[t_{1/2}]{} 12.5\%$$

For Kr-73: t = 3(27 s) = 81 s

For Kr-74: t = 3(11.5 min) = 34.5 min

For Kr-76: t = 3(14.8 h) = 44.4 h

For Kr-81: t = 3(2.1 × 10^5 yr) = 6.3 × 10^5 yr

23.   Units for N and N_o are usually the number of nuclei but can also be grams if the units are the same for both N and N_o. In this problem m = the mass of ^{32}P that remains.

$$175 \text{ mg Na}_3{}^{32}\text{PO}_4 \times \frac{32.0 \text{ mg}\,{}^{32}\text{P}}{165.0 \text{ mg Na}_3{}^{32}\text{PO}_4} = 33.9 \text{ mg}\,{}^{32}\text{P} \text{ initially; } k = \frac{\ln 2}{t_{1/2}}$$

$$\ln\left(\frac{N}{N_o}\right) = -kt = \frac{-0.6931\, t}{t_{1/2}}, \quad \ln\left(\frac{m}{33.9 \text{ mg}}\right) = \frac{-0.6931\,(35.0 \text{ d})}{14.3 \text{ d}}; \text{ Carrying extra sig. figs.:}$$

$$\ln(m) = -1.696 + 3.523 = 1.827, \quad m = e^{1.827} = 6.22 \text{ mg}\,{}^{32}\text{P remains}$$

25. $$t = 58.0 \text{ yr}; \quad k = \frac{\ln 2}{t_{1/2}}; \quad \ln\left(\frac{N}{N_o}\right) = -kt = \frac{-0.6931 \times 58.0 \text{ yr}}{28.8 \text{ yr}} = -1.40, \quad \left(\frac{N}{N_o}\right) = e^{-1.40} = 0.247$$

24.7% of the ^{90}Sr remains as of July 16, 2003.

27. $$k = \frac{\ln 2}{t_{1/2}}; \quad \ln\left(\frac{N}{N_o}\right) = -kt = \frac{-0.6931\, t}{t_{1/2}}, \quad \ln\left(\frac{N}{13.6}\right) = \frac{-0.693\,(15,000 \text{ yr})}{5730 \text{ yr}} = -1.8$$

$$\frac{N}{13.6} = e^{-1.8} = 0.17, \quad N = 13.6 \times 0.17 = 2.3 \text{ counts per minute per g of C}$$

If we had 10. mg C, we would see:

$$10. \text{ mg} \times \frac{1 \text{ g}}{1000 \text{ mg}} \times \frac{2.3 \text{ counts}}{\text{min g}} = \frac{0.023 \text{ counts}}{\text{min}}$$

It would take roughly 40 min to see a single disintegration. This is too long to wait, and the background radiation would probably be much greater than the ^{14}C activity. Thus, ^{14}C dating is not practical for very small samples.

29.   Assuming 1.000 g ^{238}U present in a sample, then 0.688 g ^{206}Pb is present. Since 1 mol ^{206}Pb is produced per mol ^{238}U decayed, then:

$$^{238}\text{U decayed} = 0.688 \text{ g Pb} \times \frac{1\,\text{mol Pb}}{206\,\text{g Pb}} \times \frac{1\,\text{mol U}}{\text{mol Pb}} \times \frac{238\,\text{g U}}{\text{mol U}} = 0.795 \text{ g } ^{238}\text{U}$$

Original mass ^{238}U present = 1.000 g + 0.795 g = 1.795 g ^{238}U

$$\ln\left(\frac{N}{N_o}\right) = -kt = \frac{-(\ln 2)\,t}{t_{1/2}}, \quad \ln\left(\frac{1.000\,\text{g}}{1.795\,\text{g}}\right) = \frac{-0.693\,(t)}{4.5 \times 10^9 \text{ yr}}, \quad t = 3.8 \times 10^9 \text{ yr}$$

Energy Changes in Nuclear Reactions

31. $\Delta E = \Delta mc^2$, $\Delta m = \dfrac{\Delta E}{c^2} = \dfrac{3.9 \times 10^{23} \text{ kg m}^2/\text{s}^2}{(3.00 \times 10^8 \text{ m/s})^2} = 4.3 \times 10^6 \text{ kg}$

The sun loses 4.3×10^6 kg of mass each second. Note: 1 J = 1 kg m^2/s^2

33.   We need to determine the mass defect, Δm, between the mass of the nucleus and the mass of the individual parts that make up the nucleus. Once Δm is known, we can then calculate ΔE (the binding energy) using $E = mc^2$. Note: 1 J = 1 kg m^2/s^2.

For $^{232}_{94}\text{Pu}$ (94 e, 94 p, 138 n):

mass of ^{232}Pu nucleus = 3.85285×10^{-22} g - mass of 94 electrons

mass of ^{232}Pu nucleus = 3.85285×10^{-22} g - $94(9.10939 \times 10^{-28})$ g = 3.85199×10^{-22} g

$\Delta m = 3.85199 \times 10^{-22}$ g - (mass of 94 protons + mass of 138 neutrons)

$\Delta m = 3.85199 \times 10^{-22}$ g - $[94(1.67262 \times 10^{-24}) + 138(1.67493 \times 10^{-24})]$ g = -3.168×10^{-24} g

For 1 mol of nuclei: $\Delta m = -3.168 \times 10^{-24}$ g/nuclei $\times 6.0221 \times 10^{23}$ nuclei/mol = -1.908 g/mol

$\Delta E = \Delta mc^2 = (-1.908 \times 10^{-3}$ kg/mol$)(2.9979 \times 10^8$ m/s$)^2 = -1.715 \times 10^{14}$ = J/mol

For $^{231}_{91}\text{Pa}$ (91 e, 91 p, 140 n):

mass of ^{231}Pa nucleus = 3.83616×10^{-22} g - $91(9.10939 \times 10^{-28})$ g = 3.83533×10^{-22} g

$\Delta m = 3.83533 \times 10^{-22}$ g - $[91(1.67262 \times 10^{-24}) + 140(1.67493 \times 10^{-24})]$ g = -3.166×10^{-24} g

$$\Delta E = \Delta mc^2 = \frac{-3.166 \times 10^{-27} \text{ kg}}{\text{nuclei}} \times \frac{6.0221 \times 10^{23} \text{ nuclei}}{\text{mol}} \times \left(\frac{2.9979 \times 10^8 \text{ m}}{\text{s}}\right)^2$$

$$= -1.714 \times 10^{14} \text{ J/mol}$$

35.   Let m_e = mass of electron; For ^{12}C (6e, 6p, 6n): mass defect = Δm = mass of ^{12}C nucleus -[mass of 6 protons + mass of 6 neutrons]. Note: the atomic masses of the elements given include the mass of the electrons.

$$\Delta m = 12.0000 \text{ amu} - 6\, m_e - [6(1.00782 - m_e) + 6(1.00866)]; \text{ Mass of electrons cancel.}$$

$$\Delta m = 12.0000 - [6(1.00782) + 6(1.00866)] = -0.0989 \text{ amu}$$

$$\Delta E = \Delta mc^2 = -0.0989 \text{ amu} \times \frac{1.6605 \times 10^{-27} \text{ kg}}{\text{amu}} \times (2.9979 \times 10^8 \text{ m/s})^2 = -1.48 \times 10^{-11} \text{ J}$$

$$\frac{BE}{\text{nucleon}} = \frac{1.48 \times 10^{-11} \text{ J}}{12 \text{ nucleons}} = 1.23 \times 10^{-12} \text{ J/nucleon}$$

For ^{235}U (92e, 92p, 143n):

$$\Delta m = 235.0439 - 92\, m_e - [92(1.00782 - m_e) + 143(1.00866)] = -1.9139 \text{ amu}$$

$$\Delta E = \Delta mc^2 = -1.9139 \text{ amu} \times \frac{1.66054 \times 10^{-27} \text{ kg}}{\text{amu}} \times (2.99792 \times 10^8 \text{ m/s})^2 = -2.8563 \times 10^{-10} \text{ J}$$

$$\frac{BE}{\text{nucleon}} = \frac{2.8563 \times 10^{-10} \text{ J}}{235 \text{ nucleons}} = 1.2154 \times 10^{-12} \text{ J/nucleon}$$

Since ^{56}Fe is the most stable known nucleus, the binding energy per nucleon for ^{56}Fe (1.408×10^{-12} J/nucleon) will be larger than that for ^{12}C or ^{235}U (see Figure 18.9 of the text).

37. $${}^{1}_{1}H + {}^{1}_{1}H \rightarrow {}^{2}_{1}H + {}^{0}_{+1}e; \quad \Delta m = (2.01410 \text{ amu} - m_e + m_e) - 2(1.00782 \text{ amu} - m_e)$$

$$\Delta m = 2.01410 - 2(1.00782) + 2(0.000549) = -4.4 \times 10^{-4} \text{ amu for two protons reacting}$$

When two mol of protons undergo fusion, $\Delta m = -4.4 \times 10^{-4}$ g.

$$\Delta E = \Delta mc^2 = -4.4 \times 10^{-7} \text{ kg} \times (3.00 \times 10^8 \text{ m/s})^2 = -4.0 \times 10^{10} \text{ J}$$

$$\frac{-4.0 \times 10^{10} \text{ J}}{2 \text{ mol protons}} \times \frac{1 \text{ mol}}{1.01 \text{ g}} = -2.0 \times 10^{10} \text{ J/g of hydrogen nuclei}$$

Detection, Uses, and Health Effects of Radiation

39.   The Geiger-Müller tube has a certain response time. After the gas in the tube ionizes to produce a "count," some time must elapse for the gas to return to an electrically neutral state. The response of the tube levels off because, at high activities, radioactive particles are entering the tube faster than the tube can respond to them.

41.   All evolved oxygen in O_2 comes from water and not from carbon dioxide.

43.   Release of Sr is probably more harmful. Xe is chemically unreactive. Strontium is in the same family as calcium and could be absorbed and concentrated in the body in a fashion similar to Ca. This puts the radioactive Sr in the bones: red blood cells are produced in bone marrow. Xe would not be readily incorporated into the body.

       The chemical properties determine where a radioactive material may be concentrated in the body or how easily it may be excreted. The length of time of exposure and what is exposed to radiation significantly affects the health hazard. (See exercise 18.44 for a specific example.)

Additional Exercises

45.   The most abundant isotope is generally the most stable isotope. The periodic table predicts that the most stable isotopes for exercises a - d are ^{39}K, ^{56}Fe, ^{23}Na and ^{204}Tl. (Reference Table 18.2 of the text for potential decay processes.)

       a.   Unstable; ^{45}K has too many neutrons and will undergo beta particle production.

       b.   Stable

       c.   Unstable; ^{20}Na has too few neutrons and will most likely undergo electron capture or positron production. Alpha particle production makes too severe of a change to be a likely decay process for the relatively light ^{20}Na nuclei. Alpha particle production usually occurs for heavy nuclei.

       d.   Unstable; ^{194}Tl has too few neutrons and will undergo electron capture, positron production and/or alpha particle production.

47. $\ln\left(\dfrac{N}{N_o}\right) = -kt = \dfrac{-(\ln 2)\, t}{12.3 \text{ yr}}, \quad \ln\left(\dfrac{0.17 \times N_o}{N_o}\right) = -5.64 \times 10^{-2}\, t, \quad t = 31.4 \text{ yr}$

       It takes 31.4 yr for the tritium to decay to 17% of the original amount. Hence, the watch stopped fluorescing enough to be read in 1975 (1944 + 31.4).

49. $20{,}000 \text{ ton TNT} \times \dfrac{4 \times 10^9 \text{ J}}{\text{ton TNT}} \times \dfrac{1 \text{ mol }^{235}U}{2 \times 10^{13} \text{ J}} \times \dfrac{235 \text{ g }^{235}U}{\text{mol }^{235}U} = 940 \text{ g }^{235}U \approx 900 \text{ g }^{235}U$

       This assumes that all of the ^{235}U undergoes fission.

Challenge Problems

51.   Assuming that the radionuclide is long lived enough such that no significant decay occurs during the time of the experiment, the total counts of radioactivity injected are:

$$0.10 \text{ mL} \times \frac{5.0 \times 10^3 \text{ cpm}}{\text{mL}} = 5.0 \times 10^2 \text{ cpm}$$

Assuming that the total activity is uniformly distributed only in the rat's blood, the blood volume is:

$$V \times \frac{48 \text{ cpm}}{\text{mL}} = 5.0 \times 10^2 \text{ cpm}, \ V = 10.4 \text{ mL} = 10. \text{ mL}$$

53.   a. ^{12}C; It takes part in the first step of the reaction but is regenerated in the last step. ^{12}C is not consumed, so it is not a reactant.

b. ^{13}N, ^{13}C, ^{14}N, ^{15}O, and ^{15}N are the intermediates.

c. $4 \, {}^1_1H \rightarrow {}^4_2He + 2 \, {}^0_{+1}e$; $\Delta m = 4.00260 \text{ amu} - 2 \, m_e + 2 \, m_e - [4(1.00782 \text{ amu} - m_e)]$

$\Delta m = 4.00260 - 4(1.00782) + 4(0.000549) = -0.02648$ amu for 4 protons reacting

For 4 mol of protons, $\Delta m = -0.02648$ g and ΔE for the reaction is:

$$\Delta E = \Delta mc^2 = -2.648 \times 10^{-5} \text{ kg} \times (2.9979 \times 10^8 \text{ m/s})^2 = -2.380 \times 10^{12} \text{ J}$$

For 1 mol of protons reacting: $\dfrac{-2.380 \times 10^{12} \text{ J}}{4 \text{ mol } {}^1H} = -5.950 \times 10^{11}$ J/mol ^1H

55. $\text{mol I} = \dfrac{33 \text{ counts}}{\text{min}} \times \dfrac{1 \text{ mol I} \cdot \text{min}}{5.0 \times 10^{11} \text{ counts}} = 6.6 \times 10^{-11}$ mol I

$[I^-] = \dfrac{6.6 \times 10^{-11} \text{ mol I}^-}{0.150 \text{ L}} = 4.4 \times 10^{-10}$ mol/L

$$Hg_2I_2(s) \rightarrow Hg_2^{2+}(aq) \ + \ 2 \, I^-(aq) \qquad K_{sp} = [Hg_2^{2+}][I^-]^2$$

Initial s = solubility (mol/L)   0   0

Equil. s $2s$

From the problem, $2s = 4.4 \times 10^{-10}$ mol/L, $s = 2.2 \times 10^{-10}$ mol/L

$K_{sp} = (s)(2s)^2 = (2.2 \times 10^{-10})(4.4 \times 10^{-10})^2 = 4.3 \times 10^{-29}$

CHAPTER NINETEEN

THE REPRESENTATIVE ELEMENTS: GROUPS 1A THROUGH 4A

Questions

1. The gravity of the earth is not strong enough to keep the light H_2 molecules in the atmosphere.

3. Ionic, covalent, and metallic (or interstitial); The ionic and covalent hydrides are true compounds obeying the law of definite proportions and differ from each other in the type of bonding. The interstitial hydrides are more like solid solutions of hydrogen with a transition metal and do not obey the law of definite proportions.

5. Hydrogen forms many compounds in which the oxidation state is +1, as do the Group 1A elements. For example, H_2SO_4 and HCl compare to Na_2SO_4 and NaCl. On the other hand, hydrogen forms diatomic H_2 molecules and is a nonmetal, while the Group 1A elements are metals. Hydrogen also forms compounds with a -1 oxidation state, which is not characteristic of Group 1A metals, e.g., NaH.

7. Group 1A and 2A metals are all easily oxidized. They must be produced in the absence of materials (H_2O, O_2) that are capable of oxidizing them.

9. In graphite, planes of carbon atoms slide easily along each other. In addition, graphite is not volatile. The lubricant will not be lost when used in a high-vacuum environment.

11. Group 3A elements have one fewer valence electron than Si or Ge. A p-type semiconductor would form.

Exercises

Group 1A Elements

13. a. $\Delta H° = -110.5 - [-75 + (-242)] = 207$ kJ; $\Delta S° = 198 + 3(131) - [186 + 189] = 216$ J/K

 b. $\Delta G° = \Delta H° - T\Delta S°$; $\Delta G° = 0$ when $T = \dfrac{\Delta H°}{\Delta S°} = \dfrac{207 \times 10^3 \text{ J}}{216 \text{ J/K}} = 958$ K

 At T > 958 K and standard pressures, the favorable $\Delta S°$ term dominates, and the reaction is spontaneous ($\Delta G° < 0$).

15. a. lithium oxide;   b.   potassium superoxide;   c.   sodium peroxide

17. a. $Li_2O(s) + H_2O(l) \rightarrow 2\ LiOH(aq)$ b. $Na_2O_2(s) + 2\ H_2O(l) \rightarrow 2\ NaOH(aq) + H_2O_2(aq)$

     c. $LiH(s) + H_2O(l) \rightarrow H_2(g) + LiOH(aq)$ d. $2\ KO_2(s) + 2H_2O(l) \rightarrow 2\ KOH(aq) + O_2(g) + H_2O_2(aq)$

19. $2\ Li(s) + 2\ C_2H_2(g) \rightarrow 2\ LiC_2H(s) + H_2(g);$ This is an oxidation-reduction reaction.

Group 2A Elements

21. a. magnesium carbonate   b. barium sulfate   c. strontium hydroxide

23. $CaCO_3(s) + H_2SO_4(aq) \rightarrow CaSO_4(aq) + H_2O(l) + CO_2(g)$

25. In the gas phase, linear molecules would exist.

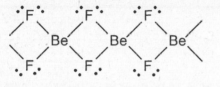

In the solid state, BeF_2 has the following extended structure:

27. $\dfrac{1\ mg\ F^-}{L} \times \dfrac{1\ g}{1000\ mg} \times \dfrac{1\ mol\ F^-}{19.00\ g\ F^-} = 5.3 \times 10^{-5}\ M\ F^- = 5 \times 10^{-5}\ M\ F^-$

$CaF_2(s) \rightleftharpoons Ca^{2+}(aq) + 2\ F^-(aq)$ $K_{sp} = [Ca^{2+}][F^-]^2 = 4.0 \times 10^{-11};$ Precipitation will occur when $Q > K_{sp}.$ Let's calculate $[Ca^{2+}]$ so that $Q = K_{sp}.$

$Q = 4.0 \times 10^{-11} = [Ca^{2+}]_o[F^-]_o^2 = [Ca^{2+}]_o(5 \times 10^{-5})^2,\ [Ca^{2+}]_o = 2 \times 10^{-2}\ M$

$CaF_2(s)$ will precipitate when $[Ca^{2+}]_o > 2 \times 10^{-2}\ M.$ Therefore, hard water should have a calcium ion concentration of less than $2 \times 10^{-2}\ M$ in order to avoid $CaF_2(s)$ formation.

29. $Ba^{2+} + 2\ e^- \rightarrow Ba;$ $6.00\ hr \times \dfrac{60\ min}{hr} \times \dfrac{60\ s}{min} \times \dfrac{2.50 \times 10^5\ C}{s} \times \dfrac{1\ mol\ e^-}{96,485\ C} \times \dfrac{1\ mol\ Ba}{2\ mol\ e^-}$

$\times \dfrac{137.3\ g\ Ba}{mol\ Ba} = 3.84 \times 10^6\ g\ Ba$

Group 3A Elements

31. a. AlN   b. GaF_3 c. Ga_2S_3

33. $B_2H_6(g) + 3\ O_2(g) \rightarrow 2\ B(OH)_3(s)$

35. $Ga_2O_3(s) + 6 H^+(aq) \rightarrow 2 Ga^{3+}(aq) + 3 H_2O(l)$

       $Ga_2O_3(s) + 2 OH^-(aq) + 3 H_2O(l) \rightarrow 2 Ga(OH)_4^-(aq)$

       $In_2O_3(s) + 6 H^+(aq) \rightarrow 2 In^{3+}(aq) + 3 H_2O(l);\ In_2O_3(s) + OH^-(aq) \rightarrow$ no reaction

37. $2 Ga(s) + 3 F_2(g) \rightarrow 2 GaF_3(s);\ 4 Ga(s) + 3 O_2(g) \rightarrow 2 Ga_2O_3(s)$

       $16 Ga(s) + 3 S_8(s) \rightarrow 8 Ga_2S_3(s);\ 2 Ga(s) + N_2(g) \rightarrow 2 GaN(s)$

       Note: GaN would be predicted, but in practice, this reaction does not occur.

       $2 Ga(s) + 6 HCl(aq) \rightarrow 2 GaCl_3(aq) + 3 H_2(g)$

Group 4A Elements

39. $CF_4, 4 + 4(7) = 32\ e^-$ $GeF_4, 4 + 4(7) = 32\ e^-$ $GeF_6^{2-}, 4 + 6(7) + 2 = 48\ e^-$

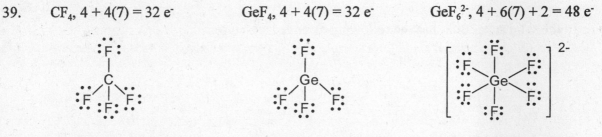

       tetrahedral; 109.5°; sp^3 tetrahedral; 109.5°; sp^3 octahedral; 90°; d^2sp^3

       In order to form CF_6^{2-}, carbon would have to expand its octet of electrons. Carbon compounds do not expand their octet because of the small atomic size of carbon and because no low energy d-orbitals are available for carbon to accommodate the extra electrons.

41.   a. $SiO_2(s) + 2 C(s) \rightarrow Si(s) + 2 CO(g)$

       b. $SiCl_4(l) + 2 Mg(s) \rightarrow Si(s) + 2 MgCl_2(s)$

       c. $Na_2SiF_6(s) + 4 Na(s) \rightarrow Si(s) + 6 NaF(s)$

43.   Lead is very toxic. As the temperature of the water increases, the solubility of lead will increase. Drinking hot tap water from pipes containing lead solder could result in higher lead concentrations in the body.

45. $C_6H_{12}O_6(aq) \rightarrow 2 C_2H_5OH(aq) + 2 CO_2(g)$

47.   The π electrons are free to move in graphite, thus giving it greater conductivity (lower resistance). The electrons in graphite have the greatest mobility within sheets of carbon atoms, resulting in a lower resistance in the basal plane. Electrons in diamond are not mobile (high resistance). The structure of diamond is uniform in all directions; thus, resistivity has no directional dependence in diamond.

Additional Exercises

49. $2 K(s) + 2 H_2O(l) \rightarrow 2 KOH(aq) + H_2(g)$, $\Delta H^\circ = 2(-481 \text{ kJ}) - 2(-286 \text{ kJ}) = -390. \text{ kJ}$

$5.00 \text{ g K} \times \dfrac{1 \text{ mol K}}{39.10 \text{ g K}} \times \dfrac{-390. \text{ kJ}}{2 \text{ mol K}} = -24.9 \text{ kJ}$ of heat released upon reaction of 5.00 g of potassium.

$24{,}900 \text{ J} = \dfrac{4.18 \text{ J}}{\text{g} \,^\circ\text{C}} \times (1.00 \times 10^3 \text{ g}) \times \Delta T$, $\Delta T = \dfrac{24{,}900}{4.18 \times 1.00 \times 10^3} = 5.96\,^\circ\text{C}$

Final temperature $= 24.0 + 5.96 = 30.0\,^\circ\text{C}$

51.   Strontium and calcium are both alkaline earth metals, so they have similar chemical properties. Since milk is a good source of calcium, strontium could replace some calcium in milk without much difficulty.

53.   The "inert pair effect" refers to the difficulty of removing the pair of s electrons from some of the elements in the fifth and sixth periods of the periodic table. As a result, multiple oxidation states are exhibited for the heavier elements of Groups 3A and 4A. In^+, In^{3+}, Tl^+ and Tl^{3+} oxidation states are all important to the chemistry of In and Tl.

55.   Major species present: $Al(H_2O)_6^{3+}$ ($K_a = 1.4 \times 10^{-5}$), NO_3^- (neutral) and H_2O ($K_w = 1.0 \times 10^{-14}$); $Al(H_2O)_6^{3+}$ is a stronger acid than water so it will be the dominant H^+ producer.

	$Al(H_2O)_6^{3+}$	$\rightleftharpoons$	$Al(H_2O)_5(OH)^{2+}$	+	H^+
Initial	0.050 M		0		~0
	x mol/L $Al(H_2O)_6^{3+}$ dissociates to reach equilibrium				
Change	$-x$	$\rightarrow$	$+x$		$+x$
Equil.	0.050 - x		x		x

$K_a = 1.4 \times 10^{-5} = \dfrac{[Al(H_2O)_5(OH)^{2+}][H^+]}{[Al(H_2O)_6^{3+}]} = \dfrac{x^2}{0.050-x} \approx \dfrac{x^2}{0.050}$

$x = 8.4 \times 10^{-4} \, M = [H^+]$; pH $= -\log(8.4 \times 10^{-4}) = 3.08$; Assumptions good.

57.   Ga(I): [Ar]$4s^2 3d^{10}$, no unpaired e^-; Ga(III): [Ar]$3d^{10}$, no unpaired e^-

Ga(II): [Ar]$4s^1 3d^{10}$, 1 unpaired e^-; Note: s electrons are lost before the d electrons.

If the compound contained Ga(II), it would be paramagnetic, and if the compound contained Ga(I) and Ga(III), it would be diamagnetic. This can be determined easily by measuring the mass of a sample in the presence and in the absence of a magnetic field. Paramagnetic compounds will have an apparent greater mass in a magnetic field.

59. $$750.\text{ mL grape juice} \times \frac{12\text{ mL } C_2H_5OH}{100.\text{ mL juice}} \times \frac{0.79\text{ g } C_2H_5OH}{\text{mL}} \times \frac{1\text{ mol } C_2H_5OH}{46.07\text{ g}} \times \frac{2\text{ mol } CO_2}{2\text{ mol } C_2H_5OH}$$

$$= 1.54\text{ mol } CO_2 \quad \text{(carry extra significant figure)}$$

$1.54\text{ mol } CO_2 = \text{total mol } CO_2 = \text{mol } CO_2(g) + \text{mol } CO_2(aq) = n_g + n_{aq}$

$$P_{CO_2} = \frac{n_gRT}{V} = \frac{n_g\left(\frac{0.08206\text{ L atm}}{\text{mol K}}\right)(298\text{ K})}{75 \times 10^{-3}\text{ L}} = 326\, n_g;\quad P_{CO_2} = \frac{C}{k} = \frac{\frac{n_{aq}}{0.750\text{ L}}}{\frac{3.1 \times 10^{-2}\text{ mol}}{\text{L atm}}} = 43.0\, n_{aq}$$

$P_{CO_2} = 326\, n_g = 43.0\, n_{aq}$ and from above $n_{aq} = 1.54 - n_g$; Solving:

$326\, n_g = 43.0(1.54 - n_g),\ 369\, n_g = 66.2,\ n_g = 0.18\text{ mol}$

$P_{CO_2} = 326(0.18) = 59\text{ atm}$ in gas phase

$$C = kP_{CO_2} = \frac{3.1 \times 10^{-2}\text{ mol}}{\text{L atm}} \times 59\text{ atm},\quad C = 1.8\text{ mol } CO_2\text{/L in wine}$$

61. $Pb(NO_3)_2(aq) + H_3AsO_4(aq) \rightarrow PbHAsO_4(s) + 2\ HNO_3(aq)$

Note: The insecticide used is $PbHAsO_4$ and is commonly called lead arsenate. This is not the correct name, however. Correctly, lead arsenate would be $Pb_3(AsO_4)_2$ and $PbHAsO_4$ should be named lead hydrogen arsenate.

Challenge Problems

63.

	Pb^{2+}	+ H_2EDTA^{2-}	$\rightleftharpoons$	$PbEDTA^{2-}$	+ $2\ H^+$	
Before	0.0010 *M*	0.050 *M*		0	1.0×10^{-6} *M*	(buffer, $[H^+]$ constant)
Change	-0.0010	-0.0010	→	+0.0010	No change	Reacts completely
After	0	0.049		0.0010	1.0×10^{-6}	New initial conditions
	x mol/L $PbEDTA^{2-}$ dissociates to reach equilibrium					
Change	+*x*	+*x*	←	-*x*	---	
Equil.	*x*	0.049 + *x*		0.0010 - *x*	1.0×10^{-6}	(buffer)

$$K = 1.0 \times 10^{23} = \frac{[PbEDTA^{2-}][H^+]^2}{[Pb^{2+}][H_2EDTA^{2-}]} = \frac{(0.0010 - x)(1.0 \times 10^{-6})^2}{(x)(0.049 + x)}$$

$$1.0 \times 10^{23} \approx \frac{(0.0010)(1.0 \times 10^{-12})}{x(0.049)},\quad x = [Pb^{2+}] = 2.0 \times 10^{-37}\ M \quad \text{Assumptions good.}$$

65. Carbon cannot form the fifth bond necessary for the transition state because of the small atomic size of carbon and because carbon doesn't have low energy d orbitals available to expand the octet.

CHAPTER TWENTY

THE REPRESENTATIVE ELEMENTS: GROUPS 5A THROUGH 8A

Questions

1. $N_2(g) + 3 H_2(g) \rightleftharpoons 2 NH_3(g) + heat$

    a.   This reaction is exothermic, so an increase in temperature will decrease the value of K (see Table 13.3 of text.) This has the effect of lowering the amount of $NH_3(g)$ produced at equilibrium. The temperature increase, therefore, must be for kinetics reasons. When the temperature increases, the reaction reaches equilibrium much faster. At low temperatures, this reaction is very slow, too slow to be of any use.

    b.   As $NH_3(g)$ is removed, the reaction shifts right to produce more $NH_3(g)$.

    c.   A catalyst has no effect on the equilibrium position. The purpose of a catalyst is to speed up a reaction so it reaches equilibrium more quickly.

    d.   When the pressure of reactants and products is high, the reaction shifts to the side that has fewer gas molecules. Since the product side contains 2 molecules of gas compared to 4 molecules of gas on the reactant side, the reaction shifts right to products at high pressures of reactants and products.

3.   The pollution provides nitrogen and phosphorous nutrients so the algae can grow. The algae consume oxygen, causing fish to die.

5.   Plastic sulfur consists of long S_n chains of sulfur atoms. As plastic sulfur becomes brittle, the long chains break down into S_8 rings.

7.   Fluorine is the most reactive of the halogens because it is the most electronegative atom and the bond in the F_2 molecule is very weak.

9.   Helium is unreactive and doesn't combine with any other elements. It is a very light gas and would easily escape the earth's gravitational pull as the planet was formed.

Exercises

Group 5A Elements

11. NO_4^{3-}

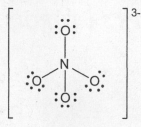

Both NO_4^{3-} and PO_4^{3-} have 32 valence electrons, so both have similar Lewis structures. From the Lewis structure for NO_4^{3-}, the central N atom has a tetrahedral arrangement of electron pairs. N is small. There is probably not enough room for all 4 oxygen atoms around N. P is larger, thus, PO_4^{3-} is stable.

PO_3^-

PO_3^- and NO_3^- each have 24 valence electrons so both have similar Lewis structures. From the Lewis structure for PO_3^-, PO_3^- has a trigonal planar arrangement of electron pairs about the central P atom (two single bonds and one double bond). $P=O$ bonds are not particularly stable, while $N=O$ bonds are stable. Thus, NO_3^- is stable.

13.   a. $NH_4NO_3(s) \xrightarrow{heat} N_2O(g) + 2 H_2O(g)$

  b. $2 N_2O_5(g) \rightarrow 4 NO_2(g) + O_2(g)$

  c. $2 K_3P(s) + 6 H_2O(l) \rightarrow 2 PH_3(g) + 6 KOH(aq)$

  d. $PBr_3(l) + 3 H_2O(l) \rightarrow H_3PO_3(aq) + 3 HBr(aq)$

  e. $2 NH_3(aq) + NaOCl(aq) \rightarrow N_2H_4(aq) + NaCl(aq) + H_2O(l)$

15.   Unbalanced equation:

$$CaF_2 \cdot 3Ca_3(PO_4)_2(s) + H_2SO_4(aq) \rightarrow H_3PO_4(aq) + HF(aq) + CaSO_4 \cdot 2H_2O(s)$$

Balancing Ca^{2+}, F^-, and PO_4^{3-}:

$$CaF_2 \cdot 3Ca_3(PO_4)_2(s) + H_2SO_4(aq) \rightarrow 6 H_3PO_4(aq) + 2 HF(aq) + 10 CaSO_4 \cdot 2H_2O(s)$$

On the righthand side, there are 20 extra hydrogen atoms, 10 extra sulfates, and 20 extra water molecules. We can balance the hydrogen and sulfate with 10 sulfuric acid molecules. The extra waters came from the water in the sulfuric acid solution. The balanced equation is:

$$CaF_2 \cdot 3Ca_3(PO_4)_2(s) + 10 H_2SO_4(aq) + 20 H_2O(l) \rightarrow 6 H_3PO_4(aq) + 2 HF(aq) + 10 CaSO_4 \cdot 2H_2O(s)$$

17. $2 \, NaN_3(s) \rightarrow 2 \, Na(s) + 3 \, N_2(g)$

$$n_{N_2} = \frac{PV}{RT} = \frac{1.00 \, atm \times 70.0 \, L}{\dfrac{0.08206 \, L \, atm}{mol \, K} \times 273 \, K} = 3.12 \, mol \, N_2 \text{ needed to fill air bag.}$$

$$\text{mol } NaN_3 \text{ reacted} = 3.12 \, mol \, N_2 \times \frac{2 \, mol \, NaN_3}{3 \, mol \, N_2} = 2.08 \, mol \, NaN_3$$

19.

$\qquad$ H$\quad$N$-$N$\quad$H (l) + O$=$O (g) $\longrightarrow$ N$\equiv$N (g) + 2 H$-$O$-$H (g)

Bonds broken: $\qquad\qquad\qquad\qquad$ Bonds formed:

$\qquad$ 1 N $-$ N (160. kJ/mol) $\qquad\qquad\qquad$ 1 N$\equiv$N (941 kJ/mol)
$\qquad$ 4 N $-$ H (391 kJ/mol) $\qquad\qquad\qquad$ 2 $\times$ 2 O $-$ H (467 kJ/mol)
$\qquad$ 1 O $=$ O (495 kJ/mol)

$\Delta H = 160. + 4(391) + 495 - [941 + 4(467)] = 2219 \, kJ - 2809 \, kJ = -590. \, kJ$

21. $1/2 \, N_2(g) + 1/2 \, O_2(g) \rightarrow NO(g)$ $\Delta G° = \Delta G°_{f, NO} = 87 \, kJ/mol$; By definition, $\Delta G°_f$ for a compound equals the free energy change that would accompany the formation of 1 mol of that compound from its elements in their standard states. NO (and some other oxides of nitrogen) have weaker bonds as compared to the triple bond of N_2 and the double bond of O_2. Because of this, NO (and some other oxides of nitrogen) have higher (positive) standard free energies of formation as compared to the relatively stable N_2 and O_2 molecules.

23.   M.O. model:

$\quad$ NO^+: $(\sigma_{2s})^2(\sigma_{2s}{}^*)^2(\pi_{2p})^4(\sigma_{2p})^2$, Bond order = (8 - 2)/2 = 3, 0 unpaired e$^-$ (diamagnetic)

$\quad$ NO: $(\sigma_{2s})^2(\sigma_{2s}{}^*)^2(\pi_{2p})^4(\sigma_{2p})^2(\pi_{2p}{}^*)^1$, B.O. = 2.5, 1 unpaired e$^-$ (paramagnetic)

$\quad$ NO^-: $(\sigma_{2s})^2(\sigma_{2s}{}^*)^2(\pi_{2p})^4(\sigma_{2p})^2(\pi_{2p}{}^*)^2$, B.O. = 2, 2 unpaired e$^-$ (paramagnetic)

$\quad$ Lewis structures: NO^+: $\left[\, :N \equiv O: \, \right]^+$

$\qquad\qquad\qquad\qquad$ NO: $\ddot{N} = \ddot{O}$ $\longleftrightarrow$ $:\ddot{N} = \ddot{O}$ $\longleftrightarrow$ $\dot{N} = \ddot{O}$

$\qquad\qquad\qquad\qquad$ NO^-: $\left[\, :\ddot{N} = \ddot{O}: \, \right]^-$

The two models give the same results only for NO^+ (a triple bond with no unpaired electrons).

Lewis structures are not adequate for NO and NO⁻. The M.O. model gives a better representation for all three species. For NO, Lewis structures are poor for odd electron species. For NO⁻, both models predict a double bond, but only the MO model correctly predicts that NO⁻ is paramagnetic.

25.   a. $H_3PO_4 > H_3PO_3$; The strongest acid has the most oxygen atoms.

b. $H_3PO_4 > H_2PO_4^- > HPO_4^{2-}$; Acid strength decreases as protons are removed.

27.   The acidic protons are attached to oxygen.

$H_4P_2O_6$ (50 valence e⁻):

$H_4P_2O_5$ (44 valence e⁻):

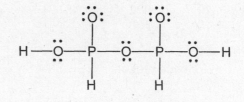

Group 6A Elements

29. $O=O-O \rightarrow O=O + O$

Break O – O bond: $\Delta H = \dfrac{146\,kJ}{mol} \times \dfrac{1\,mol}{6.022 \times 10^{23}} = 2.42 \times 10^{-22}\,kJ = 2.42 \times 10^{-19}\,J$

A photon of light must contain at least 2.42×10^{-19} J to break one O – O bond.

$E_{photon} = \dfrac{hc}{\lambda}, \ \lambda = \dfrac{hc}{E} = \dfrac{(6.626 \times 10^{-34}\,J\,s)\,(2.998 \times 10^8\,m/s)}{2.42 \times 10^{-19}\,J} = 8.21 \times 10^{-7}\,m = 821\,nm$

31.   a. $2\,SO_2(g) + O_2(g) \rightarrow 2\,SO_3(g)$ b. $SO_3(g) + H_2O(l) \rightarrow H_2SO_4(aq)$

c. $2\,Na_2S_2O_3(aq) + I_2(aq) \rightarrow Na_2S_4O_6(aq) + 2\,NaI(aq)$

d. $Cu(s) + 2\,H_2SO_4(aq) \rightarrow CuSO_4(aq) + 2\,H_2O(l) + SO_2(aq)$

33.   a. SO_3^{2-}, $6 + 3(6) + 2 = 26$ e⁻   b. O_3, $3(6) = 18$ e⁻

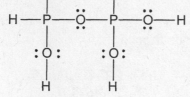

trigonal pyramid; ≈ 109.5°; sp³   V-shaped; ≈ 120°; sp²

c. SCl_2, $6 + 2(7) = 20$ e⁻

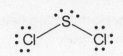

V-shaped; $\approx 109.5°$; sp^3

d. $SeBr_4$, $6 + 4(7) = 34$ e⁻

see-saw; a $\approx 120°$, b $\approx 90°$; dsp^3

e. TeF_6, $6 + 6(7) = 48$ e⁻

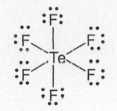

octahedral; $90°$; d^2sp^3

Group 7A Elements

35. O_2F_2 has $2(6) + 2(7) = 26$ valence e⁻; From the following Lewis structure, each oxygen atom has a tetrahedral arrangement of electron pairs. Therefore, bond angles $\approx 109.5°$ and each O is sp^3 hybridized.

Formal Charge	0	0	0	0
Oxid. Number	-1	+1	+1	-1

Oxidation numbers are more useful. We are forced to assign +1 as the oxidation number for oxygen. Oxygen is very electronegative, and +1 is not a stable oxidation state for this element.

37.   a. $BaCl_2(s) + H_2SO_4(aq) \rightarrow BaSO_4(s) + 2\ HCl(g)$

       b. $BrF(s) + H_2O(l) \rightarrow HF(aq) + HOBr(aq)$

       c. $SiO_2(s) + 4\ HF(aq) \rightarrow SiF_4(g) + 2\ H_2O(l)$

39. $ClO^- + H_2O + 2\ e^- \rightarrow 2\ OH^- + Cl^-$ E° = 0.90 V
           $2\ NH_3 + 2\ OH^- \rightarrow N_2H_4 + 2\ H_2O + 2\ e^-$ -E° = 0.10 V

       ——

           $ClO^-(aq) + 2\ NH_3(aq) \rightarrow Cl^-(aq) + N_2H_4(aq) + H_2O(l)$ $E°_{cell} = 1.00$ V

Since $E°_{cell}$ is positive for this reaction, then at standard conditions ClO^- can spontaneously oxidize NH_3 to the somewhat toxic N_2H_4.

Group 8A Elements

41.   Xe has one more valence electron than I. Thus, the isoelectric species will have I plus one extra electron substituted for Xe, giving a species with a net minus one charge.

   a. IO_4^- b. IO_3^- c. IF_2^- d. IF_4^- e. IF_6^-

43. XeF_2 can react with oxygen to produce explosive xenon oxides and oxyfluorides.

Additional Exercises

45.   As the halogen atoms get larger, it becomes more difficult to fit three halogen atoms around the small nitrogen atom, and the NX_3 molecule becomes less stable.

47. OCN^- has $6 + 4 + 5 + 1 = 16$ valence electrons.

Formal
charge   0   0   -1   -1   0   0   +1   0   -2

   Only the first two resonance structures should be important. The third places a positive formal charge on the most electronegative atom in the ion and a -2 formal charge on N.

CNO^-:

Formal
charge   -2   +1   0   -1   +1   -1   -3   +1   +1

   All of the resonance structures for fulminate (CNO^-) involve greater formal charges than in cyanate (OCN^-), making fulminate more reactive (less stable).

49. $$1.0 \times 10^4 \text{ kg waste} \times \frac{3.0 \text{ kg NH}_4^+}{100 \text{ kg waste}} \times \frac{1000 \text{ g}}{\text{kg}} \times \frac{1 \text{ mol NH}_4^+}{18.04 \text{ g NH}_4^+} \times \frac{1 \text{ mol C}_5\text{H}_7\text{O}_2\text{N}}{55 \text{ mol NH}_4^+}$$

$$\times \frac{113.12 \text{ g C}_5\text{H}_7\text{O}_2\text{N}}{\text{mol C}_5\text{H}_7\text{O}_2\text{N}} = 3.4 \times 10^4 \text{ g tissue if all NH}_4^+ \text{ converted}$$

Since only 95% of the NH_4^+ ions react:

   mass of tissue $= (0.95)(3.4 \times 10^4 \text{ g}) = 3.2 \times 10^4$ g or 32 kg bacterial tissue

51.   TeF$_5^-$ has $6 + 5(7) + 1 = 42$ valence electrons.

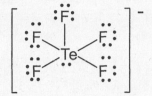

The lone pair of electrons around Te exerts a stronger repulsion than the bonding pairs, pushing the four square planar F's away from the lone pair and thus reducing the bond angles between the axial F atom and the square planar F atoms.

53.   Release of Sr is probably more harmful. Xe is chemically unreactive. Strontium is in the same family as calcium and could be absorbed and concentrated in the body in a fashion similar to Ca. This puts the radioactive Sr in the bones: red blood cells are produced in bone marrow. Xe would not be readily incorporated in the body.

The chemical properties determine where a radioactive material may be concentrated in the body or how easily it may be excreted. The length of time of exposure and what is exposed to radiation significantly affects the health hazard.

Challenge Problems

55.   For the reaction:

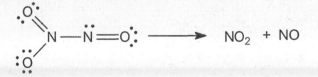

the activation energy must in some way involve breaking a nitrogen-nitrogen single bond. For the reaction:

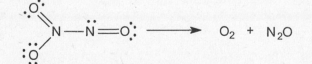

at some point nitrogen-oxygen bonds must be broken. N$-$N single bonds (160. kJ/mol) are weaker than N$-$O single bonds (201 kJ/mol). In addition, resonance structures indicate that there is more double bond character in the N$-$O bonds than in the N$-$N bond. Thus, NO$_2$ and NO are preferred by kinetics because of the lower activation energy.

57.   a.   NO is the catalyst. NO is present in the first step of the mechanism on the reactant side, but it is not a reactant since it is regenerated in the second step and does not appear in the overall balanced equation.

b. NO_2 is an intermediate. Intermediates also never appear in the overall balanced equation. In a mechanism, intermediates always appear first on the product side while catalysts always appear first on the reactant side.

c. $k = A \exp(-E_a/RT)$; $\dfrac{k_{cat}}{k_{un}} = \dfrac{A \exp[-E_a(cat)/RT]}{A \exp[-E_a(un)/RT]} = \exp\left(\dfrac{E_a(un) - E_a(cat)}{RT}\right)$

$$\frac{k_{cat}}{k_{un}} = \exp\left(\frac{2100 \text{ J/mol}}{8.3145 \text{ J/K}\cdot\text{mol} \times 298 \text{ K}}\right) = e^{0.85} = 2.3$$

The catalyzed reaction is approximately 2.3 times faster than the uncatalyzed reaction at 25 °C.

d. The mechanism for the chlorine-catalyzed destruction of ozone is:

$$O_3(g) + Cl(g) \rightarrow O_2(g) + ClO(g) \quad \text{slow}$$
$$ClO(g) + O(g) \rightarrow O_2(g) + Cl(g) \quad \text{fast}$$

$$O_3(g) + O(g) \rightarrow 2\ O_2(g)$$

e. Since the chlorine atom-catalyzed reaction has a lower activation energy, the Cl-catalyzed rate is faster. Hence, Cl is a more effective catalyst. Using the activation energy, we can estimate the efficiency that Cl atoms destroy ozone as compared to NO molecules.

At 25 °C: $\dfrac{k_{Cl}}{k_{NO}} = \exp\left(\dfrac{-E_a(Cl)}{RT} + \dfrac{E_a(NO)}{RT}\right) = \exp\left(\dfrac{(-2100 + 11{,}900) \text{ J/mol}}{(8.3145 \times 298) \text{ J/mol}}\right) = e^{3.96} = 52$

At 25 °C, the Cl catalyzed reaction is roughly 52 times faster (more efficient) than the NO catalyzed reaction, assuming the frequency factor A is the same for each reaction and assuming similar rate laws.

CHAPTER TWENTY-ONE

TRANSITION METALS AND COORDINATION CHEMISTRY

Questions

5. a. Ligand: Species that donates a pair of electrons to form a covalent bond to a metal ion. Ligands act as Lewis bases (electron pair donors).

    b. Chelate: Ligand that can form more than one bond to a metal ion.

    c. Bidentate: Ligand that forms two bonds to a metal ion.

    d. Complex ion: Metal ion plus ligands.

7. a. Isomers: Species with the same formulas but different properties. See the text for examples of the following types of isomers.

    b. Structural isomers: Isomers that have one or more bonds that are different.

    c. Stereoisomers: Isomers that contain the same bonds but differ in how the atoms are arranged in space.

    d. Coordination isomers: Structural isomers that differ in the atoms that make up the complex ion.

    e. Linkage isomers: Structural isomers that differ in how one or more ligands are attached to the transition metal.

    f. Geometric isomers: (Cis-trans isomerism); Stereoisomers that differ in the positions of atoms with respect to a rigid ring, bond, or each other.

    g. Optical isomers: Stereoisomers that are nonsuperimposable mirror images of each other; that is, they are different in the same way that our left and right hands are different.

9. Cu^{2+}: $[Ar]3d^9$; Cu^+: $[Ar]3d^{10}$; Cu(II) is d^9 and Cu(I) is d^{10}. Color is a result of the electron transfer between split d orbitals. This cannot occur for the filled d orbitals in Cu(I). Cd^{2+}, like Cu^+, is also d^{10}. We would not expect $Cd(NH_3)_4Cl_2$ to be colored since the d orbitals are filled in this Cd^{2+} complex.

11.   The d-orbital splitting in tetrahedral complexes is less than one-half the d-orbital splitting in octahedral complexes. There are no known ligands powerful enough to produce the strong-field case, hence all tetrahedral complexes are weak-field or high spin.

13. $Fe_2O_3(s) + 6\ H_2C_2O_4(aq) \rightarrow 2\ Fe(C_2O_4)_3^{3-}(aq) + 3\ H_2O(l) + 6\ H^+(aq)$; The oxalate anion forms a soluble complex ion with iron in rust (Fe_2O_3), which allows rust stains to be removed.

15.   The lanthanide elements are located just before the 5d transition metals. The lanthanide contraction is the steady decrease in the atomic radii of the lanthanide elements when going from left to right across the periodic table. As a result of the lanthanide contraction, the sizes of the 4d and 5d elements are very similar (see Exercise 7.132). This leads to a greater similarity in the chemistry of the 4d and 5d elements in a given vertical group.

17.   Advantages: cheap energy cost; less air pollution; Disadvantages: chemicals used in hydrometallurgy are expensive and sometimes toxic.

Exercises

Transition Metals and Coordination Compounds

19.   a. Ni: $[Ar]4s^23d^8$ b. Cd: $[Kr]5s^24d^{10}$

       c. Zr: $[Kr]5s^24d^2$ d. Os: $[Xe]6s^24f^{14}5d^6$

21.   Transition metal ions lose the s electrons before the d electrons.

       a. Ti: $[Ar]4s^23d^2$ b. Re: $[Xe]6s^24f^{14}5d^5$ c. Ir: $[Xe]6s^24f^{14}5d^7$

       Ti^{2+}: $[Ar]3d^2$ Re^{2+}: $[Xe]4f^{14}5d^5$ Ir^{2+}: $[Xe]4f^{14}5d^7$

       Ti^{4+}: $[Ar]$ or $[Ne]3s^23p^6$ Re^{3+}: $[Xe]4f^{14}5d^4$ Ir^{3+}: $[Xe]4f^{14}5d^6$

23. $CoCl_2(s) + 6\ H_2O(g) \rightleftharpoons CoCl_2 \cdot 6\ H_2O(s)$; If rain were imminent, there would be a lot of water vapor in the air causing the reaction to shift to the right. The indicator would take on the color of $CoCl_2 \cdot 6$ H_2O, pink.

25.   Test tube 1: added Cl^- reacts with Ag^+ to form a silver chloride precipitate. The net ionic equation is $Ag^+(aq) + Cl^-(aq) \rightarrow AgCl(s)$. Test tube 2: added NH_3 reacts with Ag^+ ions to form a soluble complex ion $Ag(NH_3)_2^+$. As this complex ion forms, Ag^+ is removed from solution, which causes the $AgCl(s)$ to dissolve. When enough NH_3 is added, all of the silver chloride precipitate will dissolve. The equation is $AgCl(s) + 2\ NH_3(aq) \rightarrow Ag(NH_3)_2^+(aq) + Cl^-(aq)$. Test tube 3: added H^+ reacts with the weak base NH_3 to form NH_4^+. As NH_3 is removed from the $Ag(NH_3)_2^+$ complex ion, Ag^+ ions are released to solution and can then react with Cl^- to reform $AgCl(s)$.   The equations are $Ag(NH_3)_2^+(aq) + 2\ H^+(aq) \rightarrow Ag^+(aq) + 2\ NH_4^+(aq)$ and $Ag^+(aq) + Cl^-(aq) \rightarrow AgCl(s)$.

27.   Since each compound contains an octahedral complex ion, the formulas for the compounds are
       $[Co(NH_3)_6]I_3$, $[Pt(NH_3)_4I_2]I_2$, $Na_2[PtI_6]$ and $[Cr(NH_3)_4I_2]I$. Note that in some cases, the I^- ions are
       ligands bound to the transition metal ion as required for a coordination number of 6, while in other
       cases the I^- ions are counter ions required to balance the charge of the complex ion. The $AgNO_3$
       solution will only precipitate the I^- counter ions and will not precipitate the I^- ligands. Therefore, 3
       mol of AgI will precipitate per mol of $[Co(NH_3)_6]I_3$, 2 mol of AgI will precipitate per mol of
       $[Pt(NH_3)_4I_2]I_2$, 0 mol of AgI will precipitate per mol of $Na_2[PtI_6]$, and 1 mol of AgI will precipitate
       per mol of $[Cr(NH_3)_4I_2]I$.

29.   To determine the oxidation state of the metal, you must know the charges of the various common
       ligands (see Table 21.13 of the text).

       a. pentaamminechlororuthenium(III) ion   b. hexacyanoferrate(II) ion

       c. tris(ethylenediamine)manganese(II) ion   d. pentaamminenitrocobalt(III) ion

31.   a. hexaamminecobalt(II) chloride   b. hexaaquacobalt(III) iodide

       c. potassium tetrachloroplatinate(II)   d. potassium hexachloroplatinate(II)

       e. pentaamminechlorocobalt(III) chloride   f. triamminetrinitrocobalt(III)

33.   a. $K_2[CoCl_4]$ b. $[Pt(H_2O)(CO)_3]Br_2$

       c. $Na_3[Fe(CN)_2(C_2O_4)_2]$ d. $[Cr(NH_3)_3Cl(H_2NCH_2CH_2NH_2)]I_2$

35.   a.

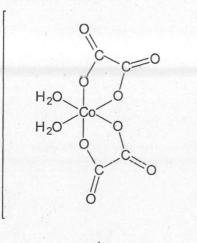

                           cis

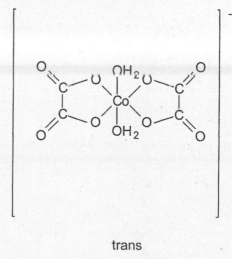

                                                              trans

       Note: $C_2O_4^{2-}$ is a bidentate ligand. Bidentate ligands bond to the metal at two positions that are
       90° apart from each other in octahedral complexes. Bidentate ligands do not bond to the metal
       at positions 180° apart.

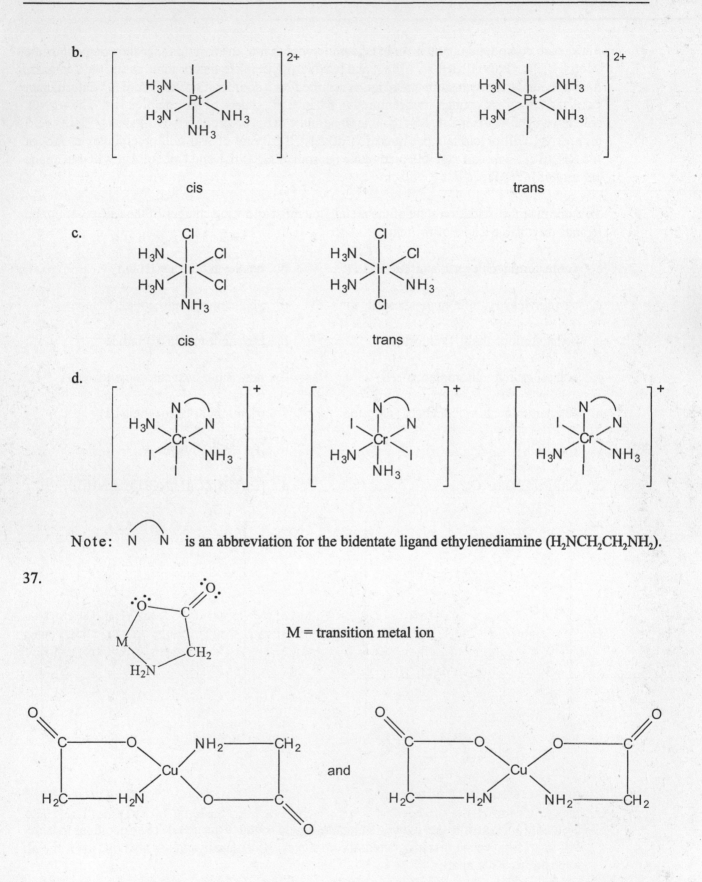

b.

cis

trans

c.

cis

trans

d.

Note: N⌒N is an abbreviation for the bidentate ligand ethylenediamine ($H_2NCH_2CH_2NH_2$).

37.

M = transition metal ion

and

39.   Linkage isomers differ in the way the ligand bonds to the metal. SCN^- can bond through the sulfur or through the nitrogen atom. NO_2^- can bond through the nitrogen or through the oxygen atom. OCN^- can bond through the oxygen or through the nitrogen atom. N_3^-, $NH_2CH_2CH_2NH_2$ and I^- are not capable of linkage isomerism.

41.   Similar to the molecules discussed in Figures 21.16 and 21.17 of the text, $Cr(acac)_3$ and cis-$Cr(acac)_2(H_2O)_2$ are optically active. The mirror images of these two complexes are nonsuper-imposable. There is a plane of symmetry in trans-$Cr(acac)_2(H_2O)_2$, so it is not optically active. A molecule with a plane of symmetry is never optically active because the mirror images are always superimposable. A plane of symmetry is a plane through a molecule where one side reflects the other side of the molecule.

Bonding, Color, and Magnetism in Coordination Compounds

43.   a. Fe^{2+}: $[Ar]3d^6$

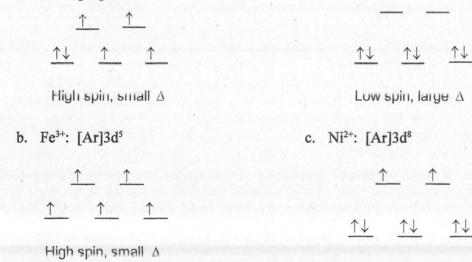

High spin, small Δ Low spin, large Δ

b. Fe^{3+}: $[Ar]3d^5$ c. Ni^{2+}: $[Ar]3d^8$

High spin, small Δ

45.   Since fluorine has a -1 charge as a ligand, chromium has a +2 oxidation state in CrF_6^{4-}. The electron configuration of Cr^{2+} is: $[Ar]3d^4$. For four unpaired electrons, this must be a weak-field (high-spin) case where the splitting of the d-orbitals is small and the number of unpaired electrons is maximized. The crystal field diagram for this ion is:

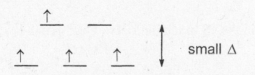

small Δ

47.   To determine the crystal field diagrams, you need to determine the oxidation state of the transition metal, which can only be determined if you know the charges of the ligands (see Table 21.13). The electron configurations and the crystal field diagrams follow.

a. Ru^{2+}: $[Kr]4d^6$, no unpaired e^-

___ ___

$\uparrow\downarrow$ $\uparrow\downarrow$ $\uparrow\downarrow$

Low spin, large Δ

b. Ni^{2+}: $[Ar]3d^8$, 2 unpaired e^-

$\uparrow$ $\uparrow$

$\uparrow\downarrow$ $\uparrow\downarrow$ $\uparrow\downarrow$

c. V^{3+}: $[Ar]3d^2$, 2 unpaired e^-

___ ___

$\uparrow$ $\uparrow$ ___

Note: Ni^{2+} must have 2 unpaired electrons, whether high-spin or low-spin, and V^{3+} must have 2 unpaired electrons, whether high-spin or low-spin.

49. From Table 21.16 of the text, the violet complex ion absorbs yellow-green light ($\lambda \sim 570$ nm), the yellow complex ion absorbs blue light ($\lambda \sim 450$ nm), and the green complex ion absorbs red light ($\lambda \sim 650$ nm). The spectrochemical series shows that NH_3 is a stronger-field ligand than H_2O which is a stronger-field ligand than Cl^-. Therefore, $Cr(NH_3)_6^{3+}$ will have the largest d-orbital splitting and will absorb the lowest wavelength electromagnetic radiation ($\lambda \sim 450$ nm) since energy and wavelength are inversely related ($\lambda = hc/E$). Thus, the yellow solution contains the $Cr(NH_3)_6^{3+}$ complex ion. Similarly, we would expect the $Cr(H_2O)_4Cl_2^+$ complex ion to have the smallest d-orbital splitting since it contains the weakest-field ligands. The green solution with the longest wavelength of absorbed light contains the $Cr(H_2O)_4Cl_2^+$ complex ion. This leaves the violet solution, which contains the $Cr(H_2O)_6^{3+}$ complex ion. This makes sense as we would expect $Cr(H_2O)_6^{3+}$ to absorb light of a wavelength between that of $Cr(NH_3)_6^{3+}$ and $Cr(H_2O)_4Cl_2^+$.

51. $CoBr_6^{4-}$ has an octahedral structure, and $CoBr_4^{2-}$ has a tetrahedral structure (as do most Co^{2+} complexes with four ligands). Coordination complexes absorb electromagnetic radiation (EMR) of energy equal to the energy difference between the split d-orbitals. Since the tetrahedral d-orbital splitting is less than one-half of the octahedral d-orbital splitting, tetrahedral complexes will absorb lower energy EMR, which corresponds to longer wavelength EMR ($E = hc/\lambda$). Therefore, $CoBr_6^{2-}$ will absorb EMR having a wavelength shorter than 3.4×10^{-6} m.

53. Since the ligands are Cl^-, then iron is in the +3 oxidation state. Fe^{3+}: $[Ar]3d^5$

$\uparrow$ $\uparrow$ $\uparrow$

$\uparrow$ $\uparrow$

Since all tetrahedral complexes are high-spin, there are 5 unpaired electrons in $FeCl_4^-$.

Metallurgy

55.   a.   To avoid fractions, let's first calculate ΔH for the reaction:

$$6\ FeO(s) + 6\ CO(g) \rightarrow 6\ Fe(s) + 6\ CO_2(g)$$

$6\ FeO + 2\ CO_2 \rightarrow 2\ Fe_3O_4 + 2\ CO$	$\Delta H° = -2(18\ kJ)$
$2\ Fe_3O_4 + CO_2 \rightarrow 3\ Fe_2O_3 + CO$	$\Delta H° = -(-39\ kJ)$
$3\ Fe_2O_3 + 9\ CO \rightarrow 6\ Fe + 9\ CO_2$	$\Delta H° = 3(-23\ kJ)$

$$6\ FeO(s) + 6\ CO(g) \rightarrow 6\ Fe(s) + 6\ CO_2(g) \qquad \Delta H° = -66\ kJ$$

So for: $FeO(s) + CO(g) \rightarrow Fe(s) + CO_2(g)$ $\Delta H° = \dfrac{-66\ kJ}{6} = -11\ kJ$

b. $\Delta H° = 2(-110.5\ kJ) - [-393.5\ kJ + 0] = 172.5\ kJ$

$\Delta S° = 2(198\ J/K) - [214\ J/K + 6\ J/K] = 176\ J/K$

$\Delta G° = \Delta H° - T\Delta S°$, $\Delta G° = 0$ when $T = \dfrac{\Delta H°}{\Delta S°} = \dfrac{172.5\ kJ}{0.176\ kJ/K} = 980.\ K$

Due to the favorable $\Delta S°$ term, this reaction is spontaneous at $T > 980.\ K$. From Figure 21.36 of the text, this reaction takes place in the blast furnace at temperatures greater than 980. K as required by thermodynamics.

Additional Exercises

57.   i. $0.0203\ g\ CrO_3 \times \dfrac{52.00\ g\ Cr}{100.0\ g\ CrO_3} = 0.0106\ g\ Cr$; $\%\ Cr = \dfrac{0.0106}{0.105} \times 100 = 10.1\%\ Cr$

ii. $32.93 \times 10^{-3}\ L\ HCl \times \dfrac{0.100\ mol\ HCl}{L} \times \dfrac{1\ mol\ NH_3}{mol\ HCl} \times \dfrac{17.03\ g\ NH_3}{mol} = 0.0561\ g\ NH_3$

$\%\ NH_3 = \dfrac{0.0561\ g}{0.341\ g} \times 100 = 16.5\%\ NH_3$

iii. $73.53\%\ I + 16.5\%\ NH_3 + 10.1\%\ Cr = 100.1\%$; The compound must be composed of only Cr, NH_3, and I.

Out of 100.00 g of compound:

$10.1\ g\ Cr \times \dfrac{1\ mol}{52.00\ g} = 0.194$ $\qquad\qquad$ $\dfrac{0.194}{0.194} = 1.00$

$16.5\ g\ NH_3 \times \dfrac{1\ mol}{17.03\ g} = 0.969$ $\qquad\qquad$ $\dfrac{0.969}{0.194} = 4.99$

$73.53\ g\ I \times \dfrac{1\ mol}{126.9\ g} = 0.5794$ $\qquad\qquad$ $\dfrac{0.5794}{0.194} = 2.99$

$Cr(NH_3)_5I_3$ is the empirical formula. Cr(III) forms octahedral complexes. So, compound A is made of the octahedral $[Cr(NH_3)_5I]^{2+}$ complex ion and two I^- counter ions; the formula is $[Cr(NH_3)_5I]I_2$. Let's check this proposed formula using the freezing point data.

iv. $\Delta T_f = iK_f m$; For $[Cr(NH_3)_5I]I_2$, $i = 3.0$ (assuming complete dissociation).

$$m = \frac{0.601 \text{ g complex}}{1.000 \times 10^{-2} \text{ kg H}_2\text{O}} \times \frac{1 \text{ mol complex}}{517.9 \text{ g complex}} = 0.116 \text{ molal}$$

$\Delta T_f = 3.0 \times 1.86°C/\text{molal} \times 0.116 \text{ molal} = 0.65°C$

Since ΔT_f is close to the measured value, then this is consistent with the formula $[Cr(NH_3)_5I]I_2$.

59.   a. 2; Forms bonds through the lone pairs on the two oxygen atoms.

b. 3; Forms bonds through the lone pairs on the three nitrogen atoms.

c. 4; Forms bonds through the two nitrogen atoms and the two oxygen atoms.

d. 4; Forms bonds through the four nitrogen atoms.

61.   a. $Ru(phen)_3^{2+}$ exhibits optical isomerism [similar to $Co(en)_3^{3+}$ in Figure 21.16 of the text].

b. Ru^{2+}: $[Kr]4d^6$; Since there are no unpaired electrons, Ru^{2+} is a strong-field (low-spin) case.

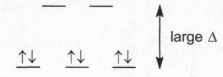

63.   Octahedral Cr^{2+} complexes should be used. Cr^{2+}: $[Ar]3d^4$; High-spin (weak-field) Cr^{2+} complexes have 4 unpaired electrons and low-spin (strong-field) Cr^{2+} complexes have 2 unpaired electrons. Ni^{2+}: $[Ar]3d^8$; Octahedral Ni^{2+} complexes will always have 2 unpaired electrons, whether high or low-spin. Therefore, Ni^{2+} complexes cannot be used to distinguish weak from strong-field ligands by examining magnetic properties. Alternatively, the ligand field strengths can be measured using visible spectra. Either Cr^{2+} or Ni^{2+} complexes can be used for this method.

65.   We need to calculate the Pb^{2+} concentration in equilibrium with $EDTA^{4-}$. Since K is large for the formation of $PbEDTA^{2-}$, let the reaction go to completion; then solve an equilibrium problem to get the Pb^{2+} concentration.

$$Pb^{2+} \quad + \quad EDTA^{4-} \quad \rightleftharpoons \quad PbEDTA^{2-} \qquad K = 1.1 \times 10^{18}$$

Before   0.010 M 0.050 M 0

0.010 mol/L Pb^{2+} reacts completely (large K)

Change   -0.010   -0.010 $\rightarrow$ +0.010   Reacts completely

After   0   0.040   0.010   New initial condition

x mol/L $PbEDTA^{2-}$ dissociates to reach equilibrium

Equil. x 0.040 + x 0.010 - x

$$1.1 \times 10^{18} = \frac{(0.010 - x)}{(x)(0.040 + x)} \approx \frac{(0.010)}{x(0.040)}, \quad x = [Pb^{2+}] = 2.3 \times 10^{-19} \, M \text{ Assumptions good.}$$

Now calculate the solubility quotient for $Pb(OH)_2$ to see if precipitation occurs. The concentration of OH^- is 0.10 M since we have a solution buffered at pH = 13.00.

$$Q = [Pb^{2+}]_o [OH^-]_o^2 = (2.3 \times 10^{-19})(0.10)^2 = 2.3 \times 10^{-21} < K_{sp} \, (1.2 \times 10^{-15})$$

$Pb(OH)_2(s)$ will not form since Q is less than K_{sp}.

67. $HbO_2 \quad \rightarrow Hb + O_2 \qquad \Delta G° = -(-70 \text{ kJ})$

      $Hb \; + \; CO \rightarrow HbCO \qquad \Delta G° = -80 \text{ kJ}$

$$HbO_2 + CO \rightarrow HbCO + O_2 \qquad \Delta G° = -10 \text{ kJ}$$

$$\Delta G° = -RT \ln K, \quad K = \exp\left(\frac{-\Delta G°}{RT}\right) = \exp\left(\frac{-(-10 \times 10^3 \text{ J})}{(8.3145 \text{ J/K} \bullet \text{mol})(298 \text{ kJ})}\right) = 60$$

Challenge Problems

69.   a.   Consider the following electrochemical cell:

$$Co^{3+} + e^- \rightarrow Co^{2+} \qquad\qquad E° = 1.82 \text{ V}$$
$$Co(en)_3^{2+} \rightarrow Co(en)_3^{3+} + e^- \qquad -E° = ?$$

$$Co^{3+} + Co(en)_3^{2+} \rightarrow Co^{2+} + Co(en)_3^{3+} \qquad E°_{cell} = 1.82 - E°$$

The equilibrium constant for this overall reaction is:

$$Co^{3+} + 3 \text{ en} \rightarrow Co(en)_3^{3+} \qquad\qquad K_1 = 2.0 \times 10^{47}$$
$$Co(en)_3^{2+} \rightarrow Co^{2+} + 3 \text{ en} \qquad\qquad K_2 = 1/1.5 \times 10^{12}$$

$$Co^{3+} + Co(en)_3^{2+} \rightarrow Co(en)_3^{3+} + Co^{2+} \qquad K = K_1 K_2 = \frac{2.0 \times 10^{47}}{1.5 \times 10^{12}} = 1.3 \times 10^{35}$$

From the Nernst equation for the overall reaction:

$$E^{\circ}_{cell} = \frac{0.0591}{n} \log K = \frac{0.0591}{1} \log (1.3 \times 10^{35}), \; E^{\circ}_{cell} = 2.08 \text{ V}$$

$$E^{\circ}_{cell} = 1.82 - E^{\circ} = 2.08 \text{ V}, \; E^{\circ} = 1.82 \text{ V} - 2.08 \text{ V} = -0.26 \text{ V}$$

b. The stronger oxidizing agent will be the more easily reduced species and will have the more positive standard reduction potential. From the reduction potentials, Co^{3+} ($E^{\circ} = 1.82$ V) is a much stronger oxidizing agent than $Co(en)_3^{3+}$ ($E^{\circ} = -0.26$ V).

c. In aqueous solution, Co^{3+} forms the hydrated transition metal complex, $Co(H_2O)_6^{3+}$. In both complexes, $Co(H_2O)_6^{3+}$ and $Co(en)_3^{3+}$, cobalt exists as Co^{3+} which has 6 d electrons.

Assuming a strong-field case for each complex ion, the d-orbital splitting diagram for each is:

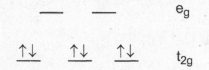

When each complex gains an electron, the electron enters a higher energy e_g orbital. Since en is a stronger field ligand than H_2O, the d-orbital splitting is larger for $Co(en)_3^{3+}$, and it takes more energy to add an electron to $Co(en)_3^{3+}$ than to $Co(H_2O)_6^{3+}$. Therefore, it is more favorable for $Co(H_2O)_6^{3+}$ to gain an electron than for $Co(en)_3^{3+}$ to gain an electron.

71. No; In all three cases, six bonds are formed between Ni^{2+} and nitrogen, so ΔH values should be similar. ΔS° for formation of the complex ion is most negative for 6 NH_3 molecules reacting with a metal ion (7 independent species become 1). For penten reacting with a metal ion, 2 independent species become 1, so ΔS° is least negative of all three of the reactions. Thus, the chelate effect occurs because the more bonds a chelating agent can form to the metal, the more favorable ΔS° is for the formation of the complex ion and the larger the formation constant.

73.

The d_{z^2} orbital will be destabilized much more than in the trigonal planar case (see Exercise 21.72). The d_{z^2} orbital has electron density on the z-axis directed at the two axial ligands. The $d_{x^2-y^2}$ and d_{xy} orbitals are in the plane of the three trigonal planar ligands and should be destabilized a lesser amount as compared to the d_{z^2} orbital; only a portion of the electron density in the $d_{x^2-y^2}$ and d_{xy} orbitals is directed at the ligands. The d_{xz} and d_{yz} orbitals will be destabilized the least since the electron density is directed between the ligands.

CHAPTER TWENTY-TWO

ORGANIC AND BIOLOGICAL MOLECULES

Questions

1. There is only one chain of consecutive C-atoms in the molecule. They are not all in a true straight line since the bond angles at each carbon are the tetrahedral angles of $109.5°$.

3. Resonance: All atoms are in the same position. Only the positions of π electrons are different.

    Isomerism: Atoms are in different locations in space.

    Isomers are distinctly different substances. Resonance is the use of more than one Lewis structure to describe the bonding in a single compound. Resonance structures are **not** isomers.

5. Substitution: An atom or group is replaced by another atom or group.

    e.g., H in benzene is replaced by Cl. $C_6H_6 + Cl_2 \xrightarrow{\text{catalyst}} C_6H_5Cl + HCl$

    Addition: Atoms or groups are added to a molecule.

    e.g., Cl_2 adds to ethene. $CH_2 = CH_2 + Cl_2 \rightarrow CH_2Cl - CH_2Cl$

7. A thermoplastic polymer can be remelted; a thermoset polymer cannot be softened once it is formed.

9. Crosslinking makes a polymer more rigid by bonding adjacent polymer chains together.

11. In nylon, hydrogen bonding interactions occur due to the presence of N–H bonds in the polymer. For a given polymer chain length, there are more N–H groups in Nylon-46 as compared to Nylon-6. Hence, Nylon-46 forms a stronger polymer compared to Nylon-6 due to the increased hydrogen bonding interactions.

13. Primary: The amino acid sequence in the protein. Covalent bonds (peptide linkages) are the forces that link the various amino acids together in the primary structure.

    Secondary: Includes structural features known as α-helix or pleated sheet. Both are maintained mostly through hydrogen bonding interactions.

Tertiary: The overall shape of a protein (long and narrow or globular) maintained by hydrophobic and hydrophilic interactions, such as salt linkages, hydrogen bonds, disulfide linkages and dispersion forces.

15.   All amino acids can act as both a weak acid and a weak base; this is the requirement for a buffer. The weak acid is the carboxylic end of the amino acid, and the weak base is the amine end of the amino acid.

17.   Hydrogen bonding interactions occur among the numerous –OH groups of starch and the water molecules.

19.   They all contain nitrogen atoms with lone pairs of electrons.

21.   A deletion may change the entire code for a protein, thus giving an entirely different sequence of amino acids. A substitution will change only one single amino acid in a protein.

Exercises

Hydrocarbons

23.   i.

CH_3–CH_2–CH_2–CH_2–CH_2–CH_3

ii.

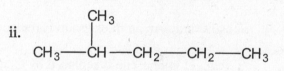

iii.

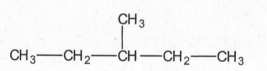

iv.

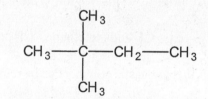

v.

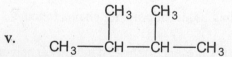

All other possibilities are identical to one of these five compounds.

25.   A difficult task in this problem is recognizing different compounds from compounds that differ by rotations about one or more C–C bonds (called conformations). The best way to distinguish different compounds from conformations is to name them. Different name = different compound; same name = same compound, so it is not an isomer, but instead, is a conformation.

a.

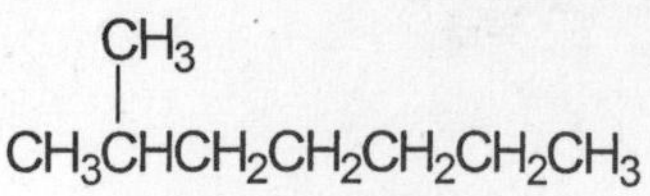

$CH_3CHCH_2CH_2CH_2CH_2CH_3$ with CH_3 branch

2-methylheptane

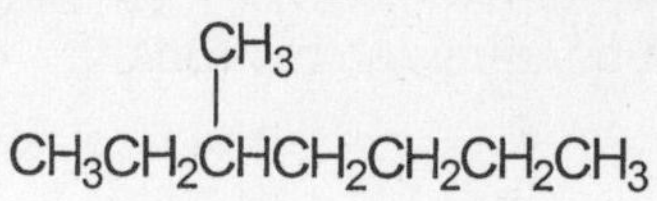

$CH_3CH_2CHCH_2CH_2CH_2CH_3$ with CH_3 branch

3-methylheptane

$CH_3CH_2CH_2CHCH_2CH_2CH_3$ with CH_3 branch

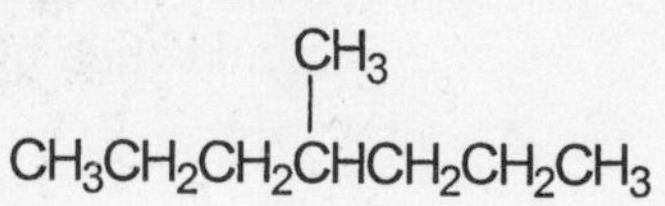

4-methylheptane

b.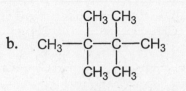

2,2,3,3-tetramethylbutane

27.   a.

CH₃—CH—CH₂—CH₂CH₃
        |
        CH₃

b.

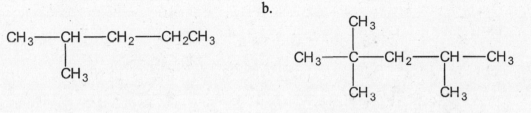

c.

CH₃—CH—CH₂CH₂CH₃

CH₃—C—CH₃
        |
        CH₃

d.   The longest chain is 6 carbons long.

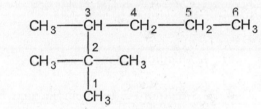

2,2,3-trimethylhexane

29.   a. 2,2,4-trimethylhexane   b. 5-methylnonane   c. 2,2,4,4-tetramethylpentane

d. 3-ethyl-3-methyloctane

Note: For alkanes, always identify the longest carbon chain for the base name first, then number the carbons to give the lowest overall numbers for the substituent groups.

31.   a. 1-butene   b. 4-methyl-2-hexene   c. 2,5-dimethyl-3-heptene

Note: The multiple bond is assigned the lowest number possible.

33.   a. CH₃–CH₂–CH=CH–CH₂–CH₃   b. CH₃–CH=CH–CH=CH–CH₂CH₃

c.
            CH₃
            |
CH₃—CH—CH═CH—CH₂CH₂CH₂CH₃

35.   a.

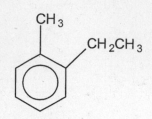

b.

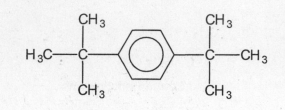

c.

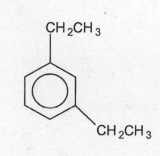

d.

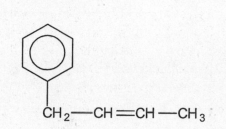

37.   a. 1,3-dichlorobutane

b. 1,1,1-trichlorobutane

c. 2,3-dichloro-2,4-dimethylhexane

d. 1,2-difluoroethane

Isomerism

39.   To exhibit *cis-trans* isomerism, each carbon in the double bond must have two structurally different groups bonded to it. In Exercise 22.31, this occurs for compounds b and c.   The *cis* and *trans* isomers for 31b and 31c are:

31 b.

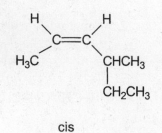

cis

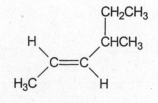

trans

31 c.

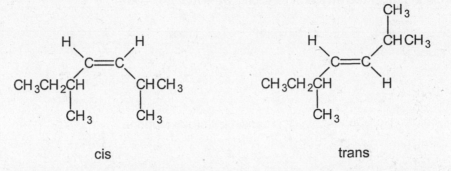

cis   trans

Similarly, all the compounds in Exercise 22.33 also exhibit *cis-trans* isomerism.

In compound a of Exercise 22.31, the carbon in the double bond does **not** contain two different groups. The first carbon in the double bond contains two H atoms. To illustrate that this compound does not exhibit *cis-trans* isomerism, lets look at the potential *cis-trans* isomers.

These are the same compounds; they only differ by a simple rotation of the molecule. Therefore, they are **not** isomers of each other, but instead are the same compound.

41. C_5H_{10} has the general formula for alkenes, C_nH_{2n}. To distinguish the different isomers from each other, we will name them. Each isomer must have a different name.

$CH_2\!\!=\!\!CHCH_2CH_2CH_3$ $CH_3CH\!\!=\!\!CHCH_2CH_3$

1-pentene   2-pentene

$CH_2\!\!=\!\!CCH_2CH_3$ $CH_3C\!\!=\!\!CHCH_3$
$\quad\quad\;\;\; |$ $\quad\quad\; |$
$\quad\quad\; CH_3$ $\quad\quad CH_3$

2-methyl-1-butene   2-methyl-2-butene

$CH_3CHCH\!\!=\!\!CH_2$
$\quad\; |$
$\quad CH_3$

3-methyl-1-butene

43.   To help distinguish the different isomers, we will name them.

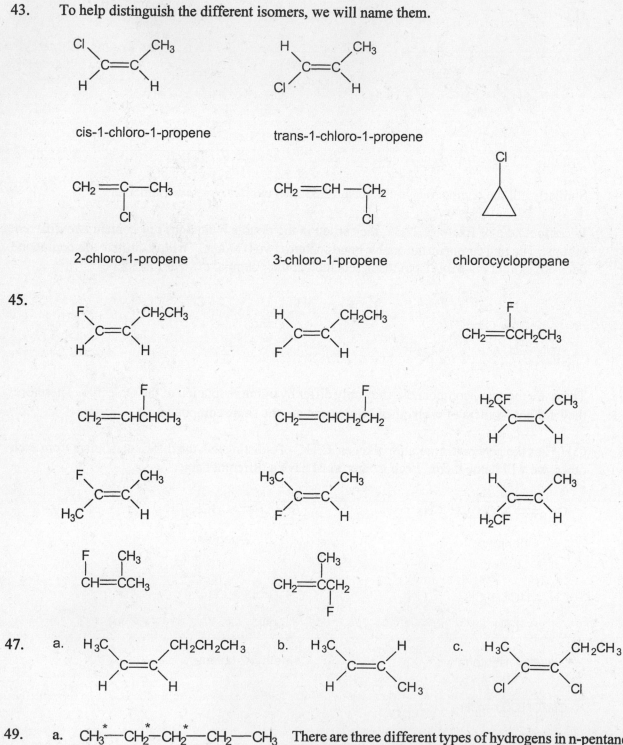

cis-1-chloro-1-propene   trans-1-chloro-1-propene

2-chloro-1-propene   3-chloro-1-propene   chlorocyclopropane

45.

47.   a.   b.   c.

49.   a. $CH_3^* \!-\!CH_2^* \!-\!CH_2^* \!-\!CH_2 \!-\!CH_3$ There are three different types of hydrogens in n-pentane (see asterisks). Thus there are three mono-chloro isomers of n-pentane (1-chloropentane, 2-chloropentane and 3-chloropentane).

b.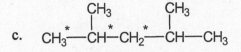

There are four different types of hydrogens in 2-methyl-butane, so four monochloro isomers of 2-methylbutane are possible.

c.   There are three different types of hydrogens, so three monochloro isomers are possible.

d.   There are four different types of hydrogens, so four mono-chloro isomers are possible.

Functional Groups

51.   Reference Table 22.5 for the common functional groups.

   a. ketone   b. aldehyde   c. carboxylic acid   d. amine

53.   a.

   b. 5 carbons in the ring and the carbon in –CO₂H: sp²; the other two carbons: sp³

   c. 24 sigma bonds; 4 pi bonds

55.   a. 3-chloro-1-butanol: Since the carbon containing the OH group is bonded to just 1 other carbon (1 R group), this is a primary alcohol.

   b. 3-methyl-3-hexanol; Since the carbon containing the OH group is bonded to three other carbons (3 R groups), this is a tertrary alcohol

   c. 2-methylcyclopentanol; Secondary alcohol (2 R groups bonded to carbon containing the OH group); Note: In ring compounds, the alcohol group is assumed to be bonded to C₁, so the number designation is commonly omitted for the alcohol group.

57.

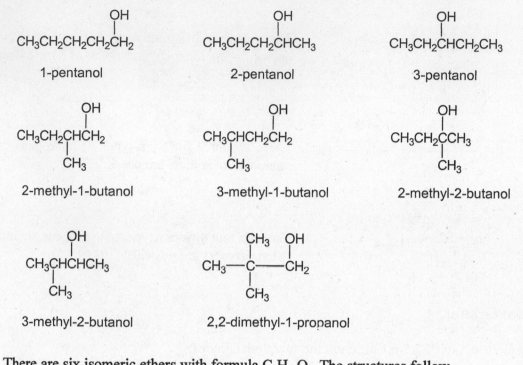

OH
CH₃CH₂CH₂CH₂CH₂

1-pentanol

OH
CH₃CH₂CH₂CHCH₃

2-pentanol

OH
CH₃CH₂CHCH₂CH₃

3-pentanol

OH
CH₃CH₂CHCH₂
      |
     CH₃

2-methyl-1-butanol

OH
CH₃CHCH₂CH₂
   |
  CH₃

3-methyl-1-butanol

OH
CH₃CH₂CCH₃
        |
       CH₃

2-methyl-2-butanol

OH
CH₃CHCHCH₃
   |
  CH₃

3-methyl-2-butanol

CH₃   OH
CH₃—C—CH₂
      |
     CH₃

2,2-dimethyl-1-propanol

There are six isomeric ethers with formula C₅H₁₂O. The structures follow.

CH₃—O—CH₂CH₂CH₂CH₃

          CH₃
           |
CH₃—O—CHCH₂CH₃

          CH₃
           |
CH₃—O—CH₂CHCH₃

          CH₃
           |
CH₃—O—C—CH₃
           |
          CH₃

CH₃CH₂—O—CH₂CH₂CH₃

          CH₃
           |
CH₃CH₂—O—CH
           |
          CH₃

59.   a. 4,5-dichloro-3-hexanone   b. 2,3-dimethylpentanal

c. 3-methylbenzaldehyde or m-methylbenzaldehyde

Reactions of Organic Compounds

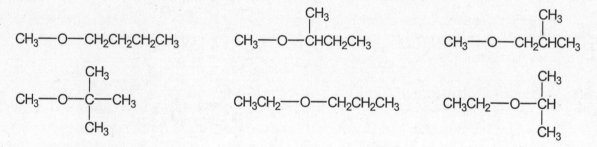

61.   a.
        H   H
        | |
   CH₃CH—CHCH₃

b.
   Cl   Cl   Cl   Cl
   | | | |
  CH₂—CHCHCH—CH
          | |
         CH₃   CH₃

c. ⬡—Cl + HCl

d. C₄H₈(g) + 6 O₂(g) → 4 CO₂(g) + 4 H₂O(g)

63.

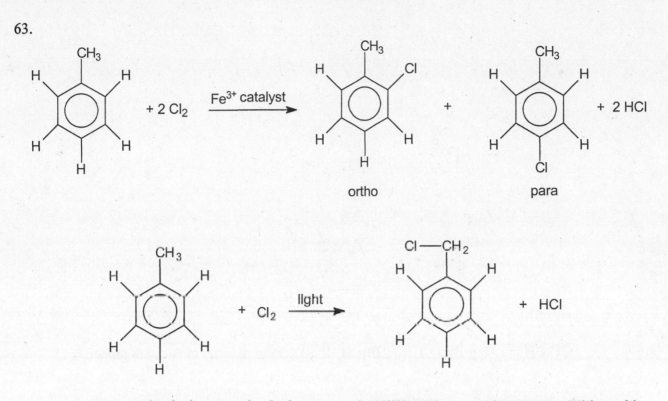

To substitute for the benzene ring hydrogens, an iron(III) catalyst must be present. Without this special iron catalyst, the benzene ring hydrogens are unreactive. To substitute for an alkane hydrogen, light must be present. For toluene, the light-catalyzed reaction substitutes a chlorine for a hydrogen in the methyl group attached to the benzene ring.

65.   Primary alcohols (a, d and f) are oxidized to aldehydes, which can be oxidized further to carboxylic acids. Secondary alcohols (b, e and f) are oxidized to ketones, and tertiary alcohols (c and f) do not undergo this type of oxidation reaction. Note that compound f contains a primary, secondary and tertiary alcohol. For the primary alcohols (a, d and f), we listed both the aldehyde and the carboxylic acid as possible products.

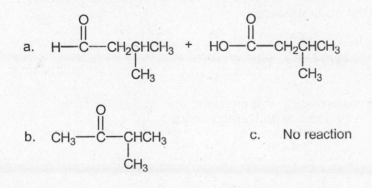

c.   No reaction

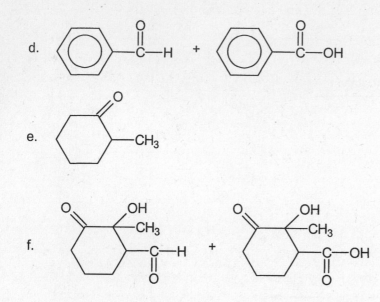

67.   a. $CH_3CH = CH_2 + Br_2 \rightarrow CH_3CHBrCH_2Br$ (Addition reaction of Br_2 with propene)

b.

Oxidation of 2-propanol yields acetone (2-propanone).

c.

Addition of H_2O to 2-methylpropene would yield tert-butyl alcohol (2-methyl-2-propanol) as the major product.

d.

Oxidation of 1-propanol would eventually yield propanoic acid. Propanal is produced first in this reaction and is then oxidized to propanoic acid.

69.   Reaction of a carboxylic acid with an alcohol can produce these esters.

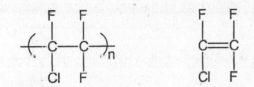

ethanoic acid   octanol   n-octylacetate
(acetic acid)

propanoic acid   hexanol

Polymers

71.   The backbone of the polymer contains only carbon atoms, which indicates that Kel-F is an addition polymer. The smallest repeating unit of the polymer and the monomer used to produce this polymer are:

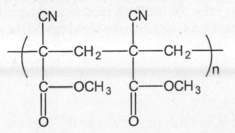

        Note: Condensation polymers generally have O or N atoms in the backbone of the polymer.

73.

        Super glue is an addition polymer formed by reaction of the C=C bond in methyl cyanoacrylate.

75. H_2O is eliminated when Kevlar forms. Two repeating units of Kevlar are:

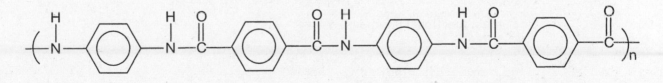

77.   This is a condensation polymer where two molecules of H_2O form when the monomers link together.

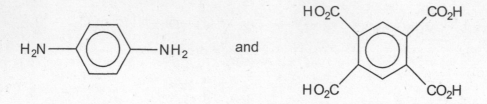

79.   Divinylbenzene has two reactive double bonds that are both used when divinylbenzene inserts itself into two adjacent polymer chains. The chains cannot move past each other because the crosslinks bond adjacent polymer chains together, making the polymer more rigid.

81   a.   The polymer formed using 1,2-diaminoethane will exhibit relatively strong hydrogen bonding interactions between adjacent polymer chains. Since hydrogen bonding is not present in the ethylene glycol polymer (a polyester polymer forms), the 1,2-diaminoethane polymer will be stronger.

      b.   The presence of rigid groups (benzene rings or multiple bonds) makes the polymer stiffer. Hence, the monomer with the benzene ring will produce the more rigid polymer.

      c.   Polyacetylene will have a double bond in the carbon backbone of the polymer.

            The presence of the double bond in polyacetylene will make polyacetylene a more rigid polymer than polyethylene. Polyethylene doesn't have C=C bonds in the backbone of the polymer (the double bonds in the monomers react to form the polymer).

Natural Polymers

83.   a.   Serine, tyrosine and threonine contain the -OH functional group in the R group.

      b.   Aspartic acid and glutamic acid contain the -COOH functional group in the R group.

      c.   An amine group has a nitrogen bonded to other carbon and/or hydrogen atoms. Histidine, lysine, arginine and tryptophan contain the amine functional group in the R group.

      d.   The amide functional group is:

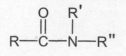

            This functional group is formed when individual amino acids bond together to form the peptide linkage. Glutamine and asparagine have the amide functional group in the R group.

85   a.   Aspartic acid and phenylalanine make up aspartame.

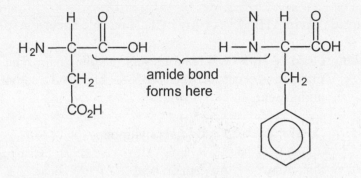

amide bond
forms here

b.   Aspartame contains the methyl ester of phenylalanine. This ester can hydrolyze to form methanol:

$$R-CO_2CH_3 + H_2O \rightleftharpoons RCO_2H + HOCH_3$$

87.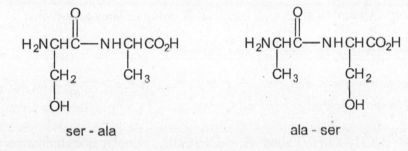

ser - ala   ala - ser

89.   a.   Six tetrapeptides are possible. From NH_2 to CO_2H end:

phe-phe-gly-gly, gly-gly-phe-phe, gly-phe-phe-gly,

phe-gly-gly-phe, phe-gly-phe-gly, gly-phe-gly-phe

b.   Twelve tetrapeptides are possible. From NH_2 to CO_2H end:

phe-phe-gly-ala, phe-phe-ala-gly, phe-gly-phe-ala,

phe-gly-ala-phe, phe-ala-phe-gly, phe-ala-gly-phe,

gly-phe-phe-ala, gly-phe-ala-phe, gly-ala-phe-phe

ala-phe-phe-gly, ala-phe-gly-phe, ala-gly-phe-phe

91.   a. Ionic: Need NH_2 on side chain of one amino acid with CO_2H on side chain of the other amino acid. The possibilities are:

NH_2 on side chain = His, Lys or Arg; CO_2H on side chain = Asp or Glu

b. Hydrogen bonding: Need N–H or O–H bond present in side chain. The hydrogen bonding interaction occurs between the X–H bond and a carbonyl group from any amino acid.

$$X–H \cdots\cdots O = C \text{ (carbonyl group)}$$

Ser Asn   Any amino acid
Glu Thr
Tyr Asp
His Gln
Arg Lys

c. Covalent: Cys – Cys (forms a disulfide linkage)

d. London dispersion: All amino acids with nonpolar R groups. They are:

Gly, Ala, Pro, Phe, Ile, Trp, Met, Leu and Val

e. Dipole-dipole: Need side chain with OH group. Tyr, Thr and Ser all could form this specific dipole-dipole force with each other since all contain an OH group in the side chain.

93.   Glutamic acid: $R = –CH_2CH_2CO_2H$; Valine: $R = –CH(CH_3)_2$; A polar side chain is replaced by a nonpolar side chain. This could affect the tertiary structure of hemoglobin and the ability of hemoglobin to bind oxygen.

95.   See Figures 22.29 and 22.30 of the text for examples of the cyclization process.

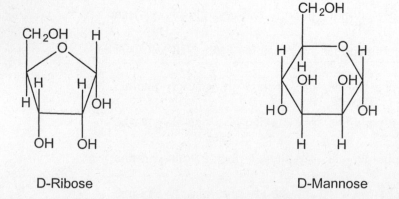

D-Ribose   D-Mannose

97.   The aldohexoses contain 6 carbons and the aldehyde functional group. Glucose, mannose and galactose are aldohexoses. Ribose and arabinose are aldopentoses since they contain 5 carbons with the aldehyde functional group. The ketohexose (6 carbons + ketone functional group) is fructose and the ketopentose (5 carbons + ketone functional group) is ribulose.

99.   The α and β forms of glucose differ in the orientation of a hydroxy group on one specific carbon in the cyclic forms (see Figure 22.30 of the text). Starch is a polymer composed of only α-D-glucose, and cellulose is a polymer composed of only β-D-glucose.

101.   A chiral carbon has four different groups attached to it. A compound with a chiral carbon is optically active. Isoleucine and threonine contain more than the one chiral carbon atom (see asterisks).

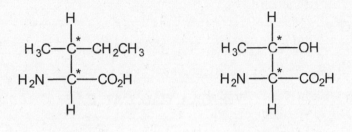

isoleucine   threonine

103.   Only one of the isomers is optically active. The chiral carbon in this optically active isomer is marked with an asterisk.

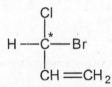

105.   The complimentary base pairs in DNA are cytosine (C) and guanine (G), and thymine (T) and adenine (A). The complimentary sequence is: C-C-A-G-A-T-A-T-G

107.   Uracil will hydrogen bond to adenine.

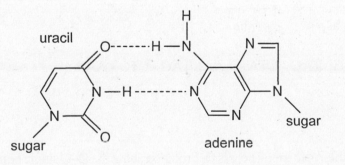

109.   Base pair:

      RNA   DNA

       A . . . . T

       G C

       C . . . . G

       U . . . . A

a. Glu: CTT, CTC   Val: CAA, CAG, CAT, CAC

    Met: TAC   Trp: ACC

    Phe: AAA, AAG   Asp: CTA, CTG

b. DNA sequence for trp-glu-phe-met:

      ACC - CTT - AAA - TAC
          or   or
         CTC   AAG

c. Due to glu and phe, there is a possibility of four different DNA sequences. They are:

    ACC - CTT - AAA - TAC or ACC - CTC - AAA - TAC or

    ACC - CTT - AAG - TAC or ACC - CTC - AAG - TAC

d. $\underbrace{T-A-C}_{met}-\underbrace{C-T-G}_{asp}-\underbrace{A-A-G}_{phe}$

e. TAC - CTA - AAG; TAC - CTA - AAA; TAC - CTG - AAA

Additional Exercises

111. CH_2Cl-CH_2Cl, 1-2-dichloroethane: There is free rotation about the C–C single bond that doesn't lead to different compounds. $CHCl=CHCl$, 1,2-dichloroethene: There is no rotation about the C=C double bond. This creates the cis and trans isomers, which are different compounds.

113.

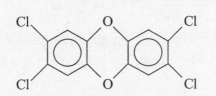

There are many possibilities for isomers. Any structure with four chlorines replacing four hydrogens in any four of the numbered positions would be an isomer, i.e., 1,2,3,4-tetrachloro-dibenzo-p-dioxin is a possible isomer.

115.   Alcohols consist of two parts, the polar OH group and the nonpolar hydrocarbon chain attached to the OH group. As the length of the nonpolar hydrocarbon chain increases, the solubility of the alcohol decreases in water, a very polar solvent. In methyl alcohol (methanol), the polar OH group overrides the effect of the nonpolar CH_3 group, and methyl alcohol is soluble in water. In stearyl alcohol, the molecule consists mostly of the long nonpolar hydrocarbon chain, so it is insoluble in water.

117.   The structures, the types of intermolecular forces exerted, and the boiling points for the compounds are:

$$CH_3CH_2CH_2\overset{\displaystyle O}{\overset{\|}{C}}OH$$

butanoic acid, 164°C
LD + dipole + H bonding

$CH_3CH_2CH_2CH_2CH_2OH$

1-pentanol, 137°C
LD + H bonding

$$CH_3CH_2CH_2CH_2\overset{\displaystyle O}{\overset{\|}{C}}H$$

pentanal, 103°C
LD + dipole

$CH_3CH_2CH_2CH_2CH_2CH_3$

n-hexane, 69°C
LD only

All these compounds have about the same molar mass. Therefore, the London dispersion (LD) forces in each are about the same. The other types of forces determine the boiling point order. Since butanoic acid and 1-pentanol both exhibit hydrogen bonding interactions, these two compounds will have the two highest boiling points. Butanoic acid has the highest boiling point since it exhibits H bonding along with dipole-dipole forces due to the polar C = O bond.

119. $85.63 \text{ g C} \times \dfrac{1 \text{ mol C}}{12.01 \text{ g C}} = 7.130 \text{ mol C}; \ \ 14.37 \text{ g H} \times \dfrac{1 \text{ mol H}}{1.008 \text{ g H}} = 14.26 \text{ mol H}$

Since the mol H to mol C ratio is 2:1 (14.26/7.130 = 2.000), the empirical formula is CH_2. The empirical formula mass $\approx 12 + 2(1) = 14$. Since $4 \times 14 = 56$ puts the molar mass between 50 and 60, the molecular formula is C_4H_8.

The isomers of C_4H_8 are:

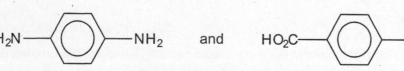

CH_2=$CHCH_2CH_3$ CH_3CH=$CHCH_3$ CH=$CHCH_3$ (with CH_3 above)

1-butene   2-butene   2-methyl-1-propene

cyclobutane   methylcyclopropane

Only the alkenes will react with H_2O to produce alcohols, and only 1-butene will produce a secondary alcohol for the major product and a primary alcohol for the minor product.

CH_2=$CHCH_2CH_3$ + H_2O $\longrightarrow$ CH_2—$CHCH_2CH_3$ (with H and OH above)

2° alcohol, major product

CH_2=$CHCH_2CH_3$ + H_2O $\longrightarrow$ CH_2—$CHCH_2CH_3$ (with OH and H above)

1° alcohol, minor product

2-butene will produce only a secondary alcohol when reacted with H_2O, and 2-methyl-1-propene will produce a tertiary alcohol as the major product and a primary alcohol as the minor product.

121.   a. H_2N—⬡—NH_2 and HO_2C—⬡—CO_2H

b.   Repeating unit:

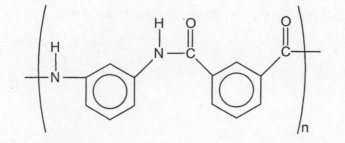

The two polymers differ in the substitution pattern on the benzene rings. The Kevlar chain is straighter, and there is more efficient hydrogen bonding between Kevlar chains than between Nomex chains.

123.   a.   The bond angles in the ring are about 60°. VSEPR predicts bond angles close to 109°. The bonding electrons are closer together than they prefer, resulting is strong electron-electron repulsions. Thus, ethylene oxide is unstable (reactive).

   b.   The ring opens up during polymerization; the monomers link together through the formation of O–C bonds.

$$\left(\!-O\!-\!CH_2CH_2\!-\!O\!-\!CH_2CH_2\!-\!O\!-\!CH_2CH_2\!-\!\right)_n$$

125.   Glutamic acid: Monosodium glutamate:

$$H_2N\!-\!\underset{\underset{CH_2CH_2CO_2H}{|}}{CH}\!-\!CO_2H \qquad\qquad H_2N\!-\!\underset{\underset{CH_2CH_2CO_2^-Na^+}{|}}{CH}\!-\!CO_2H$$

One of the two acidic protons in the carboxylic acid groups is lost to form MSG. Which proton is lost is impossible for you to predict.

In MSG, the acidic proton from the carboxylic acid in the R group is lost, allowing formation of the ionic compound.

127. $\Delta G = \Delta H - T\Delta S$; For the reaction, we break a P–O and O–H bond and form a P–O and O–H bond, so $\Delta H \approx 0$. ΔS for this process is negative since order increases. Thus, $\Delta G > 0$ and the reaction is not spontaneous.

129.   Alanine can be thought of as a diprotic acid. The first proton to leave comes from the carboxylic acid end with $K_a = 4.5 \times 10^{-3}$. The second proton to leave comes from the protonated amine end (K_a for $R-NH_3^+ = K_w/K_b = 1.0 \times 10^{-14}/7.4 \times 10^{-5} = 1.4 \times 10^{-10}$).

In $1.0\ M\,H^+$, both the carboxylic acid and the amine end will be protonated since H^+ is in excess. The protonated form of alanine is below. In $1.0\ M\,OH^-$, the dibasic form of alanine will be present since the excess OH^- will remove all acidic protons from alanine. The dibasic form of alanine follows.

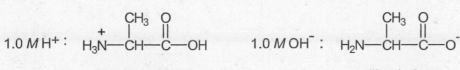

1.0 $M\,H^+$: protonated form   1.0 $M\,OH^-$: dibasic form

131.   The Cl⁻ ions are lost upon binding to DNA. The dimension is just right for cisplatin to bond to two adjacent bases in one strand of the helix, which inhibits DNA synthesis.

Challenge Problems

133.   For the reaction:

$$^+H_3NCH_2CO_2H \rightleftharpoons 2\ H^+ + H_2NCH_2CO_2^- \qquad K_{eq} = 7.3 \times 10^{-13} = K_a\ (\text{-}CO_2H) \times K_a\ (\text{-}NH_3^+)$$

$$7.3 \times 10^{-13} = \frac{[H^+]^2[H_2NCH_2CO_2^-]}{[^+H_3NCH_2CO_2H]} = [H^+]^2, \quad [H^+] = (7.3 \times 10^{-13})^{1/2}$$

$[H^+] = 8.5 \times 10^{-7}; \quad pH = -\log [H^+] = 6.07 =$ isoelectric point

135.   a.   Even though this form of tartaric acid contains 2 chiral carbon atoms (see asterisks in the following structure), the mirror image of this form of tartaric acid is superimposable. Therefore, it is not optically active. An easier way to identify optical activity in molecules with two or more chiral carbon atoms is to look for a plane of symmetry in the molecule. If a molecule has a plane of symmetry, then it is never optically active. A plane of symmetry is a plane that bisects the molecule where one side exactly reflects on the other side.

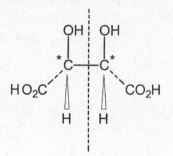

symmetry plane

b.   The optically active forms of tartaric acid have no plane of symmetry. The structures of the optically active forms of tartaric acid are:

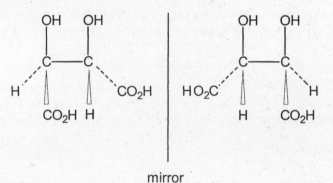

mirror

These two forms of tartaric acid are nonsuperimposable.

137.

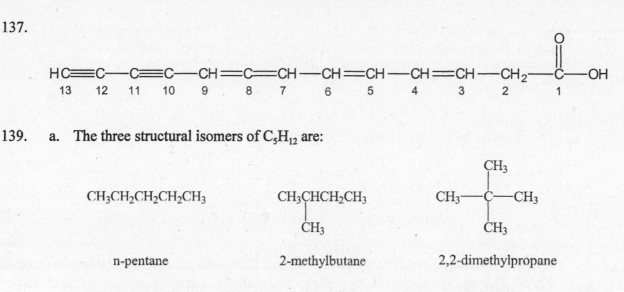

139.   a.   The three structural isomers of C_5H_{12} are:

CH₃CH₂CH₂CH₂CH₃   CH₃CHCH₂CH₃   CH₃—C—CH₃
                                          | |
                                         CH₃   CH₃   (with CH₃ on top)

      n-pentane   2-methylbutane   2,2-dimethylpropane

n-pentane will form three different monochlorination products: 1-chloropentane, 2-chloro-pentane and 3-chloropentane (the other possible monochlorination products differ by a simple rotation of the molecule; they are not different products from the ones listed). 2-2,dimethylpropane will only form one monochlorination product: 1-chloro-2,2-dimethyl-propane. 2-methylbutane is the isomer of C_5H_{12} that forms four different monochlorination products: 1-chloro-2-methylbutane, 2-chloro-2-methylbutane, 3-chloro-2-methylbutane (or we could name this compound 2-chloro-3-methylbutane), and 1-chloro-3-methylbutane.

   b.   The isomers of C_4H_8 are:

CH₂=CHCH₂CH₃   CH₃CH=CHCH₃   CH₂=CCH₃
                                                           |
                                                          CH₃   (CH₃ on top)

      1-butene   2-butene   2-methyl-1-propene or
                                                   2-methylpropene

      cyclobutane   methylcyclopropane

The cyclic structures will not react with H_2O; only the alkenes will add H_2O to the double bond. From Exercise 22.62, the major product of the reaction of 1-butene and H_2O is 2-butanol (a 2° alcohol). 2-butanol is also the major (and only) product when 2-butene and H_2O react. 2-methylpropene forms 2-methyl-2-propanol as the major product when reacted with H_2O; this product is a tertiary alcohol. Therefore, the C_4H_8 isomer is 2-methylpropene.

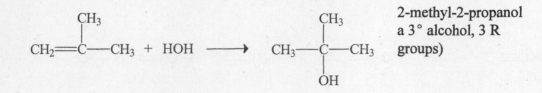

2-methyl-2-propanol
a 3° alcohol, 3 R
groups)

c. The structure of 1-chloro-1-methylcyclohexane is:

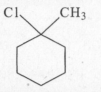

The addition reaction of HCl with an alkene is a likely choice for this reaction (see Exercise 22.62). The two isomers of C_7H_{12} that produce 1-chloro-1-methylcyclohexane as the major product are:

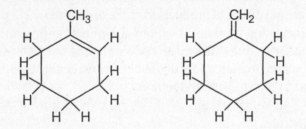

d. Working backwards, 2° alcohols produce ketones when they are oxidized (1° alcohols produce aldehydes, then carboxylic acids). The easiest way to produce the 2° alcohol from a hydrocarbon is to add H_2O to an alkene. The alkene reacted is 1-propene (or propene).

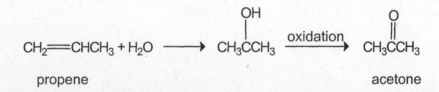

e. The $C_5H_{12}O$ formula has too many hydrogens to be anything other than an alcohol (or an unreactive ether). 1° alcohols are first oxidized to aldehydes, then to carboxylic acids. Therefore, we want a 1° alcohol. The 1° alcohols with formula $C_5H_{12}O$ are:

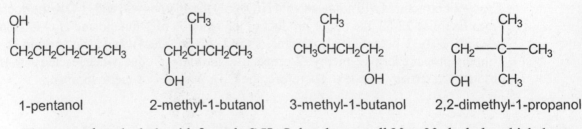

1-pentanol   2-methyl-1-butanol   3-methyl-1-butanol   2,2-dimethyl-1-propanol

There are other alcohols with formula $C_5H_{12}O$, but they are all 2° or 3° alcohols, which do not produce carboxylic acids when oxidized.

141.

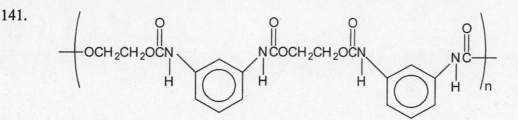

143.   a.   The temperature of the rubber band increases when it is stretched.

    b.   Exothermic since heat is released.

    c.   As the chains are stretched, they line up more closely together, resulting in stronger London dispersion forces between the chains. Heat is released as the strength of the intermolecular forces increases.

    d.   Stretching is not spontaneous so, ΔG is positive. $\Delta G = \Delta H - T\Delta S$; Since ΔH is negative then ΔS must be negative in order to give a positive ΔG.

    e.

unstretched   stretched

The structure of the stretched polymer is more ordered (lower S).